Facebook

ALSO BY STEVEN LEVY

In the Plex
Crypto
The Perfect Thing
Insanely Great
Artificial Life
The Unicorn's Secret
Hackers

Facebook

THE INSIDE STORY

Steven Levy

BLUE RIDER PRESS
New York

blue
rider
press

An imprint of Penguin Random House LLC
penguinrandomhouse.com

LIBRARY OF CONGRESS CATALOGING-IN-PUBLICATION DATA
Names: Levy, Steven, author.
Title: Facebook: the inside story / Steven Levy.
Description: [New York] : Blue Rider Press, [2020] | Includes bibliographical references and index.
Identifiers: LCCN 2019047909 (print) | LCCN 2019047910 (ebook) |
ISBN 9780735213159 (hardcover) | ISBN 9780735213166 (ebook)
Subjects: LCSH: Facebook (Firm)—History. |
Facebook (Electronic resource)—Social aspects.
Classification: LCC HM743.F33 L48 2020 (print) | LCC HM743.F33 (ebook) |
DDC 302.30285—dc23
LC record available at https://lccn.loc.gov/2019047909
LC ebook record available at https://lccn.loc.gov/2019047910

International Edition ISBN: 9781524746834

Printed in the United States of America
1 3 5 7 9 10 8 6 4 2

In memory of Lester Levy, 1920–2017.
Sorry you didn't see that Super Bowl, Dad.

Contents

Introduction 1

PART ONE

1. ZuckNet 19
2. Ad-Boarded 36
3. Thefacebook 56
4. Casa Facebook 79
5. Moral Dilemma 100
6. The Book of Change 118

PART TWO

7. Platform 147
8. Pandemic 177
9. Sheryl World 190
10. Growth! 208
11. Move Fast and Break Things 237
12. Paradigm Shift 275
13. Buying the Future 299

PART THREE

14. Election 333
15. P for Propaganda 368
16. Clown Show 399

17. The Ugly 432

18. Integrity 462

19. The Next Facebook 487

Epilogue 522

Acknowledgments 529

Notes 533

Index 565

Introduction

"HI, I'M MARK!"

The introduction is unnecessary. Mark Zuckerberg is one of the world's most recognizable faces. He's the CEO of Facebook, the world's largest social network—the world's largest human network of *any* kind, ever—approaching 2 billion members, more than half of whom log in every day. It's made him, in today's reckoning, the sixth-richest person in the world. And because he founded Facebook at such a tender age—nineteen, in his Harvard dorm room—he has become the go-to thumbnail in charts representing the mind-boggling opportunities that advanced technologies offer to even the young and obscure.

He's beyond famous. And he's *here*.

In Lagos, Nigeria.

If there were any doubt who he is, this pleasant, brown-haired young man with a goofy smile and an apparent aversion to blinking is dressed exactly like . . . Mark Zuckerberg! The signature T-shirt that signifies geek proletariat but is actually a Brunello Cucinelli creation ($325 each—he's got a closet full of them, liberating him from having to make daily decisions on couture). Blue jeans and Nikes. Exactly what you'd expect to see if Facebook's founder and CEO were to amble in. What is unusual is that no one expected him to walk into this room, in this city, in this country, on this continent.

The people in the sixth-floor loftlike studio of the Co-Creation

Hub (CcHUB)—young entrepreneurs striving to buck the enormous odds against building successful tech companies in Lagos, Nigeria—have been told only that an unnamed executive from Facebook would appear today, August 30, 2016, in advance of a Facebook-run boot camp event for tech start-ups. They had anticipated the mystery guest might be one of Zuckerberg's lieutenants, Ime Archibong, a North Carolina–raised son of Nigerian immigrants who had visited his heritage country previously. An appearance by Zuckerberg himself was too earth-shattering to contemplate.

Indeed, Facebook planned the trip with CIA-level stealth, largely because of security concerns, but also with an eye toward milking the surprise and delight his appearance would generate. Zuckerberg had never set foot on this continent, and a visit was overdue. Zuckerberg has flown in from Italy, where he and his wife attended the wedding of his friend Daniel Ek, the CEO of Spotify. After the Lake Como nuptials, he and his retinue spent a few days in Rome, meeting with the prime minister and the pope. Straight from the airport, he headed to the gritty Yaba neighborhood and CcHUB.

The Lagos start-up culture careens between an improbable optimism and a gallows humor regarding the monumental obstacles to success or even survival. But these were the people Zuckerberg wanted to meet: nerds with dreams. In the giant headquarters he built in Menlo Park, California, among the posters festooning the walls like a giant confetti blast of techno-propaganda were dozens that read BE THE NERD. So while other tech magnates devoted their initial African venture to philanthropic themes, Zuckerberg scheduled no time to hug undernourished infants in remote villages. Instead, he would meet the software strivers.

For a moment, the young tech entrepreneurs freeze in place, as if suspicious that they are viewing an apparition, or some sort of hoax. Then, convinced by the evidence of their senses, they cry out in a burst of joy and rush forward as one to converge around their famous visitor, pumping his hand, posing for Zuck selfies, and hurriedly blurting out elevator pitches.

Zuckerberg patiently interacts with them, maintaining his smile, looking each one in the eye, maybe holding the stare a bit too long. He's clearly happy. *These are my people*, he says to me as we head down the steps to talk to more entrepreneurs.

Oh yes, I am trailing Mark Zuckerberg on this trip. It is the first reporting I am doing for a book about the rise of Facebook.

On the bottom floor, there is a program called Summer of Code, where youngsters ages five to thirteen are training on computers. He goes up to a pair of boys who are sharing a PC. They look to be around seven or eight years old. "Can you tell me what you've built?" he asks, bending down to their level so he is, as they are, looking up at a screen showing blinking dots purposefully moving in formation.

"A game," says one.

Zuckerberg's eyes, which are wide at cruising level, open even farther, like the plastic oculi of a large stuffed animal: *That's what he was doing at that age!*

"Can you tell me how you built it?" he asks.

After a few more technical consultations with the children ("Can you show me the code?"), he's off to the next stop on his agenda, a start-up that trains central African engineers to do technical labor for big corporations. Zuckerberg has helped fund the effort through his foundation, which he has designated to ultimately receive 99 percent of his Facebook shares. Business visitors in Lagos rarely travel by foot in neighborhoods like Yaba, but Zuckerberg wants to walk through the streets. The sidewalks are aspirational, the dirt and concrete rutted with potholes and dotted with puddles. Cars and motorbikes buzz by. The pace is quick, to zip by the people in the shanties and storefronts before they realize what's happening. One kid manages to scoot ahead of the group to shoot a selfie. Zuckerberg seems oblivious, chatting with Archibong as he heads his retinue.

The scene is captured by his house photographer, a former *Newsweek* photojournalist who has traveled with multiple presidents. When those images hit the net, that two-block stroll among the people will make Zuckerberg a beloved figure in Nigeria. ("I thought it must be photoshopped!" said a local engineer when he first encountered it.) The next day, a run across a bridge, similarly documented on social media, will cement the image of this tech billionaire as a man of the people.

The last stop of his first day in Africa is a tiny storefront at a busy intersection. It is one of a number of "Express Wi-Fi" mom-and-pop franchises Facebook has helped fund, where locals beam Internet from shops and businesses for small fees. This particular stand, which also takes sports bets, is run by Rosemary Njoku, a woman in a black spotted dress and head scarf. She has a friend with her, who wears a long yellow print dress.

Providing Internet access to "the next few billion"—that is, people in underserved regions or who can't afford connections—has been Zuckerberg's passion for the past few years. He has promoted a variety of means to spread the Internet, from exotic technologies like self-piloting drones to a controversial plan to give people free data plans that limit their access to a subset of popular applications, including Facebook. Express Wi-Fi is a small but promising aspect of this dream, called Internet.org.

The women greet Zuckerberg in the back of a narrow space that is very warm and close—there is barely room for the three to talk. Zuckerberg, his T-shirt showing spots of perspiration, begins to quiz Njoku.

"I'd like you to help me," says the world's then sixth-richest man to a woman running a tiny business on a street corner in one of the world's poorest countries. "What advice can you give me to make this better for you?"

The question momentarily startles Njoku, but she recovers quickly. "More meters." Zuckerberg looks at her quizzically. "More meters for the Wi-Fi, to reach a bigger space, more people," she says.

Zuckerberg is momentarily silent. *What else?* he wants to know. "Hashtags," she says. "Hashtag #itsup so people know the Wi-Fi is working."

Zuckerberg brightens. "That's something we can do," he says. "The first one is pretty hard." He explains a bit about the technical difficulties and quickly loses his audience.

The next day, Zuckerberg holds a town hall for local software developers. He loves this format, much preferring it to lecturing or being interviewed fireside-style by pesky journalists. He takes special pride in telling the crowd that Facebook has built a satellite that will expand Internet coverage to many unserved areas of Africa, including Nigeria. The results will soon be delivered, as the bird is on the launchpad now, on a SpaceX rocket ship. Elon Musk's company.

One of the pre-submitted questions the moderator has on hand asks how easy—or hard—it had been for Zuckerberg to move from the total control of a software developer to the fuzzier domain of running a company. Did he miss just coding?

"I'm an engineer, like a lot of you guys," he says. "And for me engineering comes down to two real principles: The first is that you think of every problem as a system. And every system can be better. No matter how good or bad it is, you can make anything better—and that goes for you whether you're writing code or you're building hardware, or your system is a company."

Facebook, he says, attacks problems of business and culture in the same way a coder solves problems. "Running [a company is] not so different from writing code where you're writing different functions and in subroutines. . . . I do think there's something really fundamental to this engineering mindset."

Later that day, Zuckerberg visits an entertainment studio in a district called Nollywood. A collection of Nigerian celebrities—actors, DJs, musicians, comedians—has been gathered to meet him. As he tours the facilities and meets people, Zuckerberg makes sure to pursue what seems to be a current obsession: whether creators and celebrities prefer Facebook or Instagram, the mobile photo-sharing site he bought in 2012 and has so far allowed its founders to run. They all seem to prefer Instagram to Facebook. "But Facebook is bigger," he says, not at all excited about the replies.

At one point, with everyone gathered in a room to informally ask him questions, one of the Instagram stars brings up *The Social Network*, the 2010 movie that purports to tell Facebook's origin story. Since the portrait of Zuckerberg is of a conniving idiot savant who created the company because he couldn't get into a fancy Harvard eating club, and women spurned him, one might suspect that he wouldn't welcome that subject, especially since the questioner asks him if he really founded Facebook because a woman dumped him.

"My wife hates that part of the movie," he says with his aw-shucks smile. "The reality is that we were already dating. So the idea that I started Facebook to get girls makes her angry." Pause. "It's also not true."

On his fourth, and last, night in Nigeria, Zuckerberg invites me to hang out in his hotel room with some of his traveling party. The group has moved from Lagos to a fortified hotel complex in the nation's capital, Abuja, where Zuckerberg will meet the president before leaving the country. (The young CEO has so frequently met with world leaders—as sort of a peer, as Facebook's large global audience gives him a hefty constituency in many lands—that people commonly refer to the company's "foreign policy.") It had been a long day, up at 4 a.m. to board a private plane to Kenya, where he'd done a two-hour safari, met with entrepreneurs, and lunched with officials. By late afternoon, he and his entourage were back on the plane. It was then that Zuckerberg learned that the SpaceX rocket, the one carrying the satellite he had been gleefully touting as an Internet savior for the struggling continent, had blown up on the launchpad, a day before the scheduled blastoff. Facebook's satellite had been in place during the test—a time-saving measure—and was lost in the conflagration.

Zuckerberg was furious with Musk. (Facebook's own motto, "Move fast and break things," does not apply to space launches.) He turned to a natural outlet for his anger, a medium that by his own hand he had made available to a sizable plurality of the human race: Facebook. Overruling the advice of his PR people, he rage-posted a story that would wind up in the news feeds of many of his 118 million followers:

> *As I'm here in Africa, I'm deeply disappointed to hear that SpaceX's launch failure destroyed our satellite that would have provided connectivity to so many entrepreneurs and everyone else across the continent.*

In the hotel room that evening, though, Zuckerberg is jolly. He is much more relaxed and playful when surrounded by people he knows well. On the table are huge portions of local food. He takes deep pulls from a can of Nigerian beer. He playfully teases Archibong and the Facebook photographer. But upon the mention of Musk's name, he goes silent for a second. Well, closer to a minute.

"I've think I've gone through the five stages of grief," he says. Another pause. "Well, maybe not acceptance." Has he spoken to Musk? Yet another pause, this one longer and darker. "No," he finally says.

Then the conversation turns back to the trip, and he brightens. I ask him about his riff on Facebook and the engineering mindset. He's eager to expound on it, explaining how his own predilection to see things through an engineering lens has been a key to the firm's approach. "There's this fundamental thing that at an early age you looked at something and felt like: *This can be better. I can break down this system and make it better.* I remember thinking about that when I was young; it didn't dawn on me until I was older that this isn't the way everyone thinks of things. I do think that's the engineering mindset—it may even be more a value set than a mindset."

Zuckerberg is all about sharing, and he often says that the world is a better place when people share their experiences with one another. So far, the world has eaten up his philosophy, celebrating Facebook for its historic membership numbers, its ability to pull people together, and even its potential to help people address serious problems in a grassroots fashion. Facebook was credited as the driver of the liberating Arab Spring. Its privacy practices, despite steady criticism from activists and regulators, haven't punctured the Facebook narrative. And despite the dark portrait depicted in *The Social Network*, Zuckerberg is generally viewed as the plucky, egalitarian founder who likes to take recreational runs on the streets, be it in Lagos or even a smog-shrouded Tiananmen Square in Beijing.

"One of the things I'm happy about is on this trip I got to talk to real people," he says. "I went to Rome and got to talk to the pope and the prime minister—I mean, they're people and awesome—but I was happy here that I got to talk to a lot of developers and engineers."

He loved Nigeria, and Nigeria loved him. The president offered this statement:

> *In our culture, we are not used to seeing successful people appear like you. We are not used to seeing successful people jogging and sweating on the streets. We are more used to seeing successful people in air-conditioned places. We are happy you are well-off and simple enough to always share.*

One might argue that Nigeria trip was peak Zuckerberg. How could life have been any better for him? He was well on his way to connecting the world as no other human being had ever done, not even the Roman emperors he admired. The company he had founded in his dorm room was minting money; he had never worked anywhere else but now had total voting control of one of the world's most valuable corporations. His face was on countless magazine covers. He had been *Time*'s Person of the Year. A survey early in the year ranked him "Tech's most popular CEO." He was happily married, and after a series of disheartening miscarriages (news of which he would share on Facebook), his wife gave birth to their adorable daughter. Even his pet, a shaggy Hungarian sheepdog whose white fur looked twisted into dreadlocks, had a fan club. As problems go, Elon Musk's exploding satellite and concerns about Internet.org weren't so insurmountable. In short, Facebook had taken its place as one of the great American success stories. Mark Zuckerberg's world seemed perfect.

What could go wrong?

. . .

BARELY TWO MONTHS after Mark Zuckerberg returned from Nigeria, Donald Trump was elected the president of the United States. It was a shock to many, many people who supported the other candidate, Hillary Rodham Clinton.

For Facebook, the shock was compounded by something else: a huge collective finger pointed toward Menlo Park, California, where the company had its sprawling headquarters. Almost from the minute that the *New York Times* needle indicating a victor crossed over from the Clinton side to the Trump side, political observers cited the "Facebook Effect" as one possible explanation for the seemingly impossible outcome. In the weeks leading up to the election, there had been reports of so-called fake news, or misinformation intentionally spread through Facebook's algorithms, being circulated widely on Facebook's News Feed, which had become the major source of news for millions of users. The false stories—or exaggerations of trivial missteps into a narrative of evil conspiracies—overwhelmingly seemed to discourage voters from pulling the lever on Hillary Clinton.

Still, almost no one at Facebook, including the surprising number of former Republican operatives whom the company had hired to work in communications and policy, had believed that Trump had a chance to win. Facebook's rock-star chief operating officer, Sheryl Sandberg, a true-blue Clinton supporter, sent her daughter to bed that evening, promising to wake her so she could witness history as the first woman president of the United States made her acceptance speech.

The little girl slept through the night undisturbed. Sandberg still gets choked up when she talks about it.

At Facebook headquarters the next day, people were shaken. At an all-hands meeting people were in tears. Internal discussion groups popped up on the platform, wondering whether—or how much—Facebook had been a factor in the results. Still, in the immediate aftermath, the idea that Facebook was *responsible* for this outcome seemed preposterous.

Two days after the election, Mark Zuckerberg appeared at a conference at Half Moon Bay, about thirty miles north of the Facebook campus. He would be interviewed, fireside chat–style, by David Kirkpatrick, an author turned conference organizer who had himself published a book about Facebook six years earlier. Naturally, he asked Zuckerberg about the charges that Donald Trump might have benefited from misinformation circulated on Facebook via the personalized news feeds of its users.

Zuckerberg brushed off the thought. "I've seen some of the stories they're talking about around this election," Zuckerberg said. "Personally, I think the idea that fake news on Facebook, of which it's a very small amount of the content, influenced the election in any way, I think is a pretty crazy idea."

As I sat in that hotel ballroom attending the conference, the remark did not feel as if Zuckerberg made a giant blunder. The "crazy idea" remark was part of a longer, thoughtful reply, and Kirkpatrick did not challenge him on it. It remains uncertain whether anything that happened on Facebook made a significant difference in the 2016 election.

But over the next two years, as people learned more about Facebook and the way it operated, dire concerns emerged about Facebook's role, not just in the election but in the body politic at large, and the *world* at large. Time and again, they would point to the "crazy" statement as an indication that Zuckerberg was clueless—or lying—about the damage his company was doing. After months of criticism, Zuckerberg would finally apologize for the remark.

And much more.

While the election was a turning point for Facebook, many would say it was a long-overdue reckoning. To its critics, the things that Facebook had boasted about as its most cherished accomplishments were now unmasked as liabilities. The huge user base once seen as a world-changing kumbaya was now alarming evidence of excessive power. The ability to give voices to the unheard was identified as a means to bequeath an earsplitting sound system to hate groups. The ability to organize political movements of liberation was now a deadly tool of oppressors. The joyful metrics that spurred smile-inducing memes to entertain and uplift us now were fingered as an algorithmic boost to misinformation.

For the next year, Facebook's reputation tumbled.

Facebook is racist. . . . Facebook aids genocide. . . . Facebook is an outrage machine. . . . Facebook is destroying our attention span. . . . Facebook is killing the news business. . . .

And the dam burst in 2018, when news came that Facebook had allowed personal information of up to 87 million users to end up in the hands of a company called Cambridge Analytica, which allegedly used the data to target vulnerable voters with misinformation. Facebook bit-flipped from Most Admired Company to Most Reviled.

Governments on three continents would probe the company, becoming increasingly hostile at what they saw as foot-dragging or outright intransigence on Facebook's part. Investigative journalists around the world focused their attention on Facebook, and hardly a day went by that another example of its misbehavior did not emerge. The privacy worries that had led the FTC to impose a huge fine and a 2011 Consent Decree would be regurgitated; Facebook would be fined an astounding $5 billion for violating the decree. The charge that Facebook was a toxic attention sponge would be leveled in Congress and on television talk shows. Even worse were the reports that Facebook (and its subsidiary WhatsApp) spread intentional misinformation that aided genocide in Myanmar and other regions.

By 2019, government bodies around the world would be characterizing Facebook using language usually reserved for terrorist organizations and drug rings. A report issued by a parliamentary study in Great Britain called Facebook "digital gangsters." New Zealand Privacy Commissioner John Edwards tweeted that the company's leaders were "morally bankrupt pathological liars." Salesforce CEO Marc Benioff compared Facebook's toxic consequences to those of the cigarette industry.

Meanwhile, revenues and profits kept rising. This whetted the appetite for sanctions and regulation even more. As the 2020 presidential election shaped up, multiple candidates joined the calls of legislators and regulators for Facebook to be broken up. It wasn't just that Facebook might have had a hand in the result of a controversial election: the charge was that Facebook was destroying democracy itself!

In the three years since the 2016 election, the Facebook reputational meltdown has been epic. There have been other spectacular train wrecks of companies that once were the darlings of the press and beloved by investors: Enron and Theranos come to mind.

But the Facebook crisis was something unique. It began with a gloriously idealistic goal: to connect the world. But the assumption was overly optimistic, and the company pursued its naïvely utopian—and undeniably self-serving—goal with a tragic disregard for consequences. To critics, Facebook was a twenty-first-century corporate Gatsby, careless in its privileges, self-involved in serving its own needs and pleasures.

Yet there is still something to the company's insistence that the good it does outweighs what it now admits is the bad it foments. Billions of people are still on Facebook, as well as its sister companies Instagram and WhatsApp. It is still part of our lives, perhaps as much as ever.

In Nigeria, I already saw Facebook as one of the most interesting companies in the annals of business and technology. But over the next three years of reporting, I found myself documenting the most complicated, dramatic, and polarizing story that I had ever covered.

Fortunately, Facebook kept talking to me.

I FIRST MET Mark Zuckerberg in March 2006. At the time, I was *Newsweek*'s chief technology writer, working on a cover story about a phenomenon called Web 2.0, a stage in the Internet business where new companies emerged whose plan was to connect people with one another. The companies we were writing about included Flickr, YouTube—still an independent start-up—and MySpace, the leading company in a relatively new field known as social networking. I had also heard of a hot new company that was having remarkable success by focusing on colleges, and wanted to hear a bit more about it, perhaps to give it a name-check and quote its founder. He was scheduled to appear that month at PC

Forum, a conference I regularly attended. So I contacted Facebook to see if I could meet Zuckerberg.

We agreed that I'd meet Mark at lunch, just as he arrived, and we'd chat then. I didn't know much about him and was not prepared for what happened.

When we were introduced, I took it in stride that he looked even younger than his twenty-one years. I'd met a few other peach-fuzz magnates in my years covering hackers and tech companies. But what did shake me was his reaction when I asked him what seemed to me to be a few softball questions about what the company was up to.

He just stared at me. And said nothing. Time seemed to freeze as the silence continued.

I was flummoxed. This guy is the CEO, isn't he? Is he having some sort of episode? On the spectrum, as some would later speculate? Was there something I'd written that somehow made him hate me?

I didn't know then that this was common behavior for Zuckerberg. Though I was unaware at the time, I had joined the fraternity of people who'd been stunned by Mark Zuckerberg's trancelike silences.

In subsequent years, Zuckerberg seems to have addressed this issue, and actually will conduct fairly personable interviews. (On occasion, though, the frigid stare still surfaces. One of his executives refers to it as "the eye of Sauron." Others who know him well say that at those instances he's just thinking, apparently at such a high level that the world stops for him.) But at the time, I found it baffling and unnerving. I looked across the table at the companion he'd arrived with, a former venture capitalist named Matt Cohler who was now working for the company. Pleasant smile. No lifeline.

In this case, I finally stumbled on a way to break the silence, changing the subject and asking Zuckerberg if he knew anything about PC Forum. He said no, and I explained its roots as the key industry gathering in the personal-computer era, where Gates and Jobs would go at each other with smiles on their faces and shivs in their fists. After hearing that bit of history he seemed to thaw, and for the rest of the lunch he was able to talk about the company he started in a dorm room, though he did not share with me the game-changing developments that his team were building at that very moment in their second-floor office in Palo Alto, features called Open Registration and the News Feed, that would supercharge his company and put the name Zuckerberg alongside those earlier legends of the PC Forum.

I covered Zuckerberg and his company as it grew from start-up to star. In August 2007, I wrote a *Newsweek* cover story about Facebook, focusing on its shift from a college website to a service that aspired to connect the entire world. When I moved full-time to *Wired* in 2008, Facebook was one of the focuses of my reporting. I engineered a cover shoot with Zuckerberg and his role model, Bill Gates. We did an interview for *Wired*'s twentieth-anniversary issue. When Facebook introduced new products, I was often offered a preview and a sit-down with the CEO. I talked to him about search, about virtual reality, about the star-crossed Facebook phone, about the NSA's hijacking of tech-company data, about his dream of providing low-cost Internet to the developing world. When I started a publication called Backchannel on the Medium platform, I stuck to the Facebook beat, writing about the News Feed algorithm and Facebook's AI team.

But it was a simple announcement from Facebook's communications team that made me realize that the breadth of the company's ambitions could only be fully captured in a book-length study. The news was that a billion people had logged on to Facebook on the same day.

It stopped me cold. In the space of twenty-four hours, a sizable chunk of the world's population had been active on Mark Zuckerberg's network.

That was new. Occasionally a global audience of that size might be gathered for a broadcast of the World Cup final or some other major event. But in those cases, the masses were spectators. These were people logging on to a single interactive network. And this tranche of a billion people was not a spike but a baseline, as Facebook was on pace to draw in more and more of the world's population.

Zuckerberg had been talking about connecting the world for a while.

But with this milestone, it appeared that his claims had to be taken seriously. On a daily basis, Facebook would be setting records for gathering the largest collection of people ever, who would frolic with friends, relatives, contacts, and some people they couldn't pick out of a police lineup—commenting, posting news articles, selling and buying, organizing political movements, and, in some cases, bullying peers, spreading idiotic memes, and recruiting terrorists.

How did that happen? I wondered. What were the implications? Was Facebook's still-youthful leader at all equipped to manage this unprecedented phenomenon, handling all the complications that may come from fulfilling his goal to connect the world? Was anyone, let alone this odd person prone to going silent in conversations?

I decided then to dive deep into Facebook, ideally with the company's cooperation. After some months of discussion, the company, including Zuckerberg and Sandberg, agreed, giving me unprecedented access to its employees, and encouraging former Facebookers to talk as well. And, of course, I spoke to many people who never worked at Facebook but interacted with the company as cohorts, competitors, critics, clients, developers, regulators, users, or funders.

Despite the epic PR problems the company endured after the 2016 election, Facebook kept its commitment to me, and I continued my frequent visits to its campus, eventually being recognized by receptionists in multiple buildings as they checked my ID and issued me a visitor's badge. I interviewed Zuckerberg himself six times after Nigeria: in his glass-walled office; walking the roof of the headquarters building; in Lawrence, Kansas; and at his home in Palo Alto.

Obviously, after the 2016 election and crises like fake news, state-sponsored manipulation, live-streaming of suicides and massacres, rampant hate speech, Cambridge Analytica, data breaches, privacy violations, untimely employee departures, and Mark Zuckerberg allegedly serving Twitter CEO Jack Dorsey an undercooked goat, the Facebook narrative was drastically altered.

But here's what I found: the troubled post-election version of

Facebook was by no means a different company from the one it was before, but instead very much a continuation of what started in Mark Zuckerberg's dorm room fifteen years earlier. It is a company that both benefits from and struggles with the legacy of its origin, its hunger for growth, and its idealistic and terrifying mission. Its audaciousness—and that of its leader—led it to be so successful. And that same audaciousness came with a punishing price.

Virtually every problem that Facebook confronted during its post-election woes had been a consequence of two things: the unprecedented nature of the mission to connect the world, and the consequences of its reckless haste to do so. The troubles that plagued Facebook in the past three years were almost all rooted in decisions it had made in its earlier years, mostly between 2006 and 2012, when key choices were made that favored moving with lightning speed to connect the world, with implicit intent to repair any damage at a later time. Facebook now admits that the damage turned out far more extensive than expected, and is not easily repaired. All the while Mark Zuckerberg and his team insist, despite the scandals, that Facebook is still overwhelmingly a force for good in the world.

In a sense, the story of Facebook tracks with the larger story of how digital technology has changed our lives over the past few decades. Not only Facebook but all the tech giants transforming our daily lives have come under intense and skeptical scrutiny. Those great tech companies were very much based on the idealism of their founders, but now are viewed as part of a Faustian bargain: the wonders they deliver have come at a cost, to our attention, our privacy, our comity. And now we fear their power.

None so much as Facebook, which followed its leader's dictate to Move Fast and Break Things . . . and broke them. By my last interviews with him, he was explaining how he'd fix things.

This is Facebook's story. It begins, of course, with Mark Zuckerberg.

PART ONE

1

ZuckNet

ON A CHILLY night in January 1997, a twenty-eight-year-old lawyer recently turned entrepreneur named Andrew Weinreich addressed a small crowd of investors, journalists, and friends at the Puck Building in New York City's SoHo district and tried to explain what online social networking was, why the product he was announcing was the first example, and how the concept would change the world. It was a heavy lift.

Weinreich had come up with the concept as his contribution to a weekly meeting of would-be start-up founders who got together soon after the first wave of Internet companies like Yahoo!, Amazon, and eBay appeared. They would try to identify business ideas that were possible for the first time ever because of the net. Weinreich came up with an idea based around the concept of people volunteering information about their interests, their jobs, and their connections. He asked himself: *What if I could get everyone to index their relationships in a single place?*

He called his company sixdegrees, based on a concept that everyone on the planet was only six connections away from anyone else. Weinreich thought it was something Guglielmo Marconi had first stated, but actually it was a Hungarian writer named Frigyes Karinthy. In a short story called "Chain-Links," the writer assessed this huge shift.

> *Planet Earth has never been as tiny as it is now. It shrunk—relatively speaking of course—due to the quickening pulse of both physical and*

verbal communication. This topic has come up before, but we had never framed it quite this way. We never talked about the fact that anyone on Earth, at my or anyone's will, can now learn in just a few minutes what I think or do, and what I want or what I would like to do.

Hard to believe he wrote this in 1929! Karinthy's characters in this short piece tried an experiment—to see if a chain of connections could connect them to any random human among the (then) world population of 1.5 billion with only five personal introductions, beginning with one's personal network of friends and then proceeding to the next person's introduction. In the story, one of the subjects—a Hungarian intellectual like the author—met the challenge of making the connection to a random riveter at the Ford Motor Company. Karinthy's concept kicked around the social-science world for some decades until some researchers in the 1960s and '70s tried to prove it with the limited computer power of their time. In 1967, sociologist Stanley Milgram published a *Psychology Today* article on what was then called the "small world problem." In a study published two years later, he and his coauthor tried to connect random people in Nebraska with those in Boston and found that "the mean number of intermediaries between starters and targets is 5.2." In 1990, the concept would gain wide cultural currency when playwright John Guare used it to illuminate his eponymous play, *Six Degrees of Separation*, adapted for film in 1993.

Weinreich's implementation, though inspired by the Six Degrees theory, actually concentrated on two or three degrees of separation. "More often than not, I can meet the people I don't know through those I do know," he told the crowd at the Puck Building. For centuries, people have been using their friends and acquaintances to make such connections, but it had always been hit or miss. "Today we hope to change that," he promised, "with a free, web-based networking service." He compared it to putting your Rolodex online—and connecting to everyone else's

Rolodex. "If everyone uploads their Rolodex, you should be able to traverse the world," he gushed.

On that cold night in January, Weinreich expressed a mission that was astounding to consider: connecting the world in a single network. "Imagine for a moment that we had not just you in the database but every Internet user in the world," he asked his audience. (Of course there were only, he guessed, 40 to 60 million Internet users at that point.)

Weinreich assumed, as a matter of course, that connecting the world would be a boon to humanity. Why would it be otherwise?

Sixdegrees pioneered several tropes that would be part of virtually all social-networking sites. It included the viral-before-there-was-viral plan to use email invites to build the network. At the launch event, Weinreich even supplied printed-out invites inside envelopes to the attendees that were duplicates of the ones hitting their mailboxes. Then he urged them to open their browsers on computers in an adjoining room and begin submitting the emails of their friends and connections to sixdegrees. When those people got their invites they would be asked to confirm that they indeed knew those who suggested them. It was the first time that an online service used such verifications.

Sixdegrees was something new and, had it succeeded, would have been the nexus of endless studies and assessments. But it did not succeed. Weinreich's great idea was too early. At the time, most people didn't have email, let alone persistent web connections. And sixdegrees didn't let you do much besides enter your connections into the giant database. There was no temptation to relieve your boredom on sixdegrees. No way to stalk an ex-lover. No way to watch a silly cat video. You would query the database of your extended social network when you wanted a connection or recommendation. And leave.

Those who did sign up to sixdegrees quickly noted how much better the service would be if you could see pictures of people. In 1997, that was a huge hurdle because very few people had digital cameras. Weinreich even considered hiring hundreds of interns or low-paid employees to sit in a big room and do nothing but scan photographs. But he decided against it, because by then he was already considering selling the company.

While sixdegrees proved the concept of social networking—it peaked at about 3.5 million users, which was impressive at that stage of the Internet—the state of technology was still a couple of years away from nurturing the kind of connectivity that a social network would need to really thrive. Weinreich dreaded having to raise the money to wait it out. In December 1999—just in time to avoid the huge dot-com crash that would soon hit the industry—Weinreich sold sixdegrees to a company called YouthStream Media Networks for $125 million. Included in the purchase price was the pending patent, "Method and apparatus for constructing a networking database and system," which became known as the "social networking patent."

Weinreich later would say that by selling early, he never did get to implement two things he had planned all along for sixdegrees. One was allowing for users to post comments and media on the site, sneaking into the territory of other early Internet outposts for what would be called user-generated content. The other was making sixdegrees into an operating system, or platform, where third parties could create applications that would run on top of what Weinreich had dreamed would be a social network that encompassed the globe.

What Weinreich did not know was the person who would build—and surpass—his vision was only twenty-five miles from the Puck Building. And he was twelve years old.

MARK ELLIOT ZUCKERBERG was born to Karen and Ed Zuckerberg in 1984. The day was May 14, almost four months after the launch of the Apple Macintosh, which aspired to push into common use what was still seen as a device for trained experts and batty hobbyists. Not many people had personal computers then, and fewer still had modems, the noisy peripherals that connected PCs to telephone lines. The precursor to the

Internet, ARPAnet, was around, but limited to government and some computer-science students.

Ed Zuckerberg had both a computer and a modem. He had a lifelong affinity for technology in general and gadgetry in particular. When he was himself a child, his favorite subject was math.

Considering this, one might justifiably wonder whether Mark Zuckerberg's later ascension to the status of global tech idol might be a case of the son living out the father's thwarted ambitions. Ed never said as much, but he did not object when a *New York* magazine reporter writing about the family in 2012 floated the theory. "Growing up Jewish in New York City," Ed said, "if you had half a brain, your parents wanted you to be a doctor or a dentist . . . But back then, there really weren't a lot of jobs in computer programming . . . That was not the 'appropriate use of my time,' my parents would have said. It wasn't for the smart boys."

If not for the pressure it would have been different. "I would have done something in math, left to my own devices," he says now. "Absolutely. I loved math."

The Zuckerbergs lived in Dobbs Ferry, New York, twenty-five miles north of the big city. Both had grown up in working-class neighborhoods in the outer boroughs of that city. Their own parents were first-generation Americans. In 1977, while studying dentistry at NYU, Ed had gone on a blind date with a Brooklyn College coed, Karen Kempner, who hailed from Queens. He was twenty-four, she nineteen. Both had grandparents who emigrated from eastern Europe, and both were diligently studying to accomplish what was the career gold standard in each of their families: becoming a professional like a doctor or lawyer. Especially a doctor. (Ed's father was a mailman; Karen's father a precinct captain in "The 79," in Brooklyn's tough Bed-Stuy neighborhood. Her mother taught school.) Ed and Karen married in 1979, and after a couple of years living in a White Plains apartment, they moved to the Dobbs Ferry house. Among Westchester County suburbs, the town was known to be less wealthy (and less snooty) than other nearby bedroom communities, but Ed says it was simply the best house for their purposes—a sprawling, multilevel house atop a high knoll, a javelin's toss away from the busy Saw Mill River Parkway, laid out so it could accommodate a home and a dental office. "It was the only one we could afford," notes Karen. In the early '80s, Ed moved his dental practice to the ground floor, with the Zuckerberg clan basically living above the shop.

Ed brought his high-spirited personality to his work. Karen was a psychiatrist who delayed a clinical career to raise Mark and his three sisters while helping her husband run the dental practice. (Mark was the second oldest, born two years after Randi; Donna and Arielle would follow.) "My wife was a superwoman," Ed Zuckerberg said in a 2010 interview on a local radio show. "She managed to work and be home." Like many Jewish parents who had gratefully moved up a rung on the ladder to the good life, the Zuckerbergs aspired to an even higher rung for their kids, and fiercely emphasized education. (Zuckerberg once joked about it: "Good Jewish mother . . . You know, go home; get 99 percent on the test, *Why didn't you get 100?*") At one point, Karen did practice in a nearby hospital—a choice made possible because the family always had a foreign au pair helping out—but was discouraged by the failure of medical insurance to cover her patients' fees. Ed also once remarked that she thought her presence in the home might prevent her children from themselves landing on the psychiatry couch. On a Bermuda vacation, she and Ed decided she should give up the job. Her clinical skills wound up being utilized to calm nervous dental patients. Perhaps as a result of being pressured into a profession she did not formally practice, Karen Zuckerberg felt firmly that her own children should be free to pursue their passions. "You spend a lot of years working—you have to love what you do," she says. "So we always felt it was up to our children to figure that out for themselves."

Ed Zuckerberg's geeky side presented itself in a constant pursuit of exotic new dental technology. When a magazine writer visited in 2012, Ed went on at length about a $125,000 root-canal machine he'd just bought. Zuckerberg's pitch to his client was that his state-of-the-art equipment, along with a menschy compassion for patients, would make the trip of going to the dentist's office a more pleasant experience than, say, going to the dentist's office. "I was the first dentist in Westchester County who had digital X-rays, intra-oral cameras . . . all that tech stuff really got me going," he says. He billed himself as "painless Dr. Z," and his website (of course he had an early website) boasted that he "caters to cowards."

Ed bought his first personal computer in the early 1980s—an Atari 800, a "consumer" machine that was great for games but required patience, skill, and a bit of insane optimism if you wanted to actually do something useful on it. But he taught himself Atari BASIC and kept a patient database. Before Mark was born, he'd upgraded to an IBM PC, which he used to run the practice.

So it wasn't surprising to Ed Zuckerberg that his son would take to computers. From an early age, Mark had a mind attuned to logic, especially when the answer to one of his requests was no. "If you were going to say *no* to him, you had better be prepared with a strong argument backed by facts, experiences, logic, reasons," Ed Zuckerberg once told a reporter. Mark, he said, was "strong-willed and relentless," a description that many coworkers and rivals would certainly endorse.

As a tyke, Mark played with Ed's old Atari, which was a great game machine. In sixth grade, he got his own computer. "It was a Quantex 486DX," he recalled in a 2009 interview with me, and was surprised when I didn't recognize the brand name of that IBM PC clone. "I don't think it exists anymore," he explained, taking me off the hook. "But my family didn't have a lot of money, so I was lucky just to get a computer."

From the beginning, Zuckerberg used the computer to indulge a curiosity about the way people organized themselves—and how some people gained power in the process. He seems to have had this obsession since toddlerhood. "When I was a kid I had Ninja Turtles, and they would just have wars and stuff like that," he says. "What I used to do with my Ninja Turtles was create societies, and just, like, kind of model out how they'd interact with each other and things like that. I was just very interested in how systems worked like that."

So when Zuckerberg played games on computers, they indulged his world-building imagination. One of his favorites was called Civilization, a popular series in the genre of "turn-based strategy games." The idea was to build a society. He kept playing it even into adulthood.

After a few months on the computer he told himself, *All right, this is interesting—I've learned all about it, and now I want to control it.* "So I learned programming," he says. One night he demanded that his parents take him to Barnes & Noble to purchase a guide to writing C++, a key computer language for creating web applications. "He's *ten*!" recalls Ed Zuckerberg. When the acolyte coder discovered that a book explicitly targeted to "dummies" lacked key information, Dr. Z hired a tutor. For two years the tutor would visit once a week. "It was his favorite hour of the week," says his mother. The Zuckerbergs explored enrolling him in an AP computer class at the high school, but the teacher told them Mark already knew everything he'd learn in the class. The local college offered courses, but the only one Mark considered worthwhile was in the graduate department. So one night Ed Zuckerberg took Mark to the college. The teacher told Ed he had to leave his son at home during class. "He's the student!" said Ed Zuckerberg, who tells the story with pride decades later.

As Mark later told an interviewer, "I'd go to school and I'd go to class and come home. The way I'd think about it was, 'I have five whole hours to just sit and play on my computer and write software.' And then Friday afternoon would come along and it would be like, okay, now I have two whole days to sit and write software. This is amazing."

Later he would remark that from all this programming, "it reached a point where it went into my intuition. I wasn't really thinking that much about it consciously."

Zuckerberg didn't spend all his time in a bedroom lit only by a computer monitor. Teachers would later describe him as well-adjusted; though not much of a talker, when he did speak he expressed his firm opinions articulately. He was strong in math and science. He was smaller than the other kids. He played on his neighborhood Little League baseball team but didn't like it. He would later use his begrudging participation on the ballfield as an illustration of something that the company he founded might one day mitigate. "I'm not into baseball, I'm into computers," he said, suggesting that social networks would help people with disparate interests to find their tribes, as opposed to having to endure right field because it was the default activity.

Zuckerberg was much more simpatico with fencing—an individual sport that all the Zuckerberg kids would practice. The Zuckerberg family were also *Stars Wars* obsessives, and swords had the appeal of being like lightsabers. His bar mitzvah was *Star Wars*–themed. (Pre-Instagram, no photos were publicly distributed.) He and his sisters made a home movie based on *Star Wars*.

His mother called him "princely."

Though he played a lot of games, Mark wasn't satisfied to be bound by whatever rules the game creator set for the players. Being the creator was a lot better. "I wasn't into playing games—I just liked making them," he told me, skipping over the fact that he played them all the time, with a cutthroat competitiveness. One of the first games he built was a version of his favorite board game, Risk, where players attempt to conquer the world country by country, by accumulating power to make their incursions unstoppable. Zuckerberg's digital version was set in the age of the Roman Empire. You'd try to beat Julius Caesar. Zuckerberg always won.

He'd later admit his creations were terrible by any reasonable measure. But they were *his*.

"Everything was tech," his sister Randi once told a reporter about the Zuckerberg household. "We had these toys with voice changers. Mark was always thinking like, *We could get Darth Vader's voice to sound more Vaderlike if I could hack this toy*."

A more practical technology was an Internet-based intercom system that ran through the Dobbs Ferry house and allowed the dental staff to communicate with one another and the family from the downstairs office. This was dubbed "ZuckNet." Ed Zuckerberg had already hired a professional to wire the house for the T1 line, so Mark offered to write software to link the machines together. Once installed, ZuckNet proved useful not only to signal the arrival of Dr. Z's cowards but for Mark and sometimes Randi to pull an endless series of pranks, like planting a fake virus on his sister Donna's computer or tricking his mother into thinking that the Y2K bug had triggered a tech apocalypse.

In 1997, a networking product did for the young people worldwide what ZuckNet had done in the Zuckerberg house a year earlier. AOL's Instant Messenger product, or AIM, would become the software that most engaged Mark Zuckerberg in the early years of his technology life.

Zuckerberg's generation—his birth year puts him in the advance guard of the Millennials—was too late for *Bye Bye Birdie*–esque princess-phone conversations and too early for texting. But they did have computers attached by modems and, increasingly, higher-bandwidth Internet. And they had AIM, a stand-alone application with a virtual monopoly on computer chat. It was common for a kid's computer screen to have multiple chat windows open, each one an asynchronous conversation with a friend. Zuckerberg loved AIM. Because most of his high school friends lived on the other side of the busy Saw Mill River Parkway, a barrier that discouraged spontaneous visits, Zuckerberg relied on it even more than his peers.

Naturally, Zuckerberg messed with AIM's system. "If you actually talk to a lot of people who are my age, a lot of us grew up learning how to program by hacking on AOL," he says. One of the "cool things" he describes is using the Internet programming language HTML to automatically add design elements, like different color schemes, to the multiple chat boxes that populated his screen at all times. Another cool thing was hacking the program in a way that would have spurred agita in AOL chief Steve Case, had he known.

"There were all these holes in AOL where you could manipulate the service," Zuckerberg says. "Like, I could kick my friends offline because of bugs in the system."

When Zuckerberg would later build his company, the majority of those he hired were people like him, '80s kids who had spent the last years of the twentieth century submersed in the chat bubbles on their screens. "We all grew up on AIM," says Dave Morin, who would later be a key Facebook executive. "I have this whole theory that we're all not as competent at intimate communication, particularly in marriage and things like that, because we grew up on it. We didn't learn the nuance of intimate communication in person."

MARK ZUCKERBERG'S TEACHERS recognized his intelligence—and his intensity. It was clear even in nursery school, where classes would do weeklong projects on various subjects. At one point, his parents noticed that one unit, on space, had been going on longer than usual. When Ed and Karen asked about it, the teacher told them that Mark was so focused on the topic, and he'd gotten the other kids so involved, that they decided to extend the space unit to a month. After the month, Mark's space obsession continued, and the giant cardboard rocket ship the class painted wound up on his bedroom ceiling.

His parents refused numerous offers to have him skip a grade or two—he was a small kid as it was. In middle school, he had an arrangement with his teachers that after he learned the week's lessons—usually on Mondays when they were presented, he could do the work from other classes while the teachers drilled the other students. "I never saw him doing homework," says Ed Zuckerberg.

After two years at the public high school in Ardsley, which was a few miles across the Saw Mill from the Zuckerberg home, Mark clearly felt that he needed a change. He'd calculated the points he would get from taking the courses he wanted, and the mix of honors and AP courses offered at Ardsley fell short of the total he'd need to get into the top schools. And there was another reason. "Our public school didn't have any computer-science courses," he says. His parents thought that Horace Mann, an easy commute, would be best, but Mark had heard about Exeter from friends at a summer program for talented youth. Karen Zuckerberg was already sad that her oldest daughter was leaving for college and she didn't want to send off her son, too. She asked him to interview at another private school. Mark said, "I'll do it but I'm going to Phillips Exeter." As often happened, the strong-willed teenager got his way.

Phillips Exeter Academy, in Exeter, New Hampshire, was one of a cohort of haughty prep schools known as the Ten Schools Admission Organization. Modeled on their big brothers in the Ivy League, they were, as the organizational name implies, reliable feeder schools to elite colleges. Zuckerberg enrolled as an "upper" (the Exonian vernacular for juniors) in the class of 2002.

Before the school year began, Exeter held a reception in New York City for incoming students. Zuckerberg found himself chatting with another rising junior, a gangly kid with a similarly low-key demeanor whose name was Adam D'Angelo. Like Zuckerberg, D'Angelo was a suburbanite (hailing from a bedroom community in Connecticut) transferring to the tony boarding school after topping out at his public high school. They had something else in common. When Zuckerberg asked D'Angelo what he was interested in, the answer was one golden word: *programming*. Zuckerberg was thrilled—none of his public high school friends shared his passion for building things on the computer and now the first person he met at Exeter was a lot like him. "By induction [I figured] there were going to be a lot of other people here who were interested in the stuff," Zuckerberg says. "It turned out that we were actually the only two."

If Zuckerberg was intimidated by attending a private school whose students included the very wealthy—it wasn't unusual to be in a class with a Rockefeller, a Forbes, *and* a Firestone—he didn't show it. He seemed to flower at Exeter. He joined the fencing team and proved an energetic competitor, captaining the squad and winning the MVP award. He joined the team that was sent to the Math Olympiad, and though he couldn't compete at the top level, he won a secondary medal.

Outside of his own circles, he kept to himself. "I think he probably trusts very few people," says Ross Miller, who was one of his best friends at Exeter. Classes were conducted in a seminar-style participatory fashion known as the Harkness method. The school describes the method as "... a way of life ... It's about collaboration and respect, where every voice carries equal weight, even if you don't agree." Classmates recall that Zuckerberg seldom contributed to the discussions. "He was quite shy and kept to himself, usually doing work and writing code in his room," a classmate named Alex Demas later told an American Greek news website. His reputation, says Demas, was as a computer nerd. (Zuckerberg would nonetheless later comment that he admired the Harkness method: "It probably shaped my philosophy that people should be participants and not consumers.")

Thanks to a charismatic teacher at Ardsley, he had already developed a passion for the classics, and ate up Exeter's Latin program. In particular, he had a fanboy affinity with the emperor Caesar Augustus, whose legacy is a mixed one: a brilliant conqueror and empathetic ruler who also had an unseemly lust for power. The summer before his senior year, he attended a session of the Johns Hopkins program for "gifted youth" and chose a course in ancient Greek; the students worked their way through the grammar and finished by studying a speech of the Attic orator Lysias. One of his instructors, David Petrain, recalls Zuckerberg as "affable and game" and adept at memorizing forms. Zuckerberg once mentioned to Petrain that he'd started a website devoted to the love poet Catullus, but Petrain never saw it. (Petrain would later write a mildly positive recommendation for Zuckerberg's common application to colleges.)

In his senior year, he was named a dorm leader, which meant he got a bigger room. He brought up a large dental monitor recycled from his dad's office and used it as a display for Nintendo games. But his favorite game was a recent variation on Civilization, by the same creator, Sid Meier. It was a space scenario called Alpha Centauri, in which players chose one of seven different "human factions" to lead, in a complicated strategy to control the galaxy. Zuckerberg always took the role of the quasi-UN "Peacekeeping Forces." In the intricate backstory supplied with the game, the spiritual leader of the peacekeepers was a commissioner named Pravin Lal, who opined that "the free flow of information is the only safeguard against tyranny." Zuckerberg would later use a Lal quote as the signature on his Facebook profile:

> *Beware of he who would deny you access to information, for in his heart he dreams himself your master.*

Every fourth-year Exeter student took a course in Virgil's *Aeneid*, and later in life Zuckerberg would cite some key lines, using them to inspire his Facebook workforce. Recounting the plot to a reporter in 2010, he would note the resonance of Aeneas striving to build a city that "knows no boundaries in time and greatness."

Somewhere in that kid's head it all seemed to be simmering into a stew: Conquerors. Swashbuckling. Civilization. Risk. Coding. Empire-building. The recipe for Mark Zuckerberg.

ZUCKERBERG AND D'ANGELO weren't, as he later joked, the only two computer junkies at Exeter: he was part of a small cohort whose passions also lured them into spending long hours at Exeter's computer center, a recently constructed facility with state-of-the-art equipment. One was a math whiz named Tiankai Liu, who won gold in the Olympiad. Another was a fearless kid named Marty Gottesfeld, who years later would wind up in federal prison for hacking Boston Children's Hospital (he says he did it to help a fifteen-year-old patient who was being mistreated). When among his computer buddies, Zuckerberg would prance like the king of the roost.

A recent Stanford computer-science grad named Todd Perry was a teaching fellow that year, and he'd taken on extra duties since one of the regular CS instructors left early in the fall semester. He recalls Zuckerberg strolling into the computer center one evening as if he owned the place, and announced that he was about to code a certain project using Microsoft's Visual Basic. Perry felt that the task was overly complicated for someone at Zuckerberg's level—it involved techniques that Perry hadn't encountered until well into his Stanford studies—and bet a buck that Zuckerberg couldn't do it. They agreed to give Zuckerberg an hour to try to pull it off. All the nerds surrounded Zuckerberg as he coded, as if it were a gladiatorial contest. Zuckerberg collected his dollar.

On another occasion, Zuckerberg had a math teacher who promised that if his students did homework with calculators or other digital shortcuts, he'd make them do push-ups. There was no way Zuckerberg was not going to use computers to do his work, he told his friends in the computer center. Zuckerberg didn't even bother to mask his disdain for the teacher's threat—he conspicuously wrote code to do the homework and executed the push-ups as if they were a victory lap.

Exeter students are required to create a senior project before graduation, and Zuckerberg was casting around for one, listening to tunes on his computer, when the playlist he had set up went silent after the final song played. *There's really no reason why my computer shouldn't just know what I want to hear next,* he told himself. He recruited D'Angelo to partner with him in creating what would be their senior project, a personalized virtual DJ they called Synapse.

Both were big fans of an online music player called WinAmp, and they decided Synapse (which was sometimes called Synapse-ai) would ape WinAmp's functions while providing a personalized playlist. Though both Zuckerberg and D'Angelo were utter novices in artificial intelligence, they boasted about the AI in Synapse-ai, even calling the code that determined the playlist "the brain." You could either use the discrete music player they ginned up or use a plug-in they provided to AOL's WinAmp player, and Synapse would suggest songs to you based on what you had listened to before. D'Angelo, the more accomplished programmer, focused on building the brain, while Zuckerberg created the front end. "It would play songs for you based on what it knew you liked in a sequence that made sense, then we could compare different users' logs and cross-recommend stuff," Zuckerberg says. "It was cool." The pair presented Synapse as their senior project, to kudos from their instructors, who were especially impressed with D'Angelo's AI component.

But of all of the computer capers that occurred during Zuckerberg's time at Exeter, the one that proved most relevant to his future exploits would be someone else's project, with minimal participation from Zuckerberg.

It was called Facebook.

ITS CREATOR WAS a senior named Kris Tillery. Born in the Midwest, Tillery lived in West Africa and Nigeria; his parents wanted him to attend school in the United States, and so he boarded at Exeter. By his own admission, he was no computer adept, certainly nowhere as talented as D'Angelo and Zuckerberg, whose reputation was well-known throughout the academy. While Tillery was struggling with the AP course in computer science, he later recalled, he marveled at the duo who had programmed a music player with artificial intelligence!

Still, Tillery had a good vision of what the technology of the day could do. At one point he came up with a prescient idea for the turn of the century: an online grocery delivery system. The product required an automated means of getting the prices from the local store. "That was above my pay grade," says Tillery, admitting he wasn't a good enough programmer to pull that off. But he knew someone who was. "Zuckerberg built a script that would scrape prices off of the supermarket site so we could then do our grocery delivery setup," he recalls. However, the grocery service never took off.

Tillery's real legacy as an Exonian came from exporting a binder of student headshots and captions known as the Photo Address Book to the malleable and infinitely accessible digital realm. The project came about when Tillery was still a "lower"—a sophomore—who was trying to teach himself about databases. He used the student facebook in his explorations. The head of the student council lived across the hall and suggested he actually finish the project and distribute it. Tillery did so, but not without running afoul of the ever-intolerant Exeter IT Department. Using the school's servers to distribute information was verboten. Yet the administration recognized the utility of what Tillery had done. They eventually gave him permission to continue.

Thus was the Exeter Facebook sanctioned, and Tillery released it to the school's entire population, which included Mark Zuckerberg. It was devilishly useful: you could look up someone by name, of course, but users also had the ability to search other things. Phone numbers were included—every student had a landline in the dorm—and Exonians devised a game where the facebook would choose a random person, whom they would prank-call.

Tillery stopped his involvement with the facebook program after graduating from Exeter. His next stop was Harvard University. So he was present at the school in February 2004, when an online facebook suddenly appeared and swept through the school like a tornado. He wasn't surprised to see that it was created by Mark Zuckerberg. Even in his limited contact with Zuckerberg at Exeter, Tillery noticed that the intense young man had "big, big ambition." Nor was he bothered by what was arguably an appropriation of his idea. In his view, the online facebook was something he'd worked on in prep school, and he was done with it. More power to Mark.

Tillery, who now owns a vineyard in South Africa, has mixed feelings about first exposing Mark Zuckerberg to the concept of an online facebook. He is happy to have played a small part in a global phenomenon. But more recently he began to question whether that phenomenon has proved to be a good thing.

"Add up all the hours a day that all the people spend on it, and you're looking at a big number that may not be contributing positively to society's benefit or to our own personal health," he says. "The moral ambiguity of the platform—which is today the revenue based on advertising and targeting—raises big questions about how we should spend our time for our own happiness."

As for his own use, Kris Tillery deleted Facebook—the product he believes was inspired by his germ of an idea—somewhere around 2016. Zuckerberg's Facebook was, he said, making him feel bad.

2

Ad-Boarded

IN MAY 2017, Mark Zuckerberg invited me to Facebook's headquarters. Whenever Zuckerberg is working on some big speech, or an essay dictating a major change in direction for his company, he typically bounces it off a variety of people, even journalists. In this case, he was working on what he considered a personal landmark: he had been asked to give the commencement speech to Harvard's class of 2018. Sitting in the glass-walled "Aquarium" in the center of the cavernous Frank Gehry–designed Building 21 headquarters—the Facebookers who constantly buzz past us have been trained to keep eyes forward and not stare at their famous boss while he takes meetings—he outlined his speech, which he crafted after viewing a stack of commencement addresses from other business leaders. Like those august speakers, he would address weighty topics. But the exercise led him to reflect on his own time in college, leading him to venture down unfamiliar corridors of nostalgia.

"I think I'm going to be the youngest Harvard commencement speaker," he told me. His matter-of-fact tone and disquieting stare made it seem more like a data point than a boast. "It's really rare. They've been doing this for a long time, like three hundred and fifty years or something."

"What emotions are you feeling?" I asked him.

As Zuckerberg sometimes does when he needs time to process a question, he reverts to some points he had already planned to make, and circles back later. After talking for a while about issues he'd address in his talk, he paused. "I'm thinking about your question about emotions," he said. He explained that while he would be addressing weighty subjects like inequality and the lack of social cohesion in his speech, there would be personal stories as well, meaningful to him. "The emotional arc of my life," he says. "Like going from this kind of . . . I don't know what the adjective is . . . Harvard student . . ."

"What's the adjective?" I asked him, urging him to fill in the blank. I wanted to know how he viewed his younger self at Harvard, where his brief time there would be fabled and scorned.

"I don't know," he said. "I had a word, then I lost it." Pause. "Like, 'irreverent' is probably the right word."

I remarked that he was being kind to himself.

He sighed, conceding the point. "'Punk'?" he finally said.

We laugh.

The smile fades. "Do you think the punk is still there?" he asks.

"PUNK" DOESN'T BEGIN to capture Mark Zuckerberg's Harvard odyssey. But it's a start. "I am not even sure it was Mark's dream to go to Harvard," Zuckerberg's older sister, Randi, once told a CNN interviewer. Indeed, Harvard University, for all its renown, is not an obvious choice for an ambitious, computer-obsessed kid, even one who gets accepted. Consensus holds that the best schools for people like Zuckerberg are Stanford or MIT, maybe Carnegie Mellon.

But Zuckerberg had his sights set on Harvard for years. In his Exeter dorm room the only adornment on his walls was a giant banner with the school's name.

And he wasn't planning to major in computer science. He was thinking a nontechnical subject like psychology or classics. Or perhaps a science like physics. Also, Randi was already an undergraduate there. In what his parents describe as typical behavior, he didn't bother to consider an array of possible options, just an early admissions application to Harvard. If he'd been turned down, it would have been a mad scramble to apply to other schools.

His formal acceptance has been captured on video and later, via Facebook, shared with the world. The moment is strangely awkward. Home for the holidays, Zuckerberg had been in his bedroom in front of his computer—playing Civilization, of course—when he got a notification of an email from Harvard. Mark summoned his father, who promptly sprinted to the bedroom, camcorder in hand. The video recording begins with him sitting on the edge of his bunk bed in a T-shirt and flannel pajama bottoms, staring at the inbox. "Should I open it?" he asks rhetorically. He reads it in silence. "Man," he says quietly. Then, in a dead monotone: "Yay."

"What?" asks an anxious Ed Zuckerberg, unseen behind his video recorder.

"I got accepted," says Zuckerberg the younger, in a low, exuberance-free voice, with just a hint of satisfaction.

"Are you serious?"

"Yeah."

"ALL RIGHT!" screams Ed, and immediately begins narrating the moment as if he is a sportscaster, and this boy in pajamas has pulled off an astonishing athletic feat. "We're here with one of the newest members of Harvard's Class of 2006!" Mark gives a quick smile and a fist pump, but then sinks back into a computer stupor. Back to Civilization. "Don't you want to read us the email?" Ed asks.

"No, I killed it," said the newest member of the Class of 2006.

Later Ed Zuckerberg said, "I think he was happy not so much for getting into Harvard but it meant that he didn't have to waste time making other applications."

ZUCKERBERG ARRIVED AT Harvard with no intention to curb his passion for pursuing computer projects. During his very first month there, September 2002, he did a soft launch of the DJ program he and D'Angelo had created. The website was called Synapse-ai, the lowercase "ai" emphasizing the rudimentary artificial intelligence that chose the next song in the user's playlist. Zuckerberg would spend a lot of time refining Synapse in his freshman year.

He appears not to have had any problems making the transition from Exeter to Harvard. His social life revolved around Alpha Epsilon Pi, a Jewish fraternity. (He'd gotten in as a freshman in part because his older sister, Randi, was dating one of the frat brothers.) He was known as a friendly, though aloof kid who was fine with hanging out but clearly devoted to his computer.

Harvard classmate Meagan Marks, who would later work at Facebook, recalls being in a seminar-style course with Zuckerberg with only twelve students. It was about graph theory. Zuckerberg came off as exceptionally reserved. But when he did speak, he was impressive and sometimes even brilliant, as he would suggest unorthodox but winning solutions to math problems. "He would heavily disagree on something if he felt strongly about it," Marks says. "He was never afraid of being the iconoclast." Still, when she organized a dinner party for the class and had seats for only eight of the twelve students, Zuckerberg wasn't seen as someone who would be a social, fun person at the gathering, so she didn't invite him.

He made pocket money from taking contract jobs for computer work. He also took on some freelance programming jobs, like the $1,000 gig he found on Craigslist, coding a website for a Buffalo businessman named Paul Ceglia. Ceglia would later claim he owned half of Facebook, with documents that supposedly proved that Zuckerberg had agreed to this before starting the site. The courts threw out the case and Ceglia was prosecuted for forgery.

This was indicative of the weirdness of Mark Zuckerberg's residence at Harvard. From a legal standpoint, it would turn out to be a bonanza for members of multiple bar associations.

He seems to have pinned a lot of hopes on Synapse, which he no longer saw as just a class project but something that might catch on in the outside world. His partner D'Angelo would have been fine leaving it as a class project, preferring to concentrate on his studies at the college he had chosen, the California Institute of Technology. "Caltech is, like, hard—you have to do work," says D'Angelo. "Harvard, honest, it's not that much work. So I think he had a lot more time."

Synapse was slow to take off, despite Zuckerberg's attempts to promote it, including getting Synapse-ai T-shirts made that said [[my brain is better than yours]]. (The brackets were a flourish invoking programming protocol.) It wasn't until spring that things started happening with Synapse.

On April, 21, 2003, Slashdot, the premier source of news for the geek world, ran an item about an "interesting approach to digital music by students at Caltech and Harvard." It invited the millions of people in the Slashdot community to try it out, and, as was common with the site, a spirited online conversation ensued.

One of the discussion threads dealt with the program's retention of the user's music preferences. Some flagged it as a privacy violation. "I may be paranoid," one commenter wrote, "but I prefer not to have anyone, even my own computer, perform data mining on me. That's what this is, really. Personalized data mining."

On April 23, Zuckerberg himself barreled into the thread, clarifying how the program worked and touting some upcoming changes. And then he added this:

> *And a note about privacy. None of your musical listening data will be available to anyone other than you. We hope to use massive amounts of data to aid in analysis, but your individual data will never be seen by anyone else.*

IT WAS ZUCKERBERG'S first public acknowledgment of the importance of privacy in his work. Certainly not the last.

The Slashdot geeks noted another oddity involving Synapse: the callow and creepy language in the program description Zuckerberg had written. There was a rambling paragraph about all the people who would love Synapse (*"Programmers. Gangsters. Punks. Nerds. Really big nerds. Even ones from Yemen. Yeah, plenty of those . . . People who exercise to Rocky music . . . Revolutionaries. Even Canadians. Quality people. Gastroenterologists. Bums. Lots of bums. Evil geniuses. Classics professors . . ."*); a cringe-worthy *Playboy*-style passage about how the program could help in romancing "a Chinese girl"; and also some boasting about Zuckerberg's own computer prowess (*"Mark's mouse, alone, has moved enough to go around the world . . . twice."*)

D'Angelo was appalled when he saw this, and demanded Zuckerberg take it down. Of course, the Internet never forgets.

Overall, though, the Slashdot attention was a boon. Zuckerberg heard from multiple companies interested in the student project, including Microsoft and AOL. Zuckerberg and D'Angelo got an offer approaching a million dollars from one of those suitors. But the payout would be contingent on Zuckerberg and D'Angelo committing to work for that company for three years. They turned it down.

Neither was willing to leave school—at least, not for that offer. They both moved on from Synapse. "We knew that we could do something better," says Zuckerberg.

IN THE SUMMER of 2003, after completing his freshman year, Zuckerberg remained in Cambridge, interning at the David Rockefeller Center for Latin American Studies as a program analyst. He lived among a group of friends, including D'Angelo, housed in a dorm-like situation in town.

D'Angelo was interning at the MIT Media Lab, working under Professor Judith Donath, who studied social networks. It was a timely subject, because that summer the darling of the Internet was a service called Friendster, the flagship of a phenomenon dubbed social media.

"Mark thought it was interesting that I was so excited about Friendster," says D'Angelo. "He wasn't into it as a user, but it was clear to him that there was something there."

Friendster was started by a Canadian, Jonathan Abrams, who had immigrated to California in the late 1990s to work at Netscape, then the flagship start-up of the Internet revolution. He started a company of his own during the first dot-com boom in the '90s. It collapsed in the subsequent dot-com bust, but even before the sector recovered, Abrams was ready for another try. As a newcomer to California he understood he was starting from scratch in both his professional and dating life, and he'd decided to consciously map his connections as he built a new network of business contacts, potential friends, and prospective dates. What if one could do that online? In the summer of 2002, working out of his apartment, he programmed a site that would help you build and expand your network by making "friends" with people on the site. At first it was limited to people he knew and their friends, and it was focused on dating. People loved it. The instant success surprised even Abrams, who, until he examined the suddenly voluminous activity logs, had been skeptical of his own idea. "People were uploading photos, they were sending messages," he would later say on a postmortem podcast. "I mean, they were doing all the things I'd hoped they would do. And I was just watching it kind of in shock that it was working."

Tying your online persona to your true identity was a shift from other online services, where people went by fanciful or even gross nicknames, as if at a giant, messy costume ball where anonymity could let you misbehave without consequences. Knowing who you were actually dealing with, talking to, flirting with, pitching deals to, and stalking made all the difference. Anchoring people to their real names and networks forced them into portraying themselves more honestly. One feature in particular bolstered trust and lubricated social activity: once you "friended" someone, the connection could be seen from your profile. Because you could surf the site by people's interests, you could find prospective dates—or just people you might want to meet—of similar mind. One user, for example, filtered prospective dates by seeing who listed the movie *Midnight Cowboy* as a favorite. If a woman passed that dubious litmus test, she was for him, and he'd send her a message.

Abrams opened Friendster to the public in March 2003, describing it as "an online community that connects people through networks of friends for dating or making new friends." By then he had received almost half a million dollars from angel investors. People signed up in droves. Then came a big funding round from Kleiner Perkins, the Valley's prestige venture capital firm. Abrams got a $30 million buyout offer from Google—and turned it down. (A stake in 2003 Google at that size would eventually be worth more than a billion dollars.)

By Mark Zuckerberg's sophomore year of Harvard, Friendster had more than 3 million people registered, including D'Angelo and Zuckerberg.

JUST BEFORE HEADING to Cambridge for the summer, D'Angelo had decided to hack up something in that vein himself, basing it on AOL Instant Messenger. Since he and his friends used it every day, it was playing on home court. Basically, he tried to convert that chat program into something of a social network. AIM had a feature called Buddy List, which was essentially an address book for the people you chatted with. It might be more accurate to say that D'Angelo's program, which he called Buddy Zoo, actually exposed the hidden social network that you had all along when you used AIM.

It worked this way: you went to the Buddy Zoo site and submitted your Buddy List. The program would then do an analysis, yielding all sorts of insights:

- Find out which buddies you have in common with your friends.
- Measure how popular you are.
- Detect cliques you're part of.
- See a visualization of your Buddy List.
- View your Prestige, computed the way Google computes PageRank to rank web pages.
- See the degrees of separation between different screen names.

The effectiveness of the program depended in part on a lot of people submitting their lists so Buddy Zoo could garner a huge data set. To D'Angelo's astonishment, that wasn't a problem. D'Angelo had posted games he'd written before, and never gotten more than a hundred or so downloads. Synapse had done better, but that project was the result of months of work. Buddy Zoo, which took D'Angelo only a week or so to write, instantly took off. Soon after he launched it, the program had a couple hundred thousand users.

D'Angelo spent the summer adding new features to Buddy Zoo and generally trying to keep up with demand. The database of names in the giant graph he was building was well on its way to more than 10 million, and he was using it to do research in his Media Lab internship.

The combination of doing fascinating work with the instant feedback of a huge base of users transformed the way D'Angelo thought about his programming projects. After this, he told himself he would only work on projects that could have an impact on the world. "I think it had a similar effect on Mark," he says.

FOR HARVARD STUDENTS, blocking is destiny. Beginning sophomore year, the school assigns students to live in one of the twelve residential "houses," referring to a cluster of buildings that will be the center of their life outside of class. Those in the same house eat meals together, socialize with one another, and are subject to the unique customs and regulations of the house. At the end of freshman year, everyone gets a chance to join a group of eight students who form a "block" that will be assigned to one of the houses. And for the rest of their lives, Harvard graduates will mention that so-and-so had been in their block when that so-and-so's name is well-known.

But no blocking process in the annals of Harvard was as fateful as the one that ultimately determined who would room in Suite H33 at Kirkland House. To several people, the luck of dormitory geography would be worth a fortune. Those in or near that Kirkland suite would forever boast of—or, much later, admit to—witnessing history firsthand.

Zuckerberg's blocking group included his freshman roommate, who was good friends with a classmate named Sarah Goodin. She, in turn, was close to a guy named Chris Hughes. Through such loose ties a block was formed. Hughes didn't know Zuckerberg well; he'd run into him in the dorm sometimes, while visiting Goodin. He generally agreed with Goodin's assessment of Zuckerberg as a quirky computer-science kid who was nonetheless capable of being charming and funny. "He was always coding something," Hughes says.

When Zuckerberg's freshman roommate transferred out of Harvard, Hughes and Zuckerberg would end up sharing a room. They seemingly didn't have much in common. Hughes was a history and lit major, not a techie, and wasn't in the fraternity. He was gay, Zuckerberg straight. But Hughes sensed that the two of them had subtle bonds. Both were middle-class kids who leveled up by attending elite private schools (Hughes had left his hometown of Hickory, North Carolina, to attend Andover in Massachusetts, another one of the Ten Schools). And both were nose-pressed-to-glass spectators to the pageantry of Harvard as lived by those born to wealth and high privilege. As Hughes saw it, they were, each in his own way, misfits.

Their block was assigned to Kirkland House, and Zuckerberg and Hughes wound up in one of the bedrooms of Suite H33, designed for four students. The other room went to two people Zuckerberg hadn't met, Dustin Moskovitz and Billy Olson. They put all their desks in a rather cramped common room with a fireplace they never used. Zuckerberg had come with a giant whiteboard to sketch out his projects; it would sit in a narrow hallway that connected the common room to the bedrooms.

A fire door connected their suite to the one next door. It had a sign on it warning that, if opened, an alarm would blast. The alarm never went off, and the door was almost always propped open, because a frequent visitor was Joe Green, a California kid whom Zuckerberg knew through the fraternity.

Zuckerberg took a laissez-faire attitude toward classes. What seemed most important to him was working on projects. He loved building things, and the fact that he was attending one of the world's premier universities didn't distract him from spending hours and hours at his cheap wooden desk in the common room of Suite H33. He preferred communication by Instant Messenger, even with people only a few feet away from him. He didn't give much thought to the idea that by typing those words into his computer, he would create a log that would one day haunt him.

Throughout his young life, Zuckerberg has always worked on projects, jumping from one to the next. But when he returned from summer break for his sophomore year in Kirkland House, his activities seemed to become more intense than ever. As soon as he plopped down at his desk, he began churning out increasingly ambitious new ideas. But as they piled up, a theme emerged: Nearly everything he did involved connecting people in some way.

His first project that year was a program he called Course Match. Zuckerberg started by scraping the list of classes that semester from the Harvard website. ("They weren't that psyched about that," he recalls.) Students who signed up for Course Match would enter their names and email addresses and indicate what courses they had registered for. By selecting a course, a user could see who else was signed up. Or you could type in a name and see what courses that person was taking. Like future Zuckerberg projects, its simplicity was deceptive. On one hand, this preview of the classroom's dramatis personae made Course Match as compelling as a hit show on Broadway. Zuckerberg was stunned by how much time people spent with the program. "People would just spend hours clicking through," he later marveled. "Here are the courses that people are taking and, wow, isn't it interesting that this person is interested in these things? I mean, it was just text." He came to see that you could map a human network this way.

On the other hand, there were issues that Zuckerberg never considered. Very publicly exposing even something as ostensibly innocuous as one's course schedule could open up a potential rat's nest of complications. What are the implications of choosing one's courses by classmate as opposed to actual subject matter? Would women have to worry about stalkers literally breathing down their necks in the seat behind them in the lecture halls?

Course Match did not last long enough for such questions to be debated. The website—hosted on Zuckerberg's laptop, a device not up to the task—was sufficiently popular that the activity burned out the computer after a few weeks. (He later blamed the crash in part on the steam coming from the suite's bathroom, which was near his desk.) Fortunately, Zuckerberg had backed up the code. He would reuse some of it later. Course Match had taught him a very useful lesson: "People have this deep thirst to understand what's going on with people around them."

Oddly, Course Match wasn't mentioned when Zuckerberg got his first notice in the *Harvard Crimson* on October 23 of that year. Instead the story—a mini-profile of an enterprising sophomore who apparently was more interested in making software than acing his classes—focused on Zuckerberg's work with Synapse, though there was no recent news about the project. It set the table for a mutually beneficial relationship with the student newspaper, which would over the next few months sporadically cast Zuckerberg in a recurring role of campus computer wizard. The story would only get more interesting as the academic year progressed.

ZUCKERBERG'S NEXT SIGNIFICANT product was a lark that seriously got out of hand, and almost resulted in his dismissal from Harvard.

Because of Zuckerberg's predilection for documenting his activities in real time, the unflattering circumstances of this exploit would be painfully exposed at a later time, when Mark Zuckerberg was much more mature and certainly more cautious about what he documented. But his journal of making "Harvard Face Mash" provides a rare and disturbing picture of his creative process.

Peeved at an apparent romantic setback, Zuckerberg was admittedly a bit intoxicated when he sat down at his warren in the common room of the Kirkland suite a little after 8 p.m. on a Tuesday evening. A Beck's beer was at his side. After announcing that the lady in question was "a bitch," he wrote of a need for distraction. "I need to think of something to take my mind off her." For Zuckerberg, the safe space in such times was his computer.

Joe Green and his roommate, Billy Olson, were nearby, and the discussion revolved around what the distraction would be—a website with a nasty edge to it. The inspiration was the printed directory, photos included, of his fellow Kirkland House residents.

> *The Kirkland facebook is open on my computer desktop and some of these people have pretty horrendous facebook pics. Almost want to put some of these faces next to pictures of farm animals and have people vote on which is more attractive. It's not such a great idea and probably not even funny, but Billy comes up with the idea of comparing two people from the facebook, and only sometimes putting a farm animal in there. Good call Mr. Olson! I think he's onto something.*

Zuckerberg wound up making his own version of a popular website called Hot or Not, formed in 2000 by a pair of programmers who digitized wolf-whistle assessments of women by asking people to voluntarily submit their pictures to be rated by strangers. In Zuckerberg's variation people were conscripted without consent to be publicly judged by those in their own community. What's more, Zuckerberg's system varied in a way that would potentially make it more viral—and nastier. Whereas Hot or Not asked users to rate people individually on a scale of 1 to 10, Zuckerberg's system pitted two people against each other, literally a head-to-head competition.

"I thought it was more clever than Hot or Not—rating someone 1 to 10 is kind of arbitrary," says Green now, recalling the giddy circumstances of a prank that wound up in the annals of geek legendry as well as a key plot point in a major motion picture. "But you can always say which person you think is more attractive."

Zuckerberg himself would later claim that he didn't realize that the competition might actually be offensive. And of course, he had not the slightest inkling that a future creation of his might one day be cited as the source of untold cases of online bullying, which could be so punishing that in some extreme cases a target might take his or her own life. It was just another project, a fun thing to do.

Back to his beer-soaked exploit. He logged his progress by explaining the steps necessary to scrape the pictures from the online directories in each Harvard house. The process was nontrivial. Each house varied in the degree with which they protected that information. Zuckerberg went through them like a safecracker, divining each turn of the tumbler until the door swung open.

Downloading the pictures was only part of the task—his creation not only provided one-to-one comparisons but then took account of all the activity and looked Harvard-wide for winners and losers. One might argue that it foreshadowed services far in the distance, like Tinder. It took him three days of intense coding in the common room to finish the site. In the process he expanded his programming repertoire, since this project required him to deal with components of open source software like Linux and Apache and SQL that he had not previously mastered. Creating these projects was a form of training, like up-leveling an avatar in a role-playing game to take on the boss monster in some future epic clash.

He named his brainchild Facemash and secured an Internet address. In the call to action on the website's front page, Zuckerberg channeled the battlefield bravado of the classical heroes he idolized. "Were we let in for our looks?" Facemash asked its Harvard visitors. "No. Will we be judged on them? Yes."

He didn't actually launch it in any formal sense. He distributed a link to the site—he had registered the facemash.com domain on the web—to some of his friends. (He also unwisely published on the site the online journal documenting his sketchy data collection methods and the farm-animal inspiration.) Then he went to a meeting. When he returned to his dorm that Sunday evening and logged into his computer, he was astonished to find it jammed with responses to Facemash. Apparently, someone had sent it around Kirkland House. From there, it spread to all of Harvard. And it was going viral.

It was also ticking some people off. The women of Harvard in particular bristled at being rated by their looks as if they were, well, farm animals. The mailing lists of Fuerza Latina and the Association of Black Harvard Women logged a hornet's nest of outraged comments.

Between the complaints and the traffic overload, Zuckerberg concluded that Facemash wasn't worth the trouble and he began to shut it down. Soon after, the Harvard IT Department, which had been trying to deal with the unusual traffic demands, cut off Internet to all of Kirkland. Moskovitz, working on a problem set for his CS class, and Hughes, writing a paper, were annoyed at the interruption. Zuckerberg was more perturbed that Joe Green had used the distraction to bound and grab the last Hot Pocket from H33. (Hot Pockets were a culinary favorite among that crowd.)

The prank was over, but the reverberations would continue. Harvard investigated the incident and accused Zuckerberg of hacking its communications system, violating copyrights, and compromising student privacy. Green and Olson got called out, too, for lesser charges. Zuckerberg was summoned before the advisory board, a Star Chamber–esque council of deans and administrators who investigated transgressions and meted out sanctions. In Harvard parlance, he'd been ad-boarded.

Among Zuckerberg and his friends, Harvard's objections were chalked up to the school's bias against entrepreneurship. The place did not encourage building things, the activity Zuckerberg loved the most. Andrew McCollum, a fellow CS student from California who took several classes with Zuckerberg, would later explain that Harvard was self-satisfied in its own ivory tower. "You can't major in pre-med, because that's not an academic subject. You have to do biology or chemistry," he says. "You can't take an accounting course because that's much too like brass tacks. You had to go to MIT if you wanted to learn accounting. Everything had this very academic focus."

MIT was also much more tolerant of high-tech pranks. It loved its hackers. Not so much Harvard, which was regarding Zuckerberg's trick as a high offense. There was a real chance he might be sent packing.

Considering the gravity of the situation, Zuckerberg seemed oddly unaffected. His parents were largely uninvolved, though clearly unhappy at the prospect of losing a semester's worth of tuition. But they figured that he would simply explain himself and the authorities would understand. "Mark would never do anything that he knew was wrong," says his father. "He is a very ethical and fair individual," adds Karen Zuckerberg, with conviction in her voice. (As she is speaking in 2019, one suspects she is addressing more than his student behavior.) "He always was that way in the house and in school and in his dealings with people."

On the eve of the decision, he attended a "Goodbye, Mark" party thrown by his fraternity to provisionally see him off. Zuckerberg donned beer goggles for the occasion. Green had arranged for Zuckerberg to meet a friend of the girl he was dating, "It wasn't some brilliant insight," he says. The woman's name was Priscilla Chan, and the pair made their connection while in the beer line. Chan took it in stride when Zuckerberg casually mentioned that he might be kicked out of school soon. This was remarkable because her own attendance at Harvard had been an inspiring immigration story. Her determined rise to become a pediatrician could never have withstood a dismissal from college.

They agreed to go on a date. The destination was a chocolate shop called Burdick's, famed for its cake. Zuckerberg and Green concocted a scheme where Green would call him mid-date and invite him to a party, "because that would make him look cooler," says Green. When the call came Zuckerberg dramatically and quite loudly declined the invitation on the grounds he was with a great, great woman.

Chan began dating Zuckerberg regularly, and eventually became his wife. They would reenact that bit of unconvincing theater at their wedding.

On November 3, the ad board granted Zuckerberg the equivalent of a reprieve. He was put on "disciplinary probation" until May 28, 2004. There seemed to be little in the way of restrictions attached to the sentence, except he was required to see a counselor. The official transgression was "improper social behavior."

He later explained that he took the judgment as a caution not to do anything worse, or he might really get kicked out. He apologized to the women's groups and arranged to do some computer work for them. Green and Olson, who were up for more minor charges stemming from their participation, sailed through without consequences. They hadn't even bothered to attend the hearing.

An impromptu celebration erupted in Kirkland H33, with the popping of a champagne bottle. Green's father, a UCLA math professor who was visiting Cambridge to give a talk at MIT, remarked that Zuckerberg was rather cocky for a kid who almost got booted out of Harvard. "No more Zuckerberg projects for you," he told his son, an admonition that probably cost Joe Green hundreds of millions of dollars.

The most striking lesson Zuckerberg took from Facemash had little to do with transgressions and everything to do with the exhilarating spike of attention it garnered—and why that spike was so dramatic. When asked under oath about this a few years later, he replied that Facemash showed him how much people liked looking at pictures of their friends and acquaintances.

Anything else? asked the attorney.

"People are more voyeuristic than I would have thought," he said.

ZUCKERBERG WAS UNFAZED by the episode, displaying a stoicism that would distinguish him after more serious misdeeds with astronomically higher stakes. He certainly didn't curtail his activities.

"He had a real self-confidence," says Joe Green. Once Green was walking to dinner with Zuckerberg and Chan, and Zuckerberg impulsively darted into a busy street. "Watch out!" said Chan.

"Don't worry," Green told her. "His confidence force field will protect him."

Zuckerberg did take special notice of an editorial in the *Crimson*, attributed to "the *Crimson* staff" on November 6, that gave a postmortem of the Facemash debacle—with a grudging acknowledgment that it moved Harvard closer to a much-needed universal online facebook. The key, noted the editorial writer, would be building in measures that protected student privacy.

Zuckerberg took the editorial to heart, and vowed to make privacy a core component as he prepared to write the most ambitious project to date: a facebook for the school's undergrads.

The idea of putting a student facebook on the Internet was anything but novel. It was obvious and inevitable. After all, Zuckerberg had seen one at his prep school just a couple of years earlier. Students at various universities had already put directories online, some with many social features.

A full four years earlier, for instance, some Stanford undergrads running a self-styled underground website called Steamtunnels had implemented an online facebook for the school. As recounted in a September 1999 *Stanford Daily* article, the students—known only by their online handles of Drunken Master, DJ Monkey, and The Sultan—had scanned photos from the past four years of the printed directories. "We felt like we could provide a fresh, unfiltered voice and a great set of services that could liven things up for everybody," said Drunken Master. But the administration, claiming that putting the photos online without opt-in permission violated student privacy, shut down the facebook component of the system. (Years later, "Drunken Master," whose real name was Aaron Bell, would be the CEO of a start-up company specializing in ad retargeting, a practice that would become a controversial component of Mark Zuckerberg's company.)

Harvard itself had announced that it was working on an official online facebook, one it hoped to deliver within months. On December 9, the *Crimson* quoted Director of Residential Computing Kevin S. Davis as saying, "This has been on everyone's priority list for a long time." But he did not commit to a date. One reason for the delay: privacy concerns raised after one Mark Zuckerberg created outrage with his Facemash prank earlier in the semester.

To Zuckerberg's relief, Davis had said that no staff had been assigned to build the official directory yet. There was plenty of time to create something before Harvard did.

In the meantime, he was still creating smaller projects. He had spent much of the year programming and very little in class; in particular he had missed every session of a classics course called The Rome of Augustus. His passion regarding the greatest of the Romans did not extend to the art treasures associated with his hero, and he found himself unprepared to face the course final, which would be based entirely on analysis of the images and artifacts from the rule of Augustus. Especially since, in the January period during which everyone else was reading for finals, he was doing even more programming. "I was pretty screwed," he'd later admit. "There was no way I was going to cover all this material." So he decided to program his way out of it. He scraped the course website, downloaded all the images. Then he transferred them to a website of his own and sent links to the website to all his classmates taking the course and invited them to study together. "It basically was randomly showing the images and would let you contribute your notes of what you thought was important about that photo and see everyone else's," he'd later explain.

Zuckerberg's friend Andrew McCollum, who also took the course, says that it was an innovative way for Zuckerberg to disrupt the typical means of group preparation, and was nothing insidious. "Mark thought it was inefficient to organize a study group and get together in the library. Why don't we make a tool that can let people collaborate more easily to do the same thing? That was his general approach—how can you use technology to let people collaborate and remove constraints of time and space?"

A cynic might conclude Zuckerberg was creating a program to provide himself free tutoring under the guise of a study group. After all, while the site presented itself as a benefit for his classmates, its clear purpose was to help just one person: Mark Zuckerberg. But he would later argue that the Rome of Augustus hack was a virtuous effort. "I just needed to get information to study for this class, and everyone else needed information, too, so I built us this resource that would generate the information," he told me in a 2009 interview. "I just think the common thread through all this is that there is an efficient place for the world where more information is being shared and we're not in it, and there's some amount of work or products that need to get built to get people there. And I think if you built those, then you will help get the world to that place, and that's a really good thing."

It certainly was a good thing for Zuckerberg, who aced the course.

3

Thefacebook

THOUGH ZUCKERBERG WAS still not quite a campus celebrity, the *Crimson* account of the Facemash episode drew the attention of three seniors who were planning their own online project. Late in 2002, Divya Narendra had approached two of his friends, twins Cameron and Tyler Winklevoss, with an idea of a website that would deliver services, notably dating, to their classmates—and maybe beyond. They were calling it "The Harvard Connection." (Later they changed the name to the more generic ConnectU.) For much of 2003, they had been brainstorming the site, but at a seemingly leisurely pace, due to their other activities, which in the Winklevosses' case included training for the crew team (they were aspiring Olympians), attending the fancy Finals Club events, and of course their studies. Earlier in 2003, they'd hired a programmer to do the hard work of making their ideas tangible, but he bowed out with other commitments before finishing. The programmer suggested they contact the sophomore behind the Facemash caper to be the coder who could put their idea into production.

Narendra emailed Zuckerberg on November 3, and soon after he met with the ConnectU trio and agreed to do the work for them. At first he seemed enthusiastic. Over the next few weeks, though, much to the frustration of the Winklevoss twins and their partner, he would not meet his deadlines, instead offering a mix of excuses. On November 30, for instance, Zuckerberg told Cameron Winklevoss, "I forgot to bring my charger home with me for Thanksgiving so I haven't had access to my laptop since the battery ran out Wednesday evening." He promised that when he returned he'd do the work quickly.

On January 14, at a meeting at Kirkland H33, Zuckerberg finally told the ConnectU team that he was bowing out.

"It was clear to him what we wanted," Cameron Winklevoss told the *Stanford Daily* later that year. "He stalled us for months while he worked on his own idea, which he launched in February as an original idea."

Winklevoss had a legitimate gripe. As Zuckerberg said in an AIM conversation with a friend at the time:

> *Someone is already trying to make a dating site. But they made a mistake haha. They asked me to make it for them. So I'm like delaying it so it won't be ready until after the facebook thing comes out.*

This message is one of a number of damning instant messages from that time unearthed by *Business Insider* in 2010. These are a minuscule slice of Zuckerberg's output of messaging in that and other years, but bluntly address issues like dissembling and privacy that would forever haunt the adult who once wrote them. Zuckerberg would later chalk up this and other damaging instant messages in his Harvard days to immaturity, and said he regretted them. The messages, he'd argue, were out of context and did not reflect his true feelings. Later, in a text to me, he explained the impact of his being judged on casual teenage expressions that he felt gave a distorted view of his personality: "I got so frustrated that old instant messages and emails from when I was a kid kept getting surfaced out of context and making jokes or small off-handed comments seem like reflections on my core personality or values that I just decided to stop storing my older stuff." In a later interview, he'd return to the subject: "Would you want every joke you made to someone being printed and taken out of context later?"

While the conflict between Zuckerberg and the Winklevoss twins would be later memorialized in depositions and cinema, a less celebrated competitor had already launched a program at Harvard that performed some of the social functions that the others were only planning.

Aaron Greenspan was a junior that year. Like Zuckerberg, Greenspan was a builder and budding founder who inveterately created and launched small digital products. Soon after arriving at Harvard, he'd chafed at the school's implicit bias against start-ups, starting a Student Entrepreneurship Council. Greenspan had been working on his own tools to help classmates in the individual houses navigate their coursework, social life, and the general minutiae of campus life, such as textbook exchanges, or notifications that they had received packages. He linked those together in a program he called houseSYSTEM. He launched it in August 2003. One component was a student directory that he called the Universal Facebook.

To Greenspan's frustration, though, he had not been able to get much traction. He was especially miffed that he could not convince anyone at the *Crimson* to tout his efforts. After no luck with repeated emails, he marched into the newspaper's offices and managed to get someone to look at the site. Nothing came of it. In subsequent months, Greenspan seethed with envy when he read about Zuckerberg's exploits with Synapse and Facemash in the school paper. *Why did they pay him such attention?*

Greenspan didn't hold it against Zuckerberg personally, at least at the time. He had been trying to get Zuckerberg to join his student entrepreneur group. Zuckerberg professed interest but never got around to attending a meeting.

In January, their communication picked up. Zuckerberg said he was working on another project, but was "trying to keep the project on the DL" (an abbreviation for "the down-low," indicating stealth). Greenspan wondered if Zuckerberg might make his project part of houseSYSTEM, but Zuckerberg demurred, claiming that the intricacy of Greenspan's system would be a challenge for him. As with the Winklevosses, he kept his cards so close to his chest that the clubs and spades could have left imprints. "The general problem I have with these things is I don't usually have a long attention span for lots of coding," Zuckerberg messaged Greenspan, making a claim that his roommates would surely dispute. "I like coming up with ideas and implementing them quickly."

They met for dinner on January 8 at Kirkland House. Zuckerberg was with his suitemate Dustin Moskovitz and a young woman who seemed to have just randomly accompanied them. Zuckerberg struck Greenspan as very confident and incredibly laid-back. In the middle of the New England winter, he crossed the quad wearing shorts, as if he'd just stepped out of the shower. Because in his IMs Zuckerberg had been cagey about what he had been building, Greenspan relished the chance to ask him directly. Something, Zuckerberg said, about graph theory.

Is he building a Friendster for Harvard? Greenspan wondered. Zuckerberg's vagueness bothered him. But he had already made up his mind about this brash sophomore. "I didn't trust him from the moment I met him," says Greenspan now.

Greenspan felt his suspicions justified when he examined the houseSYSTEM logs from early January. Zuckerberg had left the equivalent of bread crumbs in the trail as he accessed Greenspan's website, apparently looking for ideas of his own. The logs allowed Greenspan to follow those crumbs and track Zuckerberg's activities, much as Zuckerberg's Facebook would later shadow users—and even nonusers—as they navigated their way around the World Wide Web.

But Zuckerberg wasn't hiding the fact that he was building something that might compete with Greenspan in some respects, and during the process he was messaging Greenspan to ask him how he'd done certain things with houseSYSTEM. Greenspan tacitly agreed with this dynamic, keeping his reservations about Zuckerberg to himself. That stance would change much later, when Greenspan would become the subject of a *New York Times* article claiming that Zuckerberg had plundered him as well as the Winklevosses in creating Facebook.

In truth, the idea was out there for the taking. Social media was exploding—Friendster was a phenomenon and millions of people were piling onto MySpace. And the concept of putting a school's facebook online was not quite on a par with the theory of relativity; even Kris Tillery's project at Exeter was seen by its creator simply as an obvious step in an era of digitalization.

With all that time coding, Zuckerberg's class attendance clearly suffered. He was often a no-show at the notoriously tough Operating Systems course. So much so that in January 2004, his instructor, Matt Welsh, called in the wandering student for a chat. Welsh had seen by then that Zuckerberg had no problem handling the course material without attending lectures. But he pointed out to the sophomore that in-class participation was part of the grade for CS 161. Didn't he want an A? Didn't every Harvard student want all A's?

Zuckerberg told Welsh his complicated situation, recounting his ad-board ordeal over Facemash, and explained how most of his time was being spent on writing an online facebook with social-networking aspects. Welsh was not impressed. You think you're going to compete against Friendster and Orkut? he asked the nineteen-year-old. (Orkut was a brand-new social network released by the search giant Google.) As Welsh later recounted in a blog post, Zuckerberg was "unfazed."

"It wasn't that Mark was a bad student," says his friend Andrew McCollum. "At that point Harvard had less to offer him because he basically had his path laid out; it was less and less about what happened in the classroom and more and more about these other things."

ON JANUARY 11, while he was still staving off increasingly peevish queries from the ConnectU team, and giving vague descriptions to Greenspan, Zuckerberg registered the website thefacebook.com. Facebook.com was already taken.

It's not clear how much work Zuckerberg had done on the project by then. During the winter break, in the early part of January 2004, he visited some friends in the Bay Area and was blown away by visiting the home of the big tech companies. But it is indisputable that, later that January, Zuckerberg spent one or two weeks—his own accounts vary—coding what would be known as thefacebook.com. Certainly the project was his priority that month.

He viewed this new site as the culmination of all the projects he'd been working on previously. The common thread of all those projects, Zuckerberg would later explain, was his belief that with the Internet we now had the means to more efficiently share information, but people weren't building the tools to make that happen. Building such tools would help push the world to that efficient place. "That's a really good thing," he says. "So I built these little ones like Course Match and Rome of Augustus. Facebook was kind of like the master one, because it was, like, everything about the people that you cared about."

He drew lessons from each of his previous projects. From Course Match: the ability to know the sections where your friends were enrolled. Facemash: people really wanted to see stuff about their friends. Rome of Augustus: people would gladly provide you content for free. (He would avoid the pitfalls of Facemash—and another ad-board investigation—by using only content that people voluntarily provided to the site.) In addition, he almost certainly had in mind D'Angelo's work on Buddy Zoo and how crosshatching friend lists could lead to an entire network of connections.

In that vein, Zuckerberg did one more tiny project before launching Thefacebook. One of the features he planned would allow users to include on their profiles any mentions of them in the *Crimson*, so others could match those localized news articles to real people. When he scraped the archives of the *Harvard Crimson* in preparation for this, he found he could construct a Buddy Zoo–esque graph to find out how many jumps it took to connect people on his grand network. It turned out that one person who appeared a lot in the *Crimson* was a Harvard dean and computer-science professor named Harry Lewis. As a lark, Zuckerberg decided to release an application called Six Degrees to Harry Lewis, where people could find out, through stories where multiple people were mentioned, how many leaps it took to connect them to Lewis.

In a rare moment of caution, he emailed Lewis to ask if the dean was cool with the idea.

Lewis had no problem with the idea, but, as he later told the *Atlantic*'s Alex Madrigal, he gave the young coder a caveat. "It's all public information," he wrote of the data Zuckerberg was gathering. "But there is somehow a point at which public information begins to feel like an invasion of privacy."

Zuckerberg's previous projects used minimal interface elements; basically he tossed text on the screen and had people click links to get to other pages or perform functions. But he felt this one was important enough to require some actual graphic design. McCollum had created some nice-looking pages, so Zuckerberg IMed him, saying he had created a prototype of Thefacebook and basically ordering McCollum to do the page design and the logo. McCollum objected that he was no expert in design, just a computer scientist who'd done some experiments with pirated copies of Photoshop and Adobe Illustrator. But Zuckerberg insisted, instructing him to build a page header with something like a silhouette of a person that fades into ones and zeros. (Even then, this was a tired and familiar trope to illustrate anything that deals with computers.)

McCollum wound up creating a logo with a "vector art" head shot of a young man he found online, where the edges of the picture decomposed into a digital flurry. It seemed to be based on a photo of the actor Al Pacino. Only years later did someone point out to McCollum that the original photo of what became known to millions as "Thefacebook Guy" was a picture of Peter Wolf, former lead singer of the Boston-based J. Geils Band.

Still, the site was not particularly visually arresting. And compared to its later incarnations Thefacebook was utterly primitive. It greeted visitors with the site name and the Peter Wolf graphic at the top of the page, with a block of text explaining what the site was about.

[Welcome to Thefacebook]

Thefacebook is an online directory that connects people to social networks at colleges.

We have opened up Thefacebook for popular consumption at Harvard University.

You can use Thefacebook to

- **Search for people at your school**
- **Find out who are in your classes**
- **Look up your friends' friends**
- **See a visualization of your social network**

To get started, click below to register. If you have already registered, you can log in.

At the bottom of the page—and every page of the site—Thefacebook creator made sure everyone knew who was responsible, with a line that read:

a Mark Zuckerberg production
Thefacebook © 2004

Upon signing up, you could connect with (or "friend"—the noun would quickly become a verb) classmates who were signed up—or invite ones who hadn't yet joined.

Privacy was perhaps the defining characteristic of this new website. By limiting enrollment to those who had emails on the Harvard.edu domain, he made a safe space for students to share information they volunteered about themselves. By verifying emails, he ensured that people would be interacting on the site with their real identities, a built-in safeguard against misbehavior.

Furthermore, you could restrict what you shared to certain people. By providing those protections, Thefacebook offered more privacy than any of the other social networks of its time.

The Winklevosses would later claim that using an Internet domain to ensure privacy within a community was their original, top secret idea that they shared with Zuckerberg. But that wasn't an original concept; in fact, Aaron Greenspan's houseSYSTEM, which the ConnectU team was familiar with, used the Harvard.edu domain to verify users.

Despite his apparent insouciance during the ad-board investigation of Facemash, Zuckerberg had clearly taken a lesson from the ordeal. "Facemash was probably one of the best things that could've happened to Mark and the future of Facebook because it made him extremely aware of the importance of people controlling their own data," says his classmate Meagan Marks. "When he created Thefacebook.com it was fully opt-in. He didn't scrape any data systems. You had to sign yourself up, and within a month they had more than half the student body using it. So there was no need to scrape the data."

The *users* would provide the data. Thefacebook started with no content whatsoever: just the scaffolding that allowed people to bring their own. They would do this by creating profiles of themselves. They were allotted much more space than a two-line description in a physical directory. Thefacebook urged students to upload a picture of themselves—something *they* chose, not the stiff-smiled portrait of their graduation photographer—and a host of other information, generally geared toward socializing and (one could dream) hooking up. You could put in your relationship status, and what you're "looking for." There was space for personal data like phone numbers or AIM handles, as well as your interests, political preferences and favorite books, the courses you were taking, and a "favorite quote." Though you could not have a conversation via the system, Zuckerberg did concoct a means to send a direct signal to someone—designating another user to be a recipient of a digital "poke." Exactly what that meant would be totally up to the poker and pokee, though it seemed to carry a sexual frisson. (Years later I would ask Zuckerberg if he was aware that in Larry McMurtry's celebrated book *Lonesome Dove*, characters consistently used the term "poke" as a gentle stand-in for the fornication word. It was news to him.)

But it wasn't only the profiles that provided valuable data for Thefacebook. As with Friendster, the site allowed you to "friend" other people to signal that they were in your network. But unlike Friendster, Thefacebook allowed others to browse each other's networks. "A lot of people sort of just wanted to see who other people knew," Zuckerberg would later say. "There was nothing like that that existed."

On February 4, 2004, Zuckerberg officially opened Thefacebook, and began sending emails to his friends, urging them to try it out. It was the time of year when students were shopping for courses for the new semester, and Thefacebook provided an instant utility. The feature also gave early adopters something to do when the network was sparse for them because their friends hadn't yet signed up. That sparseness didn't last long. Within minutes after Zuckerberg posted its availability, people began signing up.

The night he launched his site, Zuckerberg went for pizza with friends at a place called Pinocchio's, which they called Noch's. He'd often go there with his friend Kang-Xing Jin, also known as KX, who was his frequent partner in problem sets for the CS courses they were enrolled in. They would usually wind up speculating about what kind of earth-shattering changes technology would bring. On the evening of February 4, after watching the rapid adoption of Thefacebook, Zuckerberg and KX concluded that someone was going to connect the whole world someday. Not that they thought the glue for the human race would be the program just launched on Zuckerberg's laptop. It would probably come from Microsoft or some other giant company, they figured.

One Harvard student who knew Zuckerberg only slightly was a junior named Sam Lessin. Also living in Kirkland House, Lessin was among the cadre of Harvard students who had fiddled around with building digital services; the summer before, he had launched what he describes as "a Harvard-only eBay," called Crimson Exchange. It hadn't ended in glory.

Lessin believed Thefacebook was something extraordinary. Zuckerberg had somehow managed to bottle the lightning of social networking, a subject that Lessin had been obsessed with. Lessin's father was an East Coast tech investor who had funded sixdegrees, and Lessin had idolized Andrew Weinreich. He had watched that company's demise with sorrow and now was following the rise of Friendster, which he felt was a good product but flawed. The problem was a lack of trustworthiness, because users were not reliably identified by their true names. Now this kid from his own house had created a website that provided the security that comes with knowing whom you're dealing with, and the privacy that comes from being bounded within your community.

He immediately arranged to sit with Zuckerberg at lunch. *This could be huge!* Lessin told a poker-faced Zuckerberg. This could be worth . . . *a hundred million dollars!* It was the biggest number Lessin could think of.

What did Zuckerberg say? "He was nonchalant," Lessin later recalls. What seemed more exciting to Zuckerberg was how he could do more interesting things with Thefacebook. Not so much the money.

Still, Zuckerberg had been thinking that Thefacebook had more business potential than the other projects he'd worked on. Even before the launch he had begun talking of starting a company based on Thefacebook, and having some friends invest. He had learned from Facemash that you couldn't run a campus-wide system off a laptop and he needed some money to rent server space. Zuckerberg had first asked Joe Green, but Green, who was more into politics than computers anyway, heeded his dad's instructions to steer clear of Mark Zuckerberg's project.

Zuckerberg did manage to interest a fraternity friend named Eduardo Saverin. Saverin was a Brazilian scion to a wealthy Jewish family that left the country for Miami when Eduardo was in high school. He was involved in the Harvard Investment Club. "None of us really knew anything about business, and Eduardo was sort of the guy who seems to know about business," says Green.

Saverin kicked in $1,000, a sum matched by Zuckerberg. He would later kick in $15,000 to a joint bank account. The pair agreed to split ownership of Thefacebook. Zuckerberg would get two-thirds of the new company. Saverin, as the guy who seemed to know about business, would get the other third. "We were starting a company," Zuckerberg later explained in a deposition, when all of this became a legal matter. "It seemed like we should talk about that."

The money helped Zuckerberg rent server space. It cost $85 a month.

Over the next few days, Thefacebook began a relentless takeover of the Harvard student body. As more Harvard students signed up, the chances increased that they would find profiles of their friends, of people they might like to have as friends. In the early 1980s, computer scientist Bob Metcalfe wrote about the network effect, postulating that the value of a network increases exponentially with the number of people joining (this became known as Metcalfe's law). Hour by hour, the impetus for students to sign up began to flip from engaging in a diverting pastime to an absolute necessity, as not being on Thefacebook made you a virtual exile on the physical campus.

Later, sociologists and start-up gurus would endlessly analyze what happened at Harvard in February 2004, painstakingly deconstructing the forensics of the lightning that Zuckerberg had bottled. "In the Ivy League, where very few incoming freshmen know more than one or two people, the [physical] facebook is really a key piece of infrastructure," says danah boyd, who was a sociologist in her early twenties at the time, and one of the first to understand that a new era in social science was being born right on her computer screen. "Zuckerberg made it interactive. It had a slight social stalking element too. It was addictive. And the fact that you could see only people on your network was crucial—it let you be in public but only in the gaze of eyes you want to be public to."

A few days after launch, the *Crimson*—whose staff was beginning to regard Kirkland Suite H33 as Harvard's own Silicon Valley—weighed in on this phenomenon. "Hundreds Register for New Facebook Website," read the headline (Hundreds!), with a deck that noted that the creator of the scandalous Facemash was making a dramatic bid to restore his reputation. Zuckerberg came off as rather cocky in the story, casting Thefacebook mainly as a response to Harvard's slow-walk toward its own online facebook. "I think it's kind of silly that it would take the University a couple of years to get around to it," he said, vocalizing the first notes of a leitmotif that he would return to regularly over the next few years: this new technologic era belongs to the young. "I can do it better than they can and I can do it in a week," he boasted.

But he also took pains to note that this new project reflected how seriously he accepted the paper's recent admonition to him that students were concerned about privacy. He outlined the various ways that users could limit who saw their information. And he promised that his website would respect its users' privacy in the future. "I'm not going to sell anybody's e-mail address," he told the *Crimson*.

Zuckerberg's takeover of virtual Harvard compared favorably with the overwhelming use of force exercised by his classical heroes like Augustus, Alexander the Great, or one of the online avatars he adopted in the game Civilization. Like those ambitious warriors, he was already looking at future conquests. Instead of refining Thefacebook in the laboratory of Harvard and deliberately rolling it out to other campuses, Zuckerberg began plotting right away to colonize the nation's colleges with Thefacebook. Some of these institutions had existing online directories that would have to be overthrown. In order to do this, he needed a team to perform the chores of setting up databases for individual schools and promoting the site before the snowball of adoption would tumble down the hillside. The roles fell to those who would become Facebook's co-founders. All of them would, of course, be secondary to the single ur-founder, the one whose name appeared on every page. *A Mark Zuckerberg production.*

Dustin Moskovitz was to be Zuckerberg's key technical lieutenant. Born eight days after Zuckerberg, Moskovitz was an economics major from Gainesville, Florida. During the school year, he had kibitzed over Zuckerberg's shoulder as his roommate concocted various projects—he thought that Facemash was a stupid prank—and he became an avid participant in late-night bull sessions about how the Internet would change the world. When Facebook launched, Zuckerberg asked Moskovitz for some help to administer the site. "I didn't really pitch him," Moskovitz later told a journalist. "It was more like he was working on this thing and I was sitting next to him, and he would say, 'Hey, can you help me with this?'" But seeing Thefacebook adopted so quickly on campus, Moskovitz wanted a larger role, which would mean he'd have to actually do some coding. He undertook a crash course in programming, buying the *PERL for Dummies* book and staying up all hours to teach himself. He was unfazed when Zuckerberg informed him that the site was built not in PERL but modern languages PHP and C++. No big deal—Moskovitz would learn those too. He had an unbelievable propensity for work; eventually people would refer to him as the Ox—a nickname that gave short shrift to his intelligence and organizational skills. He quickly figured out how to mimic Zuckerberg's work, and became a master at executing the tasks required to move Thefacebook into new campuses.

Zuckerberg's roommate, Chris Hughes, immediately saw that this project wasn't a lark or a prank like Facemash. Hughes had been a big fan of Friendster but realized that a private network, limited to the digital boundary of the Harvard domain, would address the privacy concerns that Friendster raised. To Hughes, Thefacebook was the first Zuckerberg project that made him want to get involved. Though Hughes was not technically inclined, Zuckerberg asked him to take on public-facing tasks he'd rather not endure himself.

Andrew McCollum, who'd done the graphics for the page, had grown up in Idaho loving the computer, and was one of the few people at Harvard actually majoring in CS. He had been instantly impressed with Zuckerberg's intensity and his determination to do whatever it took to complete the products he dreamed up.

The co-founder with the second-largest stake, behind Zuckerberg's, was Eduardo Saverin, whose money represented skin in the game. He was charged with doing businessy things while the team planned its assaults on other campuses.

While that made five co-founders, there was no question who was alpha. Zuckerberg's self-description on the site was "Founder, Master and Commander, Enemy of the State." He was beginning to see Thefacebook as different from all his previous projects. For the first time, one of his experiments had the potential to grow into something worth all his time and attention.

Now to take over other campuses. Zuckerberg approached the task as if American colleges were nations in a giant game of Risk. Indeed, the field was not wide open—some of the schools already had some sort of facebook equivalent—and as with Risk, he would have to outsmart them.

The first was Columbia University. On the surface, that might not have seemed like the most propitious takeover site. A competitor had existed there: in mid-2003. But game master Zuckerberg was making a counterintuitive move. Instead of moving first to the schools where Thefacebook had the highest chance of working, he attacked the schools where he thought he had the lowest chance of success. Meaning those schools where students had other choices.

"That was a critical difference in Mark's personality," says McCollum. "The other people who had created these things were really happy that they were successful at the school they kind of focused on, maintaining and improving the feature set and just sort of continuing to offer a great social network. Mark wanted to see how Facebook could compete with an entrenched popular social network."

Zuckerberg had an important decision to make as he extended Thefacebook beyond Harvard: would the newcomers to the system be regarded as part of a single contiguous network, or would they be regarded as a discrete unit? Specifically, could you cruise the profiles of students from another school as you would on your home campus? He later explained the trade-off: "Would it be better for people to be able to see everyone and maybe not feel like this was a secure environment in which they can share their interests and what they thought and what they cared about? Or would it be better that more information and more expression was available, but to a smaller audience, which is probably the relevant audience for any person?" After a lot of thought, Zuckerberg decided that he would limit profile-surfing to one's own school. People would be more likely to share things like their cell-phone numbers if they knew that only others in their community could see it.

Privacy would rule. Or, as Zuckerberg later put it, "If people feel like their information isn't private, then that screws us in the long-term too."

They began to invent what would become the template for infiltrating and dominating a new campus. First, they set up a separate database. Got the Internet domain. Secured the server space. Scanned the course catalogs. Contacted the college newspaper. And finally, went live, emailing key people—friends or siblings in their own social networks, or people who had been asking when Thefacebook might come to their campus. They went live at Columbia on February 26.

One advantage that Thefacebook had over the incumbent at Columbia: privacy. Though CC Community allowed students to post more photos, and write blog posts, its contents were exposed to the general public.

One would have thought that Zuckerberg might have spent the evening of the launch monitoring Thefacebook's progress in its first extension outside Cambridge. But a special opportunity arose: Microsoft co-founder Bill Gates was on campus for a talk at the Lowell Lecture House. Gates had famously dropped out of Harvard after starting his business, a career path that he did not recommend in his lecture. Instead, he urged CS majors in the audience to graduate, and then apply for Microsoft jobs.

Gates shared what Zuckerberg would later cite as a valuable piece of information—Harvard allows students to take indefinite leaves to temporarily pursue other ventures. "If Microsoft fails, I'm going back to Harvard!" joked the billionaire. Zuckerberg would later say that had he not heard about this safety net at the lecture, he probably would not have dropped out of Harvard to work on Facebook. (But this may have been a case where his family knew him best: before he began his college adventure, his mother bet him that he would drop out; he insisted he'd get the degree. He'd later joke that he agreed to do his commencement speech in order to collect the honorary degree that would win the wager.)

In the next few days, Thefacebook launched at Stanford and Yale. The pattern was set. Over the next few months, the team would expand Thefacebook to more than one hundred campuses.

Barely six weeks after Thefacebook launched, the *Crimson* was back again, this time with a prescient think piece about how social scientists might come to view Thefacebook. Zuckerberg described himself as "just a dumb programmer" and let Hughes do the philosophizing. "It's a tool to help people improve social relations by networking with people you only know tangentially," he told the *Crimson*. Zuckerberg, despite his self-deprecating description, chimed in an objection when the reporter asked him about the resemblance to Friendster, which he dismissed as a dating site. "The type of information on the site is fundamentally different," he said. "People aren't as biased in how they represent themselves." The article concluded with kudos: "Amateur anthropologists like Zuckerberg and Hughes are changing daily lives," it read, "one poke at a time."

WHEN ZUCKERBERG HAD launched Thefacebook on February 4, the ConnectU team felt blindsided. While this guy was giving them excuses for not finishing their Harvard social-media product, he was working on his own! They set about finding another programmer, all while suffering the dread realization that they might have missed their window. Indeed, for weeks after Thefacebook appeared they had watched in horror as it not only captivated Harvard but began to colonize the rest of the Ivy League, as well as other top universities around the country. The ConnectU trio even took their grievance to Harvard's president, Larry Summers, who was clearly not thrilled that yet more budding entrepreneurs with a dream of connecting Harvard students had found their way into his office. He told them that the university had no business refereeing business disputes among students. (Summers would later call the

Winklevosses "assholes" and mock them for wearing business suits to the meeting.)

If the ConnectU team had known what Zuckerberg was boasting to his friends, they might have gone into cardiac arrest. In one of those embarrassing AIM exchanges, he confirmed the obvious conclusion: Zuckerberg was intentionally stalling the Winklevoss team while he prepared his own product.

"yea, I'm going to fuck them," he wrote, "probably in the year."

Then he corrected himself.

"Ear."

Fifteen years later, I tell Zuckerberg that it seems pretty clear he dragged his feet on fulfilling his arrangement with ConnectU.

"I'm not sure," he says. "I think I might have been conflict-avoidant. But it was . . . I don't know, I think I was pretty clear."

HE WAS MORE politic (and less candid) in a letter to a Harvard dean who had asked him to write a timeline of events of the dispute. In Zuckerberg's account, while he originally agreed to help out the ConnectU team, they kept adding on new tasks. Eventually, Zuckerberg explained, he became disenchanted with Narendra and the Winklevoss brothers, concluding that their project committed the unforgivable sin of being boring and clueless. What's more, they wanted him—were ordering him like he was some sort of backroom techie—to fix the site by the "busy work" of improving the code. Zuckerberg made it clear that such work was beneath him. During the course of that conversation, he claimed, he became appalled at the ignorance and lack of imagination of the ConnectU team. "It became apparent [they] were not as clued-in or business-savvy as they led me to believe. It almost seemed that my most socially inept friends at the school had a better idea what would attract people to a website than these guys." In the letter, Zuckerberg complained that defending himself from their charges was impinging on his classwork. "The university should be upset with them for affecting my academics by forcing me to deal with ridiculous threats."

For all of his bluster, Zuckerberg had a point. The ConnectU team had drawn its plans in the belief that success in the Internet world came from thinking up a good idea and moving it online, taking advantage of digital's superpowers. That had been the theory in the first wave of Internet start-ups, a movement that had crashed ignominiously when the inflated values of companies like Pets.com exhaled like punctured balloons. But the next wave of successes were start-ups whose founders were technically minded. They often referred to themselves as hackers. Their ideas were only starting points for a product that they would rush to release and then iterate to excellence. By the mid-2000s, the way to glory did not involve hiring people like Mark Zuckerberg as cheap labor to code up the concept you brainstormed with pals at your finals club—it was driven by the Zuckerbergs themselves.

In May, fed up with Harvard's failure to sanction Zuckerberg on grounds of insufficient honor, Cameron Winklevoss decided to go public with Zuckerberg's perfidy. Naturally, the chosen forum was the *Crimson*. Winklevoss sent a blind tip to the paper, which, not wanting to miss a single iteration in the Zuckerberg saga, assigned the story to a reporter named Tim McGinn, who interviewed the ConnectU principals. McGinn then asked Zuckerberg to come to the *Crimson* office for comment.

Zuckerberg came with computer in hand, ready to prove to McGinn and his editor, Elisabeth Theodore, that Thefacebook was an original product that owed nothing to the ConnectU idea. But first he took the odd step—for a college student who was still professing that his project was not a business—of asking the two student journalists to sign a nondisclosure agreement. (Foreshadowing, it turns out, the NDA that the thousands of visitors to Facebook HQ must sign before they are permitted to cross the threshold into its offices.) They refused to do so, but Zuckerberg proceeded anyway. He convinced them his website was in no way a clone of whatever the Winklevosses had in mind. Anyway, he admitted, "Thefacebook isn't even a very novel idea. It's taken from all these other [social networks]."

Zuckerberg was anxious, almost panicky, about the upcoming article.

In an IM exchange to Greenspan, he complained about the Winklevosses. "They blame me for stealing stuff because I helped them for like a month." He was desperate to see what the *Crimson* was up to. (The irony here is that the whole time, Greenspan was thinking, *He stole from me too!* But Greenspan did not file a complaint. "I just didn't see the logic in making a huge deal out of something that was a student project," he says. "For all I knew, it was going to disappear in a week." Besides, at the time, Greenspan considered Zuckerberg a friend.)

Zuckerberg asked Greenspan if he knew how to get on "News Talk," which was a closed mailing list for *Crimson* staffers that discussed upcoming stories and other editorial matters. Greenspan said he couldn't help.

"Oh well," Zuckerberg typed. Then he speculated what life would be like post-Harvard, when such annoyances would be behind him.

> *there are no school newspapers and ad boards after you graduate.*
> *only the new york times and the federal courts haha.*

In retrospect, Zuckerberg should have dropped the mic after making that remark, which would stand up well to Nostradamus. Instead, he tried to read the emails of the *Crimson* writer and editor.

Business Insider would later show, via Zuckerberg's IMs, how he used the users' private accounts on Thefacebook to do it. First he searched for users who identified themselves as *Crimson* staffers. Then he dove into that subset of accounts, using the site logs to see instances where someone had mistakenly entered a wrong password. He was looking specifically for a case where one of them used their email password to log into Thefacebook. Whether or not that was the scheme, he managed to access the email of at least one *Crimson* reporter. In one email that Zuckerberg obtained, Theodore discussed his visit, describing his demeanor as "sleazy." But after she viewed his demonstration of how the two sites differed, she concluded that his behavior with the ConnectU people didn't mean he stole their work.

For all of that trouble, the article itself was a fairly straightforward recitation of the charges and countercharges, concluding that, in any case, both of the sites looked like copies of Friendster. Zuckerberg wrote to the *Crimson* complaining that they should have more forcefully exonerated him, but soon dropped the matter. He had more important things to do.

That September, the ConnectU team would begin a long legal process that would eventually yield them a $65 million settlement. That would seem to be an excellent return, considering that they had no formal deal with Zuckerberg—a judge described the arrangement as "dorm chit-chat." In any case, Zuckerberg's prevarications delayed them only two months in producing a site they had been dithering on for more than a year. Nonetheless, Narendra and the Winklevoss twins later complained about how the payout was calculated, joining what would be a surprisingly large population of unhappy people who got enormously rich through Mark Zuckerberg. (houseSYSTEM's Greenspan, too, would eventually win a multimillion-dollar settlement involving the copyright status of the word "facebook," and would also harbor a contempt of Zuckerberg akin to Ahab's for the White Whale.)

But in June 2004, all of that was far in the future. Zuckerberg had big plans for the upcoming summer.

Thefacebook was going west.

THE IDEA CAME about in casual discussion. For several years, because of a family connection to Electronic Arts executive Bing Gordon, Andrew McCollum had been interning for the game company and was intending to return to Silicon Valley. Adam D'Angelo had a Google internship for the break, and was also planning to be in the Bay Area. Since Thefacebook couldn't operate from Suite H33 during the summer, and things seemed headed toward the Valley anyway, Zuckerberg figured maybe the team should get a house in California and keep working. It seemed a much better alternative than finding a summer job. "This was a cool place to spend the summer, my friends were there, and this is Silicon Valley," he later explained.

He went to Craigslist and saw an ad for a furnished house in the Barron Park section of Palo Alto, a leafy area a few miles from downtown. It had a pool. The names on the lease were Zuckerberg, Moskovitz, and McCollum. Before leaving town, he recruited two talented freshman engineers as "interns" for the company, even though they would pretty much be doing the same coding work as everyone else.

The idea of having interns was as close as his venture came to the practices of an actual company. "We didn't really think of Facebook as a start-up," says McCollum. "2004 was close to the depths of the start-up recession after the dot-com bubble burst. And so, it just wasn't really in the cultural zeitgeist, certainly not at Harvard. It was cool to see that it was successful, but it was still just this little college social network."

Before embarking on his big adventure, Zuckerberg took time to be interviewed in yet another *Crimson* article, this one a profile of "the whiz behind thefacebook.com." A reporter visited the Kirkland common room, finding clothes on the floor and half-packed boxes. Zuckerberg seemed to be either bored or impatient; everything that came out of his mouth seemed a variation of "whatever." Between the lines, you can sense a frustrated reporter enduring painful pauses between questions and barely communicative answers.

Zuckerberg resisted the *Crimson* correspondent's insinuations that he might be sitting on a gold mine. "Having [thefacebook] be wildly successful is cool, I guess," he said. "But I mean, I dunno, [money is] not the goal."

Would he ever sell the company?

"Maybe . . . if I get bored. But not anytime soon. At least not for seven or eight days."

Zuckerberg's business partner, Eduardo Saverin, is not mentioned in the story.

"My goal is to not have a job," Zuckerberg said. "Making cool things is just something I love doing, and not having someone tell me what to do or a timeframe in which to do it is the luxury I am looking for in my life."

When the reporter asked how this luxury would be paid for, Zuckerberg tossed off a verbal shrug. "I assume eventually I'll make something that is profitable," he said. "I mean, like, anyone from Harvard can get a job and make a bunch of money. Not everyone at Harvard can have a social network."

4

Casa Facebook

SEAN PARKER STUMBLED on Thefacebook by accident. He was not a college student—he'd never even *been* to college—but at twenty-six he was living with undergraduates. The house was in Portola Valley, not far from the Stanford campus, where students had recently been captivated by Mark Zuckerberg's production.

Parker's situation was kind of baffling and humiliating because he was a serious, though controversial, player in the tech community. Parker was a near legend to Gen-Y nerds—people like Mark Zuckerberg. Growing up in a Virginia suburb of Washington, DC, Parker had been an indifferent student, capable of failing courses by blatant indifference and acing the ones that engaged his intellect. He competed on the swim team until his chronic asthma put an end to that. His father, an oceanographer, got him an Atari computer, and he'd learned its tricks.

Parker's extravagant personality and nose for business surpassed his considerable, but not transcendent, programming skills. At fifteen, he'd talked his way into an internship at a start-up called Freeloader, impressing its CEO, Mark Pincus, with his enterprise and chutzpah. Later, Pincus would learn that Parker had "kept a room with press clippings about me and other people who were interesting to him."

Parker spent a lot of time on IRC—Internet Relay Chat, a bulletin board that was a hacker hangout—and it was there he met another teenager with big ideas. In 1998, Shawn Fanning had the vision to realize that it was possible to disrupt an entire industry from a dorm room. As a freshman at Northeastern University, Fanning understood that the open nature of the Internet allowed even a nineteen-year-old like him to set up a collaborative database that would let people share music files without a central server. He called it Napster. Parker volunteered to join in, and he wound up writing Napster's business plan and helped secure angel funding. Millions downloaded Napster, triggering an orgy of free music sharing that almost destroyed the recording industry.

Parker often found himself in bizarre situations like that. His email signature at Napster was "specialization is for insects," and he was indeed diligent in defying speciation. His website later boasted:

> *A girl named Nina once remarked of me, "I can't tell if you're animal or machine." This would have hurt, if I were human. Luckily I'm a chinese* [sic] *hamster that has been fitted with an experimental math co-processor wetwired into my brain by Ray Kurzweil. I also have an empathy chip.*

"His brain doesn't work like anyone else's," Parker's fiancée, now his wife, once told me. "He has five thoughts for every sentence."

None of those thoughts could save Napster. Even though the company eventually struck a deal with the media giant Bertelsmann, music executives and investors never forgot its association with copyright infringement. That, and the fat target it presented for copyright litigation, ensured its demise. Parker didn't make a dime from it, but managed to emerge with excellent connections in the music world.

His next act was Plaxo, a start-up that tried to crowdsource everyone's contact lists. (It fulfilled Andrew Weinreich's 1997 vision about a great global networked Rolodex, when he launched sixdegrees.) Napster had been viral because of great word of mouth, but Plaxo had virality built in. With a click of a button, new users would bombard those on their contact list with requests to upload their addresses and phone numbers to Plaxo. Those targeted would often become furious at the multiple solicitations in their inboxes. Dealing with that rage was a cost of doing business for Plaxo. Sooner or later, the thinking was, people would yield to the inevitable and sign on. Indeed, Plaxo looked like it was gaining traction for a while.

But Parker got outplayed. His investors became alarmed at his erratic behavior and eventually moved to toss him from his own company. Parker was bitter that he had to fight his partners to get the money he felt he was entitled to. As with Napster, he had lost control of a venture he helped start. He blamed the venture capital firm Sequoia for cutting him out, and held a fierce grudge.

Now, in 2004, he was Silicon Valley's dark prince in exile, renting a bedroom in a house full of students.

Friends told him to put his financial house in order and take a job, but he was holding out for something bigger. "They were like, you're getting deeper in debt, you've got to focus on paying your bills," he says. "I had a bank account shut down on me, I couldn't get credit, but my belief was, I'm just going to hold out for the large payday. I'm just going to keep doing what I'm doing and as long as I create something of massive value, the money will follow."

Parker had a superpower when it came to detecting the next big thing. In 2004, he sensed it would be based on the power of networks, like Napster and Plaxo were. Parker had even gotten tight with Jonathan Abrams, who headed Friendster, and he was hanging around a small group of people in San Francisco who believed that social media would take over the world.

So, when the girlfriend of one of Parker's roommates pulled up Thefacebook on her computer one day, in the spring of 2004, Parker rubbernecked. He was struck that it looked a lot like Friendster or MySpace, but used only people's real names. "It was all about identity for me when I first saw Facebook," he'd later say.

The student described how Thefacebook was going totally viral in the schools that had it. At the word "viral," Parker sprang into action. He blind-emailed the company, telling them he'd been working with

Friendster, and suggested a meeting to see if there was anything he could do with this site he clearly admired. Eduardo Saverin answered the email, and they arranged a meeting. For Zuckerberg, it would be a big deal to meet with one of the founders of Napster.

They met in a chic New York City restaurant—Parker and Zuckerberg and Saverin, along with the girlfriends of the Harvard kids. It would be nice to verify as accurate the dialogue in the scene in *The Social Network* where Justin Timberlake, doing his silver-tongued Parker imitation, told young Zuckerberg, "You know what's cool? A billion dollars." But that was a screenwriter's fantasy. The dinner was mostly Parker bonding with Zuckerberg, whom he'd instantly determined was the only Thefacebooker worth connecting with. "I don't think I said five words to Eduardo," Parker later recalled.

Perhaps the most memorable comment came when Zuckerberg said that as amazing as Thefacebook was, he had something bigger in store, referring to it as a "secret feature."

But for all the dinner's excitement—Parker overdrew his anemic bank account when he picked up the tab—the follow-up was vague. Parker and a friend he knew who worked at a start-up tried to imagine what that "secret feature" might be, but couldn't figure it out. "I thought that was kind of the end of it," says Parker's friend. "It seemed to be in the category of things that Sean is potentially interested in."

As the summer approached, things were looking grimmer for Parker. He moved out of the Los Altos house and was crashing at the home of his girlfriend's parents—never an ideal situation. One June evening, he was outside the house when he saw a group of bedraggled teenagers flopping down the street. He wondered if he should worry about getting mugged. Then one of them called out, "Parker!"

It was Mark Zuckerberg.

THE HOUSE ZUCKERBERG rented via Craigslist was a flat-roofed one-story cottage in the same Barron Park neighborhood where Parker was staying. It was on a dead-end street called La Jennifer Way. With five bedrooms and a pool in the backyard, it was perfect home, office, and party space for Zuckerberg, Moskovitz, McCollum, D'Angelo, the two interns, and an array of visiting contributors. But they mostly worked. They repurposed the living room into a suburban equivalent of the Kirkland common room, pushing tables together to fit the rows of computer monitors.

Later, an attorney would ask Mark Zuckerberg about that summer in what was dubbed Casa Facebook.

"It was fun," said Zuckerberg,

All right, what did you do every day?

"Woke up, walked from my bed to the living room, and programmed."

Okay, what time did you wake up in the morning?

"It probably wasn't the morning,"

Okay, how late did you stay up programming?

"I don't know, like, it's quiet at night."

Okay.

"You can get work done."

Did you work all night sometimes?

"Yeah. I mean, although, I guess, that's relative when you're on shifted hours like that."

The Harvard kids made *some* concessions to the California life. Using Craigslist again, Zuckerberg bought his first car. It was a beat-up Ford Explorer that had no key. "You just had to turn the ignition," he later told a reporter.

Andrew McCollum came up with the idea of stringing a zipline over the pool. He built it from about $20 worth of supplies from Home Depot. "It wasn't a very good zipline," he admits. No one could figure out how to attach the rubber handholds to the threaded metal handle, and so gripping the line could cut your palms. Ultimately, the anchors to the chimney gave way, rendering the zipline useless and causing some house damage. "I guess I should have thought that the chimney might not be that steady," says McCollum. "But it's made out of brick!"

Some of the residents smoked marijuana, but Zuckerberg stuck to beer. He had an aversion to needles that was somehow triggered by being around people smoking weed. "Even just seeing us smoke pot, he would get light-headed because he would connect pot to drugs and drugs to needles," says one early Facebooker. "He would have to leave the room because he was making himself sick, just in his own head."

Mostly, though, it was fast food, video games, and work, preparing Thefacebook for another campus and then another. A typical outing was like the one on that night in June when Zuckerberg, Moskovitz, and a couple of others were on foot, starting the half-mile trek down Matadero Lane toward El Camino Real, where there was a Happy Donuts store. They had hardly started when Zuckerberg spotted Sean Parker's familiar face.

As they conversed, Parker made a brash call to move out of his current situation and couch-surf with Zuckerberg and his team, placing himself smack in the eye of the Facebook hurricane. His possessions then were minimal: the only big-ticket items were his BMW 5 Series and some kickass speakers.

That night, his new housemates were astonished to overhear a phone call where Parker hammered out the details of his Plaxo settlement. The Harvard kids listened with awe. *This was the big league!*

PARKER BEGAN TO get involved with Thefacebook, taking Zuckerberg under his wing. He found the nineteen-year-old founder an odd duck, but their differences were complementary: Zuckerberg was prone to clamming up, and Parker would not shut up. But when Zuckerberg did talk, it was clear that all that silence was incubating sharp insights.

Parker would later comment to me about how he was struck with Zuckerberg's obsession with concepts of power and dominance, constantly citing books of ancient Greek and Roman conquerors. He could erupt in bursts of bravado, bounding around the place with his fencing gear, annoying subordinates by thrusting his foil inches from their faces.

But he also could appear shy and insecure. That summer, and for much of the next few months, he was wrestling with the question of whether he should go all in on this one project of his. He'd always skipped to the next idea. Zuckerberg would often ask Parker if he thought Thefacebook would be around in a couple of years. Parker would always assure him that it would. With every campus takeover, it was obvious that Zuckerberg had unleashed a whirlwind.

As a well-networked serial founder, Parker knew the intricacies of starting a business, a subject that Zuckerberg knew nothing about. He didn't even view his project as a serious business. "I remember driving up the 101, and seeing all these great companies, and thinking to myself, *Oh, wow. These are such amazing companies. Maybe one day I'll start a company*," Zuckerberg would later recall about that summer. "And I had already started Facebook!"

When originally incorporating Thefacebook, Eduardo Saverin made the rookie mistake of registering it as a Florida corporation. (His parents lived in Miami.) Parker helped Zuckerberg reincorporate in Delaware, a state with a business-friendly regime that allowed big companies to operate with the most impunity and least transparency. That was Business 101.

Even so, Parker was no substitute for a corporate attorney. At one point while filling out documents, he asked Zuckerberg how many vacation days employees should get. "Three weeks," said Zuckerberg, meaning fifteen business days. Parker, who never punched a clock, thought a week was a week, and wrote down twenty-one days. From that point on—and to this day—every new Facebook employee gets twenty-one days of paid vacation instead of fifteen.

Facebook needed money. Saverin's investment and Zuckerberg's own contribution were being eaten away by the increased server costs and other expenses. Parker took charge of the fundraising. One of the first people he called was Reid Hoffman.

In Silicon Valley, no one knew social networks better than Hoffman. In the wake of the dot-com bust in 2001, many investors thought the consumer Internet was dead. But Hoffman believed that a new wave was coming, based on software that deepened personal connections.

Hoffman, who had gotten rich from his involvement in PayPal, was willing to spend the money making bets on this. One of those bets was Friendster. In addition, he was starting his own social-network company, LinkedIn, designed to grease professional connections. "I was thinking there was going to be a revolution where your real identity and real relationships would be a platform for applications by which you navigate your life," he says.

Another early call went to Parker's old boss Mark Pincus, with whom he'd stayed in touch since his adolescent internship. Pincus had even put $100,000 into Napster. Despite losing his investment, Pincus had been inspired by Napster's success at gathering a huge self-organized community where everyone brought their own content.

Leave it to Parker to know the two most influential believers in the social-networking movement. Back in 2002—even before Friendster—Pincus had shared Hoffman's belief that a new wave was coming where people would use the Internet for personal connection. That concept was central to an informal brainstorming group that met in Pincus's San Francisco house, where he had taken down walls to make it more like a loft. Parker himself would become part of the scene. Through the discussions held there, Pincus envisioned a massive virtual exchange that he called the Cocktail Party. The global Internet had the potential of making the entire world into a single gathering where you could look across the room, see someone interesting, and have someone introduce you. Hoffman gave the Internet-of-people concept a name that actually stuck: Web 2.0.

Pincus had joined Hoffman as an early, though still skeptical, Friendster investor. "Nobody thought it was going to be a great idea," he says. Pincus hedged his bet: he split his $15,000 investment with a friend.

Now the social-media idea was taking off. Pincus had started a social-network site called tribe.net, devoted to local community building. He thought of it like Craigslist with pictures. It became best known as the digital glue behind the "burners" who attended the Burning Man festival in the Nevada desert. But it had yet to explode more generally, and it wasn't clear that it would.

And now Parker was telling him about a kid with what might be the best bet yet.

Some months before, well before anyone in Silicon Valley was paying attention to a site working out of a dorm room to connect college students, some interesting news had reached Hoffman: the sixdegrees patent was going up for auction. (Andrew Weinreich now admits that he gave Hoffman the tip.) YouthStream, which had shut down its acquisition, sixdegrees, in 2000, now understood that the real value of its purchase might be its intellectual property, which laid claim to how connections were made on social networking and other key aspects of such sites. The holder of that patent would have a virtual boot on the neck of any competitor.

Hoffman and Pincus did not necessarily want to plant their heels on anyone's neck: they were trying to save their own. Their worry was the winner would be Yahoo! or Friendster. The former was a competitor that did not shy away from enforcing its intellectual property. Friendster was a trickier matter. Pincus and Hoffman were both friendly with its CEO, Abrams—after all, they had stakes in the company! But neither man trusted Abrams with the patent. Hoffman in particular worried that Abrams might build a business component in Friendster that would compete with LinkedIn.

But even if that didn't happen, there were troubles with Friendster that indicated that it might not be the best partner. The company wasn't handling growth well. Its servers were overwhelmed; customers were furious about how slowly pages loaded. Its positive buzz had transitioned to a grouchy murmur, and Friendster seemed destined for a downward spiral. In Abrams's desperation, they worried, he might use the patent as a weapon to stifle competitors. "Jonathan felt like social networking was his idea," says Pincus. "He felt like nobody else should do it." (The venture capitalists on Friendster's board wound up firing Abrams the same month as the sixdegrees patent auction was held.)

Pincus and Hoffman did a lot of war gaming to figure out their bidding strategy, to no avail; they wound up the under bidder. But Yahoo!

insisted on taking a thirty-day pause before closing the deal, for due diligence. Pincus and Hoffman said they'd be happy to wire the money the next day, and the cash-hungry seller took their $700,000 rather than wait for Yahoo!'s higher bid. Pincus and Hoffman agreed that the patent should not be used to generate cash, but to protect the then-tender ecosystem of social networking.

The ultimate winner would be Mark Zuckerberg. Without paying a penny, he wound up not ever having to defend against a patent that might extinguish his young company.

PARKER BROUGHT ZUCKERBERG to Pincus's office in Potrero Hill that August. Pincus found the kid audacious. He looked fourteen, and totally owned it, wearing flip-flops and long basketball shorts. His business card read, "I'm CEO . . . bitch." But it was Thefacebook's story, at least as Parker explained it, that blew Pincus's mind. Something like 80 percent of its users signed on every day. This was unheard of, and a dramatic contrast to the metrics at Pincus's own social software, tribe.net, where a single-digit percentage of members logged on daily. *He's created the Cocktail Party!* Pincus thought, albeit one where the guests weren't old enough to drink yet.

Hoffman was eager to meet Zuckerberg, too, but was sensitive to criticism he'd gotten after his Friendster investment. People were saying that it wasn't seemly for him to be funding his potential competitors. So Hoffman decided maybe someone else might lead the investment round. He suggested that they all meet in the office of his former PayPal colleague Peter Thiel. Hoffman figured that Thiel might be sufficiently impressed to lead the round. If not, Hoffman figured, he'd step up himself. Criticism or not, he wasn't going to pass on this opportunity.

Thiel was head of the Founders Fund, an investment firm he started after leaving the company that made his own fortune, PayPal. (Besides Hoffman, other veterans of that payment company included Tesla founder Elon Musk and entrepreneur Max Levchin. They would become known as the PayPal mafia because of their undue influence—in both investing and philosophy—on Silicon Valley thereafter.) Thiel didn't name his fund by accident: he believes that the prime indicator of a successful company is a driven, iconoclastic founder, the kind of person who perseveres even when others think they are utterly insane. He also favors businesses that aim to utterly dominate the field they compete in; his dream investment is a firm shooting for monopoly power in its market.

So while others might have seen red flags in an edgy Sean Parker and a strangely intense teenager pitching a website for college kids to share their course selections and relationship news, Thiel was poised to recognize what was really important in the brief presentation: the numbers that demonstrated how thoroughly Facebook had entranced its users. It was a metric of dominance.

Before the meeting, Thiel had invited a junior member of his staff to sit in. Matt Cohler had also worked for LinkedIn; at twenty-eight he already knew the start-up game. He also had an excellent bullshit detector.

Cohler, Thiel, and Hoffman sat on one side of the conference table, and Parker and Zuckerberg on the other. While Zuckerberg maintained a mysterious silence, Parker recounted Facebook's dramatic lift-off. Then he shared those insane metrics. Thiel loved it. The magic wasn't just in the growth in users, but the way Thefacebook instantly became part of users' lives, with most of them visiting the site every single day. In contrast, most sites would be celebrating to see 15 percent of their users on a given day.

The other impressive signal was *why* Thefacebook had been created. A lot of the pitches that Cohler had heard were hatched by someone studying a market map to identify an alleged need, then building something to fill the gap. Zuckerberg hadn't approached Thefacebook that way. He was building something that he wanted for himself, and people were naturally flocking to it. That had been the story of the giants of tech—Apple, eBay, Yahoo!, Google. At that point, Cohler didn't think that this reticent kid would be joining the leaders of those companies as the Valley's next superstar CEO. But Zuckerberg did seem serious and purposeful enough that he wasn't going to screw things up.

Cohler himself wanted to be part of it. But before he did, he approached Zuckerberg. When he'd heard Thefacebook's amazing numbers, he wondered whether there had been some sort of mathematical error. "Don't take this personally—I don't think you're lying to me," he said. "But I'm just concerned that maybe you have a bug in your database, so I want to validate the claims that you're making."

Not only were the numbers correct, but when Cohler began talking to users, he found the qualitative data even more shocking than the quantitative. Are you using this facebook thing, he'd ask, and they'd be surprised anyone would even ask. It would be like asking someone if they'd checked out this running-water thing. *Using it? I'm living it.*

Thiel's investment was $500,000 for 7 percent of the company, valuing the company at $5 million. Hoffman and Pincus also participated, each agreeing to invest $37,500 in Facebook. "When they offered me and Reid the chance to invest in that round, I felt then like we had won at lotto," says Pincus.

Thiel had one message for his new protégé as the meeting ended. "Just don't fuck it up," he told Zuckerberg.

Parker started to bring some friends in. He kept telling the team about a crony of his named Aaron Sittig, who had created the original Macintosh version of Napster. Sittig, a laconic Southern Californian who had mostly grown up in Spain, had helped Parker design Plaxo too, and Parker knew he could help Thefacebook. Sittig finally showed up in August. By then he was quitting a job at a start-up he'd joined, and was burned out on start-ups in general. He'd enrolled at Berkeley to study philosophy and was looking forward to going back there in September. (He considered philosophy almost like a design degree, and considered Wittgenstein the first network theorist.)

Sittig got in the habit of hanging out at the La Jennifer house, more for fun than anything else. Usually there was a movie on the TV—the team was obsessed with movies, and often would spurt out lines from *Top Gun*, one of their shared favorites. Or they would watch the Olympics. Sittig felt he knew the type—young hackers flailing away on some mad pipe dream. One day he encountered Adam D'Angelo, who was waving his hands around in front of a tiny webcam. "I have really bad RSI, and I'm trying to build an invisible keyboard so I can type in the air," he said. Typical nerd weirdness.

As Sittig kept visiting, though, he became more impressed. These undergrads might have had zero experience, but they were super smart, super driven, super motivated. Especially Zuckerberg. Clearly the leader. When people started goofing off and talked about going out to see a movie, Zuckerberg would say, *Hold on—we're going to sit down and finish, and then we'll go to the movie.*

Sittig decided to take a close look at Thefacebook. His first thought was how ugly and clunky it seemed. But then he started to look closer, questioning why Zuckerberg had set each element in the configuration it was in, and realized he had underestimated it and its founder. Zuckerberg seemed to have hit on something brilliant. Everything was in the right place, optimized so that people would flow through it.

For instance, when you went to your profile page, a legend appeared in big letters saying, "This Is You." To a Mac designer used to a Zen minimal aesthetic, this at first seemed dumb and redundant. Isn't it obvious that it's you? Your picture is there! But when Sittig thought about it, he realized that there was something actually magical about that. "This was a brand-new type of thing," he says. "Mark was basically framing it for people—this is your page, this represents you, this is how other people see you. It needed that explicit frame for people to learn how to use it."

Sittig believed the future of social networking was up for grabs. Everybody had originally loved Friendster, but the company was blowing its lead with poor performance. Now there was a race for dominance, with a company from Southern California, MySpace, taking the lead. Right in the heart of the Valley, though, these Harvard teenagers had elbowed their way into the race. Only, no one knew it yet.

Everyone kept urging Sittig to pitch in, and he finally agreed to spend a few days helping out. But he wouldn't be concentrating so much on

Thefacebook. Instead he'd work on that "secret feature" Zuckerberg had mentioned to Parker in New York.

IT'S RIDICULOUS TO consider now, but even as Thefacebook was taking off, Zuckerberg was just as passionate about a second project. He'd even mentioned it in the meeting in Thiel's office. If Thefacebook didn't turn them on, he told the potential investors, there's this other thing he was working on, a program to help people share files with one another. It was called Wirehog.

"No, no, no," said Hoffman. "Facebook's the good idea. Abandon Wirehog." Advice that, in the short term, Zuckerberg ignored.

Wirehog's origin came from Zuckerberg's continuing obsession with taking something from AOL Instant Messenger, making it better, and busting it out on the Internet. Zuckerberg and his friends would often want to share files with one another. AIM handled this poorly.

"Wirehog was partly born out of just this frustration we had with AIM, which had this feature where you could send a file to someone else on AIM and it just never worked," he says. "Wirehog was sort of a solution to that," he says.

Wirehog was basically a means to make the files on other people's computers visible and accessible to you, via your own machine. Or if you were using a remote computer, you could use Wirehog to access your own files, which were sitting on your home computer's hard drive. Or you could share your documents and media with friends, letting them browse your virtual file cabinets and photo albums. "Mark was like, *Well, college kids seem to like this Facebook thing. But what they really would like is having access to each other's media content*," says Aaron Sittig.

Zuckerberg envisioned it as Thefacebook's twin, another iteration on all those projects he'd done that led up to Thefacebook. "Mark likes doing things quickly. He likes building prototypes and getting them out there," says Andrew McCollum, who spent much of that summer and the months after working on the project, along with Adam D'Angelo. "Wirehog was a cool thing that he'd built. They sort of coexisted." Thefacebook was built to connect people, and Wirehog would let those people share the things that interested them.

Parker, with his instinctive product genius, had taken a close look at what McCollum and D'Angelo were doing on Wirehog, and understood that the concepts were advanced. It not only let you spread your files across many devices with full access to them, but allowed people to share selected files with their friends in various galleries segregated by content—photos, documents, music. The music gallery had a player built in so you could play music from someone else's library on your own computer. One night, discussing the product with the two coders, Parker gave them a suggestion that basically foretold what would later be known as cloud computing: "You have to make it really easy for people," he said. "People should have just one thing to put files in, then everyone could put their stuff in and make it available." He even suggested a new name for the project: Dropbox.

Nonetheless, Parker realized that Wirehog was not only a distraction but a dangerous one. Because of its basic function of file sharing, without distinction as to whether the content was copyrighted, it was uncomfortably similar to the project that had brought him fame and notoriety, and, ultimately, grief: Napster. Pirating music on Wirehog wasn't anywhere near as good a way to do so as Napster had been. But that was a distinction that would not placate record labels.

Parker told Zuckerberg that Wirehog would be the end of Thefacebook. *These music guys are gonna sue you for all you're worth, and this is never going to work. We've got to shut down Wirehog!*

Zuckerberg respected Parker, but he wasn't taking orders from him. "Mark is an interesting combination—he's very strong willed but also adaptable," says Andrew McCollum. "He recognized that Sean had a lot of valuable experience and could help him learn a lot about how to kind of navigate the new world that he was in. But he didn't just do whatever Sean thought or said. It wasn't that kind of relationship."

Zuckerberg suggested that maybe they should talk to the music-business guys and see what they had to say. Parker already *knew* what they would say, but he indulged Zuckerberg and set up a meeting. Despite Napster, or maybe because of it, Parker had gotten close with some top music executives. They flew to LA and went to the house of Tom Whalley, who was then the head of Warner Bros. music. At the end of the meeting a surprise guest walked in—Seagram heir and self-styled musician Edgar Bronfman, who at the time was making a bid to buy the record company. The music honchos had a directive for Zuckerberg regarding Wirehog, one he'd heard before: *Shut this fucking thing down.*

Before the meeting ended, though, Parker and Zuckerberg had a peace offering: the opportunity for Warner Bros. or Bronfman personally to invest in Thefacebook. They turned it down. According to Parker, they felt the price was too high.

Ever stubborn, Zuckerberg incorporated Wirehog. He even gave Parker a share of the company, perhaps to quell his criticism. In January, Parker did set up a meeting with the venture capital firm Sequoia, ostensibly to pitch the new company. Indeed, VC firms who weren't invited to put money into Thefacebook were more than willing to consider another project from the same team.

The meeting was set up for the early morning, and, as usual, much of the team had stayed up until 4 a.m. or so, coding. So it might kindly be inferred that the late hours were the reason that Mark Zuckerberg showed up to the meeting in pajamas. Actually, they were players in a Sean Parker revenge pageant. Parker had blamed Sequoia for cutting him out of his share of Plaxo and was recruiting his new colleagues to humiliate his nemeses. To their discredit—and later regret—the teenage Facebookers were on board with the prank. "We really didn't care if they invested or not," says McCollum. "In fact, within that pitch there actually was a slide called 'Ten Reasons You Shouldn't Invest in Wirehog.' One of them was, 'We Work with Sean Parker and You Guys Hate Him.'"

Wirehog never did get funding. The beta version Zuckerberg released to several schools didn't take off. Wirehog's two key developers, D'Angelo and McCollum, were headed back to their respective schools. Thefacebook, meanwhile, was growing madly—it hit a million users in December 2004, and Thiel threw a big party in San Francisco for his hot start-up. Ultimately, Facebook's success killed what was left of Wirehog. "We put a bullet in that thing," Parker later would brag.

Wirehog is now an obscure footnote to its famous older sibling. And a company actually called Dropbox, founded a few months after the Wirehog episode, is now worth around $10 billion. Its first venture capital round was backed by Sequoia.

As the summer of 2004 ended, it was time for Zuckerberg and his friends to make some decisions. First, they had to move. Their landlord had gotten complaints of noise and misbehavior. On three occasions, it had been necessary to vacuum broken glass from the bottom of the pool. When the landlord sent someone to peer through the front door to investigate, the visitor reported, "The house seemed to be in total disarray and very dirty." There was also the chimney damage from the zipline fiasco.

Originally, the assumption was that everyone would head back to Harvard and keep the company going from there, maybe having someone in California handle operations. But Thefacebook would need much more attention than a caretaker could handle. One day Moskovitz took Zuckerberg aside. *We're getting a lot of users,* he told him. *And we're needing more and more services. We have no ops guy to run this*—we're *the ops guy. I don't think that we can do this and take a full course load.*

That's when the Bill Gates option kicked in. Just as Gates had talked about in that lecture hall, when he went off to Albuquerque to start Microsoft, he and his co-founders could take advantage of Harvard's liberal leave policy. So, Zuckerberg figured, why not just take one term off and try to get all that infrastructure under control? Once they got it set up, they could go back to Harvard maybe for spring semester. Thefacebook would still be growing but they'd be able to run it more autonomously. You don't want to be too old when you return, Zuckerberg figured—if you waited four years, for instance, it would be weird.

For the time being, they would stay in California. "I mean, it was never a formal decision, like, should we go back to school?" Zuckerberg later explained to the *Crimson*, which continued to keep close watch on its undergrad tech star in exile. "We all just kind of sat around one day and were, like, 'We're not going back to school, are we? Nah.'"

They found a rental house in Los Altos, not too far from the original Casa Facebook. The landlady took one look at Zuckerberg and asked how old he was. Twenty, he replied. "You think I'm going to rent you my million-dollar house?" she asked.

"Yes," said Zuckerberg. This time there was no zipline.

Peter Thiel had instructed Zuckerberg to organize the shares for his co-founders and establish a schedule for their options to vest. Zuckerberg hadn't even known what a vesting schedule was, but once he took on the tasks, he realized that Moskovitz should have a significant stake—and Eduardo Saverin was overrepresented. Moskovitz, the Ox, was becoming indispensable to keep the site open as it grew. "If Dustin had left, Facebook would have been in a very bad state," says McCollum. Though Moskovitz's 5 percent stake was more than justified, it caused some rancor. "Everyone else was like, '*What the fuck are you doing?*'" Zuckerberg later told a reporter. "And I was like, *what do you mean? This is the right thing to be doing. He clearly does a lot of work.*"

No one, really, had any idea of what those initial stakes would become. "Facebook became so successful that it's impossible to have any amount of equity in Facebook [from that time] and feel you got shortchanged," says McCollum.

Eduardo Saverin would be an exception.

IT WAS MIDNIGHT when Ezra Callahan showed up at the Los Altos house that Zuckerberg had convinced the landlady to rent him. Callahan had just gotten back from Europe, and his friend Sean Parker—whom he'd lived with in the Los Altos house earlier in the year—convinced him that joining his start-up would be a good way to kill time on a gap year before law school. The gig would pay around $30K per year, with stock options—of dubious value, he figured. He could even live at the company pad in the Los Altos Hills. Dustin Moskovitz answered the door. "I'm Ezra," said Callahan. "I work for you. Sean said I could crash here." Moskovitz shrugged and let him in.

The work routine set at the La Jennifer house had moved to the new one in Los Altos with basically the same personnel. Callahan joined Zuckerberg, McCollum, Moskovitz, and Parker as full-time residents, a roster that would continue to grow, with people sleeping several to a bedroom and often more on the floor. This house wasn't furnished, and the team bought the bare minimum of furniture from IKEA. Zuckerberg didn't bother with a dresser; he had no problem with stacking his clothes in piles. The only indication of Thefacebook's new funding was new wheels for Zuckerberg. After he showed up an hour late for a meeting, Thiel said he'd buy a new car for him. "Keep it under $50,000," he said. Zuckerberg chose an Infiniti. Even then, it was a lease, not a purchase.

That fall, the roles of Thefacebook's leadership became clearly, if implicitly, defined. Zuckerberg was the awkward genius who was the first and final product authority. Parker was the visionary who understood the significance of the product and helped Zuckerberg understand its potential, as well as schooling him on how to play the Silicon Valley game. Moskovitz was the implementer, who translated Zuckerberg's product vision to the engineers who would actually write the code to make it happen, as well as churning out reams of code himself.

Chris Hughes, who was staying there for a few weeks before returning to Harvard, did communications. Cohler provided the business heft; he was the steady hand charged with harnessing Zuckerberg's geeky brilliance and Parker's out-of-control ideas. "He was very much the adult in the room," says Callahan, who was twenty-three at the time, four years younger than Cohler. "Mentally he was forty to me and had the gravitas of that. Only later would I realize he barely knew what he was doing, either."

Before the Thiel investment, Thefacebook had been running on financial fumes. Zuckerberg was borrowing from the funds his parents set aside for the rest of his college education. Thefacebook's rate of growth had been constrained by the server space they could rent, and how quickly they could create the separate database that each college needed. By the end of the summer, they had fewer than fifty schools set up, with hundreds more clamoring for the service. After the money came, Zuckerberg could hire more people to handle expansion. (And he paid back his parents.)

In early September, they moved their servers from their East Coast company to a bigger co-location facility in California called Equinex. "We really didn't know what we were doing, because we were just college students," says McCollum. They bought dozens of new servers and spent an all-nighter unboxing them, setting them on the racks, installing Linux, and hooking them up to the network. Hundreds more colleges would come online.

As the holiday season approached—still less than a year since Zuckerberg launched Thefacebook—they all realized that the project was going to be huge. "The idea that this was going to be a billion-dollar company was there from day one," says Callahan. (He was talking about *his* day one that fall, not the first day at Harvard.) But no one had a clue how many billions. In any case, Facebook still had no serious business model. Callahan was asked to help out with this.

Six months earlier, the idea of delegating that task would have gone to Eduardo Saverin, the co-founder in charge of the company's commercial aspect. But Saverin had remained on the East Coast that summer, selling ads for Thefacebook, with plans to return to Harvard to finish his degree. He'd missed the shift from an East Coast college product to an ambitious Silicon Valley start-up. Sean Parker, the louche Yoda of La Jennifer Way, thought that Saverin was a drag on the company and would often say so. As one observer puts it, "Eduardo was talking about thousands when Sean was talking about millions."

At one point, Parker asked Callahan if he could fill in for Saverin's duties. Eventually, Zuckerberg agreed that Saverin had to go. Zuckerberg would let his lawyers and bankers break the news, by way of presenting a reformulation of the partnership that set up his friend's stake in the company to shrink. When Saverin discovered that he had signed documents that essentially cut him out of the company, he was outraged. Zuckerberg gave his justification to a friend on Instant Messenger (again, unearthed by *Business Insider*):

> *I maintain he fucked himself. First by completing none of his three assigned tasks. He was supposed to set up the company, get funding and make a business model. He failed at all three and took the offensive against me with no leverage. That just means he's dumb. And now that I'm not going back to Harvard I don't need to worry about getting beaten by Brazilian thugs.*

Saverin, though, didn't need Brazilian thugs to avenge Zuckerberg's machinations. After litigation, in 2007, he claimed an impressive 5 percent of the company—worth billions—and forced Facebook to permanently confirm in its official annals that he was a co-founder. He also enlisted an author to write a story about the betrayal from his point of view, with a movie—*that* movie—in production even before the book publication. Saverin decamped to Singapore, reportedly for tax advantages, and is now known less for what he did or did not do for Facebook than for being the real-life character portrayed by a movie star in *The Social Network*.

And of all of Facebook's founders, Eduardo Saverin had the least to answer for when the horrors befell the project once launched on his thousand-buck investment.

5

Moral Dilemma

IN MARCH 2005, Thefacebook finally moved into an office. Parker secured a second-floor space on Emerson Street in downtown Palo Alto, over a Chinese restaurant.

By then Zuckerberg had moved out of the Los Altos house. As the company was getting bigger it was less seemly that the CEO was bunking with the underlings. After crashing in different locations for a few months, Zuckerberg would move to a small apartment in downtown Palo Alto, a few blocks from the office. He had no TV, just a mattress on the floor and a few sticks of furniture. He was the CEO and biggest shareholder of a company with more than a million users and he still stacked his clothes on the floor.

In the first few weeks in the office, Thefacebook faced a financial crisis. Though it hadn't yet spent all of Thiel's angel money, the server bills and other costs were accumulating. The company still needed a new pot of cash, ideally coming from an investor who could act as an adviser to a CEO who had never even worked for a big company before, let alone run one. There would be no problem getting the money. But the choice of lead funder was fraught.

Zuckerberg had a strong preference for who he wanted to fill that role: *Washington Post* chairman and CEO Don Graham. Not a venture capitalist. Chris Ma, the father of one of Zuckerberg's Kirkland House classmates, headed business development for the *Post*, and his daughter Olivia's description of Thefacebook's conquest of the college market intrigued him. In January 2005, Parker and Zuckerberg went to Washington, DC, to explore a business relationship. Ma invited Graham to the meeting, and the *Post* CEO listened in fascination as Zuckerberg described how Thefacebook worked. He wondered, though, whether privacy was an issue. Are people convinced that their posts will be seen only by those whom they want to see them? he asked.

People were indeed comfortable with sharing, Zuckerberg told him. A third of his users, he said, share their cell-phone numbers on their profile page. "That's evidence that they trust us."

Graham was startled at how emotionless and hesitant this kid was. At times, before he'd answer a question—even something that he must have been asked thousands of times, like what percentage of Harvard kids were on Thefacebook—he would fall silent, staring into the ether for thirty seconds or so. *Does he not understand the question?* Graham wondered. *Did I offend him?*

Nonetheless, before the meeting was over, Graham became convinced that Thefacebook was the best business idea he'd heard in years, and told Zuckerberg and Parker that if they wanted an investor who was not a VC, the *Post* would be interested.

Negotiations were well along when Matt Cohler joined the company. He didn't agree with Zuckerberg's choice. Yes, he was pleased that the two CEOs had strangely bonded. And Graham would certainly be valuable as a mentor. But Cohler told Zuckerberg that you only do an "A round" once. (Thiel's money was the "seed round," and the next step in start-up life is an A round from a venture capital firm.) Ten percent of the company was at stake! Cutting a better deal than the *Post* was offering could yield a difference that would prove significant to the company's finances later on.

Word had already gotten out among the VCs that an incredibly promising start-up was seeking money. Cohler doesn't deny spreading the word himself, and Parker—while giving Graham the impression that he was all for the *Post* investment—was also eager to get a better deal from a Silicon Valley VC.

Though the benchmark firm and the top VC at the time, Kleiner Perkins, loved Thefacebook, ultimately their investments in Friendster presented a conflict. The most persistent VC was a company called Accel: one of its partners virtually camped out in front of Thefacebook's office until he got a pitch. When Accel's lead investor, Jim Breyer, finally made his offer, it was nearly twice what the *Post* was offering. He'd pay $12.7 million at a valuation of an eye-popping $98 million for that year-old company run by a twenty-year-old. What's more, he was okay with an arrangement where Zuckerberg would retain long-term control of the company—Breyer and Thiel would be on the board, but Zuckerberg would control two seats and Parker would have one. Zuckerberg would never get frozen out of his own company like Parker had.

But by then Zuckerberg had already agreed to take the *Post* investment. No papers were signed. But in spirit, the handshake had been made.

Zuckerberg was in a quandary. He had kind of a management crush on Graham, and had even followed him around for a day at the *Post* to see how a CEO operated. But Cohler and Parker were making strong arguments. The more money Facebook had, the faster it could grow, and the more ammunition it would have to take on MySpace, which had many more users. Still, Zuckerberg did not want to think of himself as someone who would go back on a deal, especially with someone he respected so much.

He called Graham. "Don, I have a moral dilemma," he said, and explained that Jim Breyer was offering him twice the money.

Graham wanted to make sure Zuckerberg understood that going to a VC meant that part of the company would be owned by a party whose goal was reaping the highest return. It might push for a sale that Zuckerberg opposed, or agitate for an IPO before Facebook was ready.

"I understand that," said Zuckerberg.

"So, it's important for you to have that additional money?" asked Graham.

"It really is," he said. "We need to grow. We need to grow fast."

Graham thought that maybe he should match Accel's offer—but figured that Accel would then just top that, and there would be another phone call asking him to top the offer. He simply said he couldn't match it, and Zuckerberg would have to make his choice.

Soon after the call, Jim Breyer took Parker, Cohler, and Zuckerberg out to the Village Pub in Woodside. It was one of the few places in the Valley where a fusty, Old World formality prevailed. Breyer was celebrating his imminent victory, and when the wine list arrived, he ordered a Quilceda Creek Washington State cabernet, raving about the legendary and legendarily expensive wine. Parker and Cohler were jazzed to drink it. Zuckerberg said he couldn't partake because he was still underage. He ordered a Sprite.

"I like Sprite," he would later explain in an onstage interview with Breyer that year.

As the dinner progressed, Zuckerberg appeared more and more uncomfortable. Cohler wondered if he wasn't used to fancy restaurants. Zuckerberg finally excused himself and went to the men's room. And did not return. Finally, Cohler went back there to see what happened.

Mark Zuckerberg was on the floor of the Village Pub men's room. And he was weeping.

What's up? asked Cohler.

Zuckerberg told him he couldn't go through with the Accel deal—it was wrong. "I gave my word to Don Graham, and that's all that matters," he said. "I'm not doing this."

The tortured moralist was a side of Zuckerberg that Cohler had not seen—certainly it would have been a shock to the Winklevoss twins—and Cohler was impressed. It also presents a contrast to Zuckerberg's stoicism in future jams. There would be plenty more moral dilemmas in the future, and most often Zuckerberg would resolve them with cold-blooded pragmatism.

Which is what happened here, after the tears dried. Cohler later explained that he wanted Zuckerberg to go forward without regrets and reservations. After the dinner, Cohler suggested Zuckerberg call Graham and ask him what he thought he should do.

"It was probably an unfair thing for me to do," says Cohler. Anyone who knew Don Graham could guess *The Washington Post* CEO's response to that question. Certainly, Mark Zuckerberg knew. Graham was not the kind of person who would compromise his values for a business advantage.

"Mark, if that's your feeling, I will release you from your moral dilemma," said Graham. He told Zuckerberg to take the Accel money. Take as much as you can get, Graham said. He hoped that the two of them would remain friends. (Zuckerberg would later ask Graham to join Facebook's board, and the two did remain close.)

But at twenty, Zuckerberg had learned something about business and also about who he was. He had made a major decision between what seemed morally right and what was right for him and Thefacebook. That month he painted the word FORSAN on the wall near his desk. It was a reference to a famous passage in the *Aeneid*, *"Forsan et haec olim meminisse iuvabit,"* where Aeneas addresses his lost and battered troops, saying, "Someday perhaps, remembering even this will be a pleasure."

WITH THE MONEY in the bank, one of the first orders of business was buying the Facebook.com domain, so the company could drop the awkward "the" from its name. It belonged to a company called AboutFace, which had nothing to do with colleges; it built employee directories for law firms and corporations. Parker was able to snag the domain for $200,000, clearing the way for Thefacebook to become simply Facebook.

More critically, with the venture money, Thefacebook could hire more people. The Harvard coders had done well to get the site off the ground, but it required people actually trained in computer science to scale the product for the huge population that used it. (Matt Welsh, Zuckerberg's Operating Systems instructor, blogged that "the original version of Facebook was a mess, technically.") But hiring those engineers was a challenge for a start-up, especially one that only college students could see. One tactic was to stand outside the CS department at Stanford and physically buttonhole geeky-looking people. Cohler's pet tactic was a bait and switch. He'd solicit top students for summer internships and then convince them to drop out of school.

One big win from that approach was Scott Marlette, a master's student who was frustrated with his professors. Even before meeting Zuckerberg, Cohler had lured him with the argument that at a small company, his impact would be huge. On his first day, he walked the two blocks to the Apple Store to buy a laptop, found a place in the office, and began to address the complicated infrastructure problems that the smart but inexperienced undergrad dropout engineers hadn't been able to fix. The process was helped along immeasurably by an experienced world-class engineer named Jeff Rothschild, who at fifty was almost like the ancient mariner. With experience at HP, he was perhaps the biggest whale Facebook landed. Marlette and Rothschild would keep Facebook's systems running as growth skyrocketed. Even with that expertise, the task was challenging: at one point, the heat in Facebook's server cages rose so high that Rothschild ran an expedition to buy the entire inventory of fans in local Walgreens stores to drive the temperature down.

The engineers coming to Facebook either were straight out of school—often leaving before they graduated—or moving after a year or so at a bigger company like Microsoft or Oracle. The latter group would be taken aback at the persistent chaos of the workspace above the Chinese restaurant on Emerson Street. Before noon, the office would be sparsely populated, and in the early afternoon people would drift in and settle at their workstations for fourteen hours or so of coding. Zuckerberg would be wandering around, often in his pajamas. "It basically felt like my dorm room at Carnegie Mellon, is the best way to describe it," says Aditya Agarwal, who chatted with Zuckerberg and had a more substantial interview with Jeff Rothschild. Agarwal got an offer that day, deciding to accept only when his girlfriend, Ruchi Sanghvi, who was still at Berkeley, told him how the product was blowing up. It paid $75,000 and "some options." Sanghvi herself joined a few months later.

That pattern happened often—an engineer would look askance at this ragtag operation and then make a call to a younger sibling in college, who would put to rest any doubts about the prospects for this dubious company. "I remember calling my brother, who was in his sophomore year at Johns Hopkins, where Facebook had just launched," says Soleio Cuervo, who had been puzzled by the mismatch between the stratospheric IQs he encountered at Facebook and its seemingly frivolous mission. *Facebook is bigger than God out here!* his brother told him. *It's blowing up!* Cuervo became Facebook's second designer, joining Sittig.

Zuckerberg actively took a hand in recruiting, particularly when trying to lure someone from an industry giant. A typical prime target was Greg Badros, who had just moved to Google's Gmail team. Also, perhaps not coincidentally, he was in charge of Google's social network Orkut. (The dual role told Zuckerberg a lot about the low priority Google assigned to social networking.) Interviewing Badros at the Facebook office would risk exposure, so they met at Zuckerberg's apartment on Ramona Street in downtown Palo Alto. Badros was struck by how sparsely Zuckerberg lived, in a one-bedroom apartment with a tiny table, a mattress in the corner with one blanket and no sheets. Sitting on the steps outside was Priscilla Chan, doing her homework there so she wouldn't disturb her boyfriend's recruitment push. (The couple had split when Zuckerberg first came to California but had gotten back together when Chan started medical school at Berkeley.) They spoke for more than an hour, and Badros, who had been wondering whether this wunderkind was out to change the world or just cash in on the start-up boom, had his answer: the former, times ten. Badros didn't take the job then, but Zuckerberg's intensity and curiosity stuck with him. Less than two years later he was at Facebook.

As the company grew, Zuckerberg's move-fast process often led him to hire people who might not fit, especially managers, and subsequent quick dismissals would lead to missing connections in the hiring process. In the spring of 2005, Sittig and Parker met a promising Stanford sophomore at an open house hosted by a Palo Alto start-up also casting for new hires. They urged him to come to Facebook, even for an internship. He met Zuckerberg at a Palo Alto noodle bar and fell into conversation about what it's like to start a company. Even as Zuckerberg was telling him it was the hardest thing ever, the student was thinking, *I want to do that someday.*

The student's peers and even a Stanford mentor advised against joining what they considered a silly college-based start-up. And he was having fun at Stanford; he had a girlfriend and was into his fraternity. Still, it was tempting. He kept corresponding with the engineering lead who was recruiting him. He was leaning toward taking the job. And then the mails went silent, and he decided Facebook didn't want him anymore. What he didn't know was that the engineer he'd been corresponding with had been fired. Had he pursued it, the job would have been his.

But Facebook was not done yet with Kevin Systrom.

WHAT CHARACTERIZED FACEBOOK'S method was the speed with which new code got pushed out. For instance, when Agarwal was at Oracle, he worked for months before he was allowed to make his first "commit" to the code base, and even then, his work went through four reviewers to make quadruple sure that the changes wouldn't affect anything. Even then, the actual change didn't appear to customers for years, because products were on a two-year release cycle.

At Facebook, they pushed out code four or five times a day. Essentially, Zuckerberg and Moskovitz were operating by the same rules as they did when Facebook was a dorm-room project. Since they never worked at any other company, they didn't realize how subversive their process was, that it essentially violated the accepted best practices of software development. Even Google rebuilt their indexes only every two weeks or so, queuing up changes for the regular updates. "We had no dogma, nothing to unlearn," says Agarwal. "So it's obvious—why would you wait?" If veterans who lived under the previous paradigm felt this to be blasphemy, they clearly weren't Facebookers. The attitude was, *We don't give a shit how long it took you to write code in your previous gig. At Facebook we want to move at light speed.*

The indoctrination came on an engineer's first day, when you'd download the code base, get your dev environment set up, and go fix a bug. If you hadn't coded in PHP before, or didn't know how an object-relationship model worked, figure it out—today. No excuses.

Maybe if Facebook had been working on something that clearly affected people's safety or well-being, it might have been different. But it was only . . . Facebook. What serious consequences could come from a glitch?

Almost to underline that point, if you introduced some horrible bug in your code that brought down the whole software stack of cards, with hundreds of thousands of Facebook users deprived of their passion, your embarrassment would be leavened by an email CCed to the entire engineering team: *Congratulations! You brought the site down—which means you're moving fast!* (Making the same mistake twice was not so celebrated.)

The process later became formalized: every employee who had anything to do with engineering, even someone at the VP level, would have to attend a "boot camp." Each attendee would be quickly exposed to the system and make a "code base commit" (making a change in the actual computer program that ran Facebook) within the first twenty-four hours—and push out the code in the next day or so. It was like being given control of a rocket ship the first time you entered the cockpit.

The main job of the engineers was "firefighting." A large part of what everyone did was fix problems to make sure the site didn't go down. Moskovitz was the central dispatcher. Later on, he would have regular meetings with Katie Geminder, an engineering manager hired in 2005, and they would try to keep track of who was working on what. But for months, the process was ad hoc.

Since the vast majority of the workforce consisted of people in their early twenties—including the boss—stepping into Facebook was like taking the SAT while playing a video-game version of beer pong. Employees were encouraged to basically consider the office their home. Zuckerberg even gave his workers a $600 a month bonus if they agreed to live within a mile of the office.

The company ethos was right there on the office wall. Only a few months after moving into the first office, Facebook opened a second one on the main downtown drag, 156 University. PayPal had started there, imbuing the space with excellent karma. In the lease negotiations, Parker offered a sweetener: in exchange for the right to take more space in the building, he'd allow the landlord to invest $50,000 in Facebook, a rare opportunity. The landlord was willing, but his partners nixed the deal, potentially costing them hundreds of millions of dollars.

Parker had the idea to hire a hot graffiti artist named David Choe to decorate the place. Choe thought that social-networking sites, including Facebook, were a big joke, and gave an outrageous quote of $60,000. Unlike the landlord, Choe gave in to Parker's offer to accept payment in Facebook stock options. Choe produced a sprawling mural with grotesque and profane images that metastasized all through the office. It was as if *Playboy* had commissioned Hieronymus Bosch to tag a subway car. Later, a girlfriend of Parker's would festoon the ladies' room with similarly non-PC images. The artwork was beloved by the nearly all-male Facebook workforce. But not so much by the (few) women working there.

Parker and a few Facebookers celebrated Choe's work at dinner at the Cheesecake Factory. They asked Choe if he understood the value of those stock options he'd received. Did he even know what stock options were? Choe admitted he didn't. But his previous job had paid him with a drum kit, so this seemed similar. Choe's stock options would eventually be worth more than $200 million.

Zuckerberg was an awkward public speaker, and at first was terrified to address even the ten or fifteen people at early all-hands meetings. "I had to sit down, I was so afraid," he later said. At another all-hands, he felt so awkward that he bailed halfway through, turning it over to Matt Cohler. But eventually he got fairly comfortable addressing his troops, and held weekly Friday meetings where anyone could ask him anything. When the session ended, he would shout out, "Domination!" in the spirit of the ancient leaders who had so long ago captured his imagination. Invoking a hero's journey for Facebook to a Kool-Aid-saturated workforce empowered a natural introvert to express the ambition he hid behind a poker face, and to do it with a bit of air-quote irony that provided a measure of deniability. "It was tongue in cheek but he also meant it," says Geminder. "And it was inspiring."

Eventually, Parker told him to cut it out, as one day it might be cited in an antitrust suit.

IN JUNE 2005, Mark Zuckerberg gathered his employees and told them what he had in mind for Facebook's second summer.

A site redesign.

A photo application.

A personalized newspaper based on users' social activity.

An events feature.

A local business product.

And a feature that he called I'm Bored, which would give people things to do on Facebook.

It was a list that would transform his site from a college directory to the world's premier social utility.

Zuckerberg's product vision was something that had been evolving since the move to California. At Harvard, he had simply pursued ideas on a whim, coding them up himself and releasing them with little thought. Even Thefacebook, which was a conscious culmination of his work to date, was quickly coded and shipped.

Now he had a growing team of engineers and a base of millions of users. While much of the effort was devoted to scaling the site and extending it to new campuses, he knew it was critical to introduce new features that would expand its powers and make it more addictive. He wanted those features to be ambitious, while the company still kept his own ethic of moving fast and fixing later. This meant not only directing projects cooked up in his own head but empowering his employees to make up their own, as long as they promised to add some value to Facebook. There was a try-anything flavor to the workflow. Moskovitz made sure that people were assigned to see that the crucial stuff got done, like keeping the site up, and doing the legwork to get the next campus rolling. But the group was always talking about new ideas. "We spent so much time together, sitting in the living room, eating meals, hanging around the pool, always talking about things we could build," says McCollum. "And the next step would be, someone would just build it." The projects had to be done fast, though. It was like you were constantly cramming for a final in a course where you hardly showed up during the semester. And the grader was Zuckerberg. "Mark would always be the final arbiter of what went in the product and how it worked," adds McCollum.

ONE PROJECT COMPLETED while the team was still at the La Jennifer house was the Wall. While Facebook still wasn't competing with AIM in real-time communications, Zuckerberg wanted Facebook to become a forum for expression. While he didn't openly promote his own politics, if he even had them, free speech was a passion for him. He had begun to see how the product he had created could be a powerful tool, maybe a historic one, to give people a voice. He would later tell me that free speech was "the founding ideal of the company." At the moment, Facebook users could only input snippets of information about themselves on the profile page. After endless internal debate, Facebook came up with a way to start actual interaction, by introducing kind of a dynamic whiteboard in the center of the profile page. In the style of *Wikipedia*, people could add to or edit parts of someone's profile. "That was when *Wikipedia* seemed like the most innovative, unbelievable, incredible thing," says Chris Hughes. "We created a spot on the profile where you could write anything, free-form, and other people could write too." It eventually evolved into a way for people to add text-only remarks on someone's profile page. The comments were organized in reverse chronology, like a blog, and appeared in the center of the page. People could comment on someone's profile, discuss what happened at the party last night, or just talk trash.

It subtly changed the character of the site, moving it from a directory to something more interactive. Facebook, via the profile information, was already hosting what was known in the industry as "user-generated content." But throwing open the gates of discourse raised a lot of issues that Facebook didn't have answers to. Who got to control the Wall? Did people own it, or did Facebook? What should be allowed there, and what was verboten? In true Facebook fashion, it released the feature with no resolution to those knotty questions.

No one thought much about how to deal with inappropriate comments.

Following the Wall was a feature called Groups, which were a variation of the profile page where the Wall could act like a bulletin board to organize people around a shared topic. Any Facebook member could launch a group, which would be accessible in the same way a profile was. Zuckerberg had lofty ideas for this: existing student groups could move their governance online; students running for college offices would establish virtual campaign headquarters; activists would petition to make changes on campus. Much of this did occur, as notices formerly posted on corkboards in dorm vestibules moved to Facebook. Still, the most popular groups seemed to be whimsical or borderline dumb. There was the Anti-Popped Collar Club and the Students for the Relocation of Harvard University to the Alternate Universe Where Kerry Won. There was even a group about loving Facebook Groups. The *Crimson* reported on a phenomenon called "facebook group whores"—people who promiscuously accepted every invite to a new group.

Now Zuckerberg's ambitious initiative marked a new thinking of the website. What all the ideas had in common was that they all seemed to dovetail with an expansion of Facebook. Also, most of them were sufficiently complicated that they would take much longer than the run-and-done projects of the past.

Indeed, of the five ideas on the Zuckerberg agenda, only one was completed before Labor Day: the redesign. Aaron Sittig, who had finally joined Facebook full-time, headed the project. One of the first things he did was question "the weird-looking creepy guy" on the top of the page. Not long after, worried about intellectual-property issues, they tracked down where the vectorized image came from. It turned out to have been in a Microsoft Office clipart collection that was freely available to anyone with an Office license. That meant Facebook could not trademark its own page. The Facebook Guy was gone.

Sittig's aesthetic was clean and modern. It was a striking contrast to what was then the dominant social network, MySpace, which had about ten times as many users but was painful on the eyes. It allowed users to customize their pages, and viewing it was like visiting Shibuya with a hangover.

Eschewing that riot of color, Sittig limited himself to a blue palette. This had the advantage of registering most clearly to Zuckerberg, who is color-blind and can't see reds or greens. After a lot of searching, Sittig finally found a shade he liked. It was the background on the website for the powerful and politically connected Carlyle Group equity fund, once described by author Michael Lewis as "the salon des refusees for the influence-peddling class." Sittig appropriated that blue, and it ultimately became synonymous with Facebook to billions of people. Inside the company, the Facebook site would eventually be referred to as the Blue app, and sometimes just called Blue.

Next to be implemented was Photos. Facebook's users were allotted a single profile photo, and that was it. So desperate were they to share photos that some would constantly update their profile images, sometimes multiple times a day. Others discovered that while Facebook limited the width of photos on the page, there was no similar vertical limitation. So they would edit a bunch of photos in a stack, as you'd see in a strip of pictures from a photo booth, and use those ad hoc photo albums as their profile shot.

MySpace, which did allow people to post some photos, had just increased the number from fifteen to fifty. Meanwhile, the hottest photo-sharing site at that time was Flickr, where people publicly shared their photos, often tagging them with labels that people could search for. That was great for pictures of things. But Facebook was a place for *people*, specifically people you knew.

Zuckerberg assigned Sittig, Marlette, and Doug Hirsch, Facebook's new product manager, to come up with something new. They started to whiteboard what the app might look like. As always, Zuckerberg would be weighing in, they'd work on it, talk to Mark, work, talk to Mark, and so on.

Marlette was a photographer himself, but he realized that Facebook users would care less about the artistic aspects of their shots. Their pictures, like the other things posted to their profiles like interests or relationship status, would be an expression of who they were. So it didn't matter if they were high-resolution—it made better sense to use lower quality images so they would load fast and not overwhelm Facebook's servers with storage issues.

One night Hirsch and Sittig were brainstorming. Hirsch suggested that Photos should have a social component. Sittig had a brilliant idea—why not tag the people in the photos? It was something our brains already did when we looked at pictures. So it was simple to execute, no AI involved—just have people click on the faces in the photos and fill in a blank text box.

Facebook didn't have the artificial intelligence for facial recognition yet. But it did have fanatic users hugely motivated to share. Sittig set up a system where people could quickly note who was in the photo—if the person was on your social graph, just typing a few letters would automatically fill out the full name. The whole thing was engineered to encourage you to tag the people in the photo. Then a flywheel would take over, to extend the experience to others. When you were tagged in a photo, you'd get notified, and of course you would go to that person's profile page to see the picture. If the person with the photo wasn't your friend yet, you might friend the person right then. And you might be more likely to then post photos of your own.

At least, that was the theory: No one was sure. "We knew there was a desire for more photos," says Sittig, "but we weren't sure what people would do with this."

When it was time to launch, in October 2005, they put up a big display, setting it up with a big grid that showed what people were uploading, and whether they were tagging. It was around eight or nine at night. The first photos that came up were someone's desktop wallpaper from Windows. Not promising. But the next set was a group of girls at a party. Terrible photos—you couldn't even see the background, just bodies and faces lit by a flash. But they were tagging!

"That's when we knew it was going to work," says Sittig. "A photo like that is entirely about telling a story publicly that you are friends with these people, it's a way of expressing public support for the people you are in photos with, and reinforcing a relationship that you already have."

Within a few months, Facebook was the most popular photo-sharing site in the world. It was all they could do to keep the servers running to store all those photos.

SEAN PARKER WAS not there to see the Photos launch. On August 27, just after midnight, Parker had been on vacation, accompanied by a college-age Facebook employee he was dating, when the police entered the house he was renting in North Carolina. They found what they assumed was cocaine, and arrested Parker for possession. Parker undoubtedly dreaded the consequences, correctly suspecting that even if the charges were eventually dropped (as indeed they were), it opened the possibility that once again, he would be robbed of a continuing role in a company he helped build.

His fears were well founded. Parker was arrested late on a Friday, and the board held an emergency meeting that weekend. Zuckerberg reorganized the company in a way that essentially demoted Parker. "I wanted him not to be president anymore," he would say not long afterward, under oath in the ConnectU case deposition. It wasn't just the fact that Parker's arrest had put Facebook in legal jeopardy. By then, Zuckerberg viewed Parker more as a visionary than an operating president. He wasn't doing a very good job of managing the sales team. Moreover, "he freaked people out," said Zuckerberg.

"Sean's hours were so irregular. You'd go days without seeing him," says Ezra Callahan—Parker's friend. "He was super flaky and unreliable and hard to get ahold of. When you needed him, he'd come in at the eleventh hour and save the day, but you couldn't rely on this guy for anything."

Though he lost his job, Parker was not banned from Facebook's offices, and neither from Zuckerberg's good graces. For the next few years he would be a mercurial presence, swooping in at various times, a usually welcome presence in product meetings. (Not much different from when he was supposed to be at all the meetings.) Zuckerberg valued his opinion. And Parker made the most of that. Parker would remark that Zuckerberg owed him. He was like the guy who saved your life in Vietnam and never let you forget it.

A month after Parker left, Facebook raided Amazon to hire Owen Van Natta, a thirty-six-year-old executive known for his deal-making skills, to head business development. After a few weeks, Zuckerberg promoted him to chief operating officer.

Years later, when Parker became somewhat maligned as a con man, a characterization cemented by Justin Timberlake's manic and unflattering performance in the infamous movie, people would write Parker off as a footnote in the Facebook saga. But those who were actually there say otherwise.

"The company just would have been sold without him," says Adam D'Angelo. "It would have been taken over by venture capitalists. It was priority number one for him not to get [Facebook] burned in the way he got burned."

Parker wasn't burned by Facebook like he was at Plaxo. He'd made sure of that during the negotiations with the VCs. His own deal specified that even if he left Facebook, he would retain his percentage in the company, a cut that would eventually provide him a reliable standing in the Forbes Billionaires list.

But it was Mark Zuckerberg who kept control, with the biggest stake of all. "Whether it's Peter Thiel or Sean Parker, these people thought they were manipulating Mark," says one early Facebook employee. "But Mark saw Sean as a useful tool to do the job that sucks the most—fundraising. In hindsight, it was genius that Mark convinced Parker to raise all the money for him."

6

The Book of Change

ZUCKERBERG CARRIED A notebook. In 2006, the year that put Facebook on a course for greatness and infamy, you would have seen him in the Palo Alto office, head down, scrawling on an unruled journal in his crabbed, compact script, sketching out product ideas, diagramming coding approaches, and slipping in bits of his philosophy. Those who visited his one-bedroom apartment, with its mattress on the floor and a kitchen that never saw an egg boiled, might spot a stack of completed notebooks.

Zuckerberg was no longer doing much coding, but he'd use those notebooks to convey a detailed version of his product vision. The method compensated for his interpersonal shortcomings. When Facebook engineers and designers would roll into the office in the late morning or early afternoon they would sometimes find a few photocopied pages sketching out a design for a front end or a list of signals for a ranking algorithm. The pages didn't necessarily end the conversation, but often opened one up, as recipients would use them as a basis for collaboration with their boss.

Whiteboards appeared in abundance in every Facebook office, and employees couldn't survive without excellent dry-erase skills. But a Zuck notebook carried an almost papal sanctity. It was like glimpsing his soul.

In later years, Zuckerberg would be less enthusiastic about logging things. He'd explain that Facebook's lawyers identified his cherished notebooks as potential evidence in future intellectual-property lawsuits, and so he destroyed them. But that wasn't the only reason. After he was humiliated by the publication of the callow IMs he wrote at Harvard, he was reluctant to archive his personal thoughts anymore, even as he urged users to do so. (Years later, he would even demand a personal exception to the protocols of his messaging services, and had Facebook delete his half of private communications from the logs of the people he chatted with. Stored chats would suddenly be one-sided, with the Zuckerberg half of them gone. Facebook described it as "limiting the retention period of Mark's messages," while everyone else's retention period was infinite. Under pressure, Facebook promised to allow ex post facto "unsends" for all. The feature took almost a year to implement.)

But the notebooks aren't totally lost. Snippets left behind, presumably those he copied and shared, present a revealing window into his thinking at the time. I managed to get a seventeen-page chunk from what might be the most significant of his journals in terms of Facebook's evolution. It was called "Book of Change." Dated May 28, 2006, the first page has his address and phone number information, with a promise to pay a $1,000 reward to anyone returning the book to him if lost. He even scrawled an epigram, a message to himself:

> *"Be the change you want to see in the world."*
> —Mahatma Gandhi

The Book of Change grapples with the two projects that would transform Facebook from college network into Internet colossus.

THE FIRST WAS called Open Registration, referred to internally as Open Reg. It would flip the very essence of Facebook, from a college networking program to a general social utility.

It's unclear how long Zuckerberg had planned this. He would often recount the night Facebook launched, when he met with friends at Pinocchio's pizza parlor and talked about how someday some big company would do on a global scale what he had just done at Harvard. Over the next two years, he became emboldened to think that he was the person to do it. It's hard to pinpoint a moment when this epiphany occurred: early employees agree that this wasn't the mission in 2004 or early 2005. In June 2005, he told an interviewer that while a lot of people "are focused on taking over the world or doing the biggest thing to gain users," he was more into making a difference by concentrating on upping his service to the college world.

By 2006, though, expanding to everyone *was* the mission.

Zuckerberg now insists that once Facebook became more than a college project, he was viewing it as more than a utility for colleges. "I just think I was never particularly excited from a mission perspective of building the next MTV, which was the broad conversation that we had at the time," he says.

Open Reg was the necessary step to make it happen. But it was dangerous. Zuckerberg realized that taking the step of opening Facebook to the general public could potentially jeopardize its base of college students, who thought that Facebook was theirs. He decided to move incrementally, so people might still get a sense that private information would stay within one's community.

First came a seemingly organic move to high schools. But even that step required a massive technical rejiggering that set the stage for a wider expansion. Up to that point, Facebook had a siloed architecture. When Facebook entered a college, it created a discrete database for that institution. That set a natural boundary to the community that established a high level of privacy by way of infrastructure design. You couldn't browse the profiles of those outside your school. And you didn't miss it, because colleges, including their alumni, had thousands of people in their communities: the networks were still interesting.

But that structure wouldn't work for high schools. There were only a couple thousand possible colleges—but around 40,000 high schools. That required setting up a more open system that ultimately would supersede the siloed college networks. The technical challenge would occupy Facebook's newer engineers for weeks.

No one was sure how well Facebook would play in high schools. Facebook started the process by having college students invite their younger siblings who were still in high school, hoping that the association with older kids would imbue it with a sophistication that MySpace lacked. That worked fairly well, but there wasn't the personal immersion that college kids had. High schoolers were still well connected to networks beyond their classmates—parents, after-school activities, and friends in other schools. The best news for Facebook was that one fear didn't come to pass: that having younger teenagers on the system would make it uncool for college kids.

That spring Facebook tentatively opened up beyond education with Work Networks, using a company instead of a school as the basis for a community. Facebook selected one thousand big employers, ranging from tech companies to the armed forces, and allowed anyone with an email address on that firm's Internet domain to sign up. It flopped. Unlike college or high school, people generally wanted to keep their social lives separate from their jobs. Not every workplace was like Facebook, where people worked, played, and romanced in a tight community.

Allowing entry to all, by simply providing an email address, was the obvious solution. You would be able to network with everybody—friends, relatives, colleagues—because everybody could be on Facebook. Still, Zuckerberg initially felt that new users would have to be rooted in *some* network. If people weren't rooted to a community, it would be much harder to know whether they were who they said they were, or just tricksters like you found all the time on MySpace. That level of trust was key to Facebook's success.

So Zuckerberg decided that absent a school or workplace, new users would be organized by where they lived. But a network the size of a big city was so sprawling that it hardly provided any security at all. It wasn't as if you would be okay with someone checking out your profile because you both lived in Chicago. Other scenarios would give Facebookers the shivers. Could an adult "friend" a high school student? Wouldn't that be creepy? Or just uncool? "Facebook would be taking this service that college students considered their own, and opening it up to a bunch of randos," says one employee from the time. "The thinking of folks at the time was that older people were just going to make it lame."

Maybe older people were too lame to even *use* Facebook. That worry was underlined by user testing, a standard practice in software companies belatedly adopted by Facebook at the urging of Katie Geminder. After much politicking, she got a go-ahead to hire an outside research firm to test the kinds of users who could not be demographically located within the Facebook workforce, like people over forty. It uncovered a lot of biases in the product, many related to its college roots. One example came when a woman research subject of a certain age encountered the "poke" command. She asked the facilitator what that was and the response, out of the facilitator playbook, was, *What do you think it is?* She didn't know. *How would you find out?* She said she'd try the Help section. *Okay, why don't you try that?* And she did. The answer she found there when she asked for help on the word "poke" was "If you have to ask, you shouldn't be here." Not very welcoming.

Zuckerberg wrestled with all of this in the Book of Change. A day after he started the notebook, he began a page called "Open Registration," and asked, "So what do we need to hash out before we build this?" His focus was making it happen; the idea that Open Reg might lead to billions of users, and a mile-wide rat's nest of unintended consequences, was not part of his analysis in 2006. He diagrammed the information flow on the sign-in process, asking people to declare whether they were in college, in high school, or "in the world." He decided that people should use their zip codes to identify their geographical network. He even mused on how privacy should work. Could you see profiles of "second-degree" friends in your geographical region? Or anywhere? "Maybe this should be anywhere, as opposed to just your geo," he wrote. "That would really make the site open but probably not a good idea just yet."

That comment seemed to characterize his mindset, where he seemed at least as concerned about the *perception* of privacy as about privacy itself. He knew by then that Facebook was destined to be totally wide open, and that he was pulling back on the original deal he offered to Harvard and other schools, where only your classmates could browse your profile. But he still wanted people to feel as safe, despite that. In designing Open Reg, he posed one final question to himself.

"What makes this seem secure, whether or not it actually is?"

WHILE SOME OF Facebook's engineers worked to implement Open Reg, another team was remaking the site itself, centering it on a product that would come to be synonymous with Facebook: the News Feed. It would become Facebook's biggest boon and also the source of its future woes. The Feed was the personalized newspaper Zuckerberg had brought up in his summer of 2005 to-do list. He didn't really get to it until late in the year, brainstorming it with Adam D'Angelo, who was around for his Caltech winter break.

D'Angelo and Zuckerberg saw it as a way to reinvent Facebook. Both agreed that despite its success, Facebook was kind of broken. The home page was wasted space—people would quickly abandon it and go to their Friends page, to see who had updated their profile. Then you would painstakingly click through each of those to see what was different. The Facebook logs showed that a huge number of people actually went through their friend list in alphabetical order to make sure they were on top of new activities. "This is how all social networks worked at the time, but this felt very inefficient," says D'Angelo. "Everyone is spending so much time clicking around on those profiles."

Zuckerberg's solution was the News Feed. The information that was now buried on profiles would be distributed directly to your friends, like a newsboy tossing papers onto the front porch. With Facebook, the news would hit on your front page. One way to do that would be to put little boxes on the home page, with updates on things like events and new friends and other developments since you last logged in. The other, more ambitious, approach was to do a continuous stream of news, flowing up your screen in reverse chronology. Zuckerberg chose the latter.

D'Angelo began working on it, but then headed back to Caltech for spring semester. Ruchi Sanghvi was one of the few engineers on staff experienced enough to work on the complicated architecture, so she wound up on the project.

Another person tapped to help develop the product was a recent Stanford graduate named Chris Cox. Born in Atlanta and raised in Chicago, Cox was not a typical nerd: he had movie-star good looks with a smile that could fire up spotlights. He was a serious musician, playing multiple instruments but particularly excelling as a jazz pianist. Stanford brought out his geek side, and he majored in symbolic systems—a cult program whose alumni include Reid Hoffman and Google's Marissa Mayer—and took courses with world-renowned masters of AI. He was around the AI lab when it won the DARPA self-driving car challenge.

After graduating in 2004, he decided to take a year off before grad school, traveling the country and doing computer consulting. When he returned from his trek, he was back at Stanford and living in one of a group of houses in Palo Alto known as the Grateful Dead Houses, run by a landlord who was a serious fan of the band. (Cox had moved into the "Truckin'" house.) Also living there was Ezra Callahan, who'd moved there from the Los Altos house. Every day Callahan would come home and tell Cox how amazing Facebook was, and that he should come work there. Cox would say he wasn't interested. Why would a Stanford AI graduate with dreams of solving natural language processing work for a silly company with posts and pokes?

Callahan ultimately convinced him to come in, and he interviewed with Moskovitz, Jeff Rothschild, and Adam D'Angelo. Moskovitz explained to him that Facebook was the seed of a collaboratively created directory of people with one authentic representation for every individual. It would be interconnected, it would be live, and there were plans to expand it beyond colleges.

What really blew Cox's mind was the question they asked him in the interview: *How would you design a feed that presented the latest news of your friends to you?* As Cox stumbled to answer, he realized that there were serious computer-science obstacles to producing such a product. As the discussion progressed, Cox realized that his interlocutors were as talented as top engineers in bigger companies, especially the more experienced Rothschild.

They offered Cox the job on the spot, but he took a week to think on it, during which his friends, mentors, faculty advisers, and family all told him it was a horrible idea to leave graduate school for this weird little company. But he followed his gut and began in November as the company's twelfth engineer.

Along with Cox and Sanghvi, Zuckerberg assigned a third engineer, Dan Plummer, to work on the project. At thirty-nine, Plummer was almost twice the average age of his colleagues. Facebook had convinced him to leave a faculty job at UCSD to come there as its first research scientist. He was trained as a scientist, and had done serious research on vision problems, but was a kickass computer scientist as well. Plummer was also a champion cyclist.

The team called the project Feed, in keeping with describing its products generically (Photos, Groups, etc.). But the social-networking trademark for that word had just been secured—by Viacom, the owner of MTV, which at the time was trying to buy Facebook. So it became the News Feed instead.

From the start, everyone understood that News Feed would take months to develop—a drastic departure from Facebook's usual process of churning out something in a few late nights and releasing it immediately. Right away came a tragic setback. On January 4, 2006, Plummer was cycling in the Los Altos Hills, not far from Palo Alto, when he was fatally hit by a falling tree branch. When people got back to the office after the winter break, a few words were said in his memory. And then everybody got back to work. "It was almost as if he was washed away into the ocean without a trace," one Facebooker later wrote.

Not exactly. As would be the case for thousands of the newly departed, Plummer's memory would live on, in a sense, as an existing Facebook profile. You can check out of life, but you may never leave Facebook. (Years later, the company would develop detailed protocols for what happened to your pages when you died, explicitly making the case for Facebook presence ad infinitum.) Plummer's profile is still searchable; a month before he died, he posted some pictures of his new puppy.

Plummer's place was filled by another recent hire, Andrew Bosworth. Everyone called him Boz. His arm was tattooed with the word "*Veritas*": it was Harvard's motto but, more important, in Roman mythology it was the name of the Goddess of Truth. She was Boz's muse, as he always seemed to say what he thought, even when others wouldn't. Some considered him a truth-teller and others an obnoxious loudmouth. But he was a hard and smart worker.

Bosworth was a rare tech engineer whose blood went back generations in the Silicon Valley; his family had raised apricots and prunes in the hills looming above Sunnydale and Cupertino since the 1890s. When Bosworth grew up, the homestead had become a riding and boarding stable for the well-off inhabitants of a tech-transformed Valley.

He learned programming from an HP employee involved in his 4-H Club, and studied computer science at Harvard, which he chose because it looked like he might make the football team there. In his senior year, he was a teaching assistant in the popular Intro to Artificial Intelligence class when one of the brighter students jarred the campus with a prank called Facemash. *Hey, man*, Boz emailed Mark Zuckerberg, *this is probably not the best idea.*

Boz signed up for Thefacebook on its second day—user 1,681. But a year and a half later, when a recruiter sent him an AOL instant message trying to lure him to the company, he felt it was too small an operation to consider. Besides, he told the Facebook recruiter, he and a friend had just bought a house. *You'll be able to buy TEN houses in Silicon Valley!* the recruiter told him. *I know Silicon Valley*, Boz thought, *and nobody can afford ten houses there.*

But he figured he'd get an expenses-paid trip to see his family. Before his trip out west he went to lunch with a group of eight Harvard friends working at Microsoft at the time, and told them he was going to talk to

Facebook. They all had a good laugh. Within a year, five of them would be working at Facebook.

What sold him on Facebook was the speed with which the company pitched product. At Microsoft, when Boz had come up with a great feature idea for the release of a product, users would not see it for more than a year. At Facebook, you could put ideas into production within hours.

And then there was his former student, Zuckerberg, who was shockingly bold in his ambition. *We are going to connect the world and be this global fabric!* Zuckerberg told him. *Can you imagine what that would be like?*

Bosworth was hooked, and from that moment he became Zuckerberg's loyal lieutenant.

At Zuckerberg's request, Sanghvi became product manager on the project, although the role was new at the company and she wasn't sure what that meant. Someone dropped a few books about management on her desk and she diligently plowed through them. She was working on multiple products then, and sometimes the demands got out of hand. One time, Bosworth came up to her saying he needed her to do something for News Feed immediately, and she told him she was launching the other product the next day. He told her he really needed her to do the News Feed thing and she blew up. "Boz, if you don't leave me alone, I'm going to throw a shit fit." And she stood up and screamed. It was just another day in the Facebook office.

The team was guided by copied pages from the Book of Change. Zuckerberg was thinking hard about what would appear on the News Feed, diving deep into the criteria that would determine which stories should appear on one's News Feed, and how they might be ranked. Zuckerberg was just trying to improve Facebook. He wanted to make it easier for people to see what was important in the world of the friends they had consciously connected to. He had one word in mind as a yardstick for inclusion: "interesting-ness." It sounded innocent at the time. He had no idea then how important such a ranking would be, that democracies could fall and minds could deaden by the wrong stories appearing on one's News Feed.

As depicted in his compressed penmanship, Zuckerberg envisioned a hierarchy of what made stories compelling, with the key factor being a blend of curiosity and narcissism. There were three tiers of value to stories. The top was "stories about you." The highest priority for Facebook would be to share those times when people posted on your wall, blogged about you, tagged you in a photo, or commented on your post or photo.

The second most important category involved people you cared about—those whom Facebook understood to be in your social circles. He provided examples of the kinds of stories those might include: changes in your friends' relationships; life events of those you know. Next there were "friendship trends." These involved people who moved in and out of circles you were familiar with. Finally, he mused on a future utility that mentioned "people you've forgotten about resurfacing."

The least important stories, but still worthy of inclusion, were less about you and your social world, a category he called "stories about things you care about and other interesting things." Here is where Zuckerberg sketched out how his vision of News Feed as a personalized newspaper was much broader than the boundaries of one's connections. The News Feed might also include a stream of information that could well augment or even supplant traditional means of news and entertainment. His list of possible stories might include:

- Trends in media, interest groups, etc.
- Events that might be interesting
- External content
- Platform applications
- Paid content
- Bubbled-up content

Zuckerberg was only getting started. Over the next two days he feverishly outlined ideas about privacy, how Facebook would open to high schools and then to everybody, the design of a "Mini-Feed" that would track activities of and about individual users, and a host of other ideas. At one point, his pen seems to run out of ink and he switches writing implements. "Sweet, this pencil works better," he writes, and a page later he is sketching out what seems to be a grand vision for Facebook. He calls it "The Information Engine."

> *Using Facebook needs to feel like you're using a futuristic government-style interface to access a database full of information linked to every person. The user needs to be able to look at information at any depth . . . The user experience needs to feel "full." That is, when you click on a person in a governmental database, there is always information about them. This makes it worth going to their page or searching for them. We must make it so every search is worth doing and every link is worth clicking on. Then the experience will be beautiful.*

One way Zuckerberg felt that he could give that depth of information would be making personal profiles about people who *weren't* on Facebook. He called these "Dark Profiles," and spent several pages writing about them. He envisioned users creating these Dark Profiles of their friends—or just about anyone who didn't have a Facebook account. By giving the name and email address of the person, you could start such a profile—you'd be informed if one already exists—and you could add information to it, like a person's biographical details or interests. The owner of one of those dark accounts would be part of the Facebook conversation. Every so often email alerts might pop up in a dark-profiled person's inbox about activity involving them on Facebook. Presumably that would motivate them to sign up. Zuckerberg was aware that opening profiles for people who had no desire to be on Facebook might stir up some privacy concerns. He spent some time pondering how this could avoid being "creepy." Maybe, he wondered, dark accounts might not be searchable? It's not clear how much of this idea was implemented. Kate Losse, an employee at the time, would later write that she worked on a dark-profile project around September 2006. "The product created hidden profiles for people who were not yet Facebook users but whose photographs had been tagged on the site," she wrote in her 2012 memoir. She now explains that when those nonusers responded to an email—provided by the person tagging them in the picture—the tagged photos would be waiting for them. "It was kind of peer-to-peer marketing at Facebook directed at people who had friends on the site but hadn't signed up yet," she says. Ezra Callahan confirms this, adding that though the idea that users would be able to create and edit dark profiles of friends Wikipedia-style was brainstormed, it was not executed. (Facebook has always insisted that dark profiles do not exist.)

Aaron Sittig was in charge of the News Feed design, which would be a major part of a general Facebook redesign. He understood how it would transform the site. "The idea of a front page that was linear, chronological, and customized per individual user had never really been done before," he says.

While the News Feed showed the user what was going on with their friends, Zuckerberg also envisioned a second feed that would tell friends what was going on about *you*. That feature, the Mini-Feed, would live on the profile page, taking up as much space as the Wall. "When someone arrives at someone else's profile, they need to feel like they know what's going on with that person and who that person is," he wrote. In reverse chronological order, it would reveal all your Facebook "events"—who posted a picture about you, whom you friended, what changed in your relationship. "The idea is to present a log of each person's life, but hopefully not in a creepy way," wrote Zuckerberg in his notebook. "People should have control of what displays in their event stream and they can add and remove things, but they shouldn't be able to turn them off."

Through the spring and into summer the team worked away. One day Cox got a story to appear on the prototype News Feed page he set up for himself. It noted some activity from his boss, and since they were Facebook friends, post zero appeared:

Mark has added a photo.

Oh my God, it's alive! he said to himself. *Facebook just got ten times more useful!*

People, he thought, *will just love this.*

WHILE ZUCKERBERG WAS madly creating his private product manifesto, he was involved in a drama that did not find its way into the Book of Change. He was fending off potential buyers of Facebook who would wrest it from his control. While still at Harvard, he had seemed to welcome the idea of selling off his project, even joking that if the Winklevoss twins won a lawsuit against Thefacebook for stealing their idea, paying damages would be the problem of whoever bought the site. But now he was totally committed to his creation. He felt it could make a difference in the world. But not if someone else owned it.

After word leaked about Facebook's metrics, a conga line of suitors came calling. For other social networks, buying Facebook would neutralize a threat. To big tech companies with little presence in social networking, it was an opportunity to enter that arena. For media companies, it represented a pipeline to young customers.

Zuckerberg spent an extraordinary amount of time taking meetings with those suitors, often accompanied by Owen Van Natta, who was an experienced dealmaker. It would take a while for Van Natta to accept the fact that Zuckerberg entered—and exited—every discussion about being acquired with the intention of rejecting the offer. Van Natta felt that if those tech and media giants were interested, Zuckerberg should hear them out. In the process, Zuckerberg would learn about the business he was entering.

Sometimes, the discussions led to partnerships, as was the case with Microsoft. In 2006, the two companies reached a deal for Microsoft to sell Facebook ads to international clients, an arrangement that provided Facebook much-needed revenue.

Other times, Facebook's aim was to gather intelligence. Zuckerberg, Van Natta, and Cohler had several meetings with MySpace, their putative rival, just to get to know them better and see if they could learn something. "The point for us was to learn and better understand their team and culture and how they thought about [their] product. The point for them was to buy Facebook," says Cohler, who admits to being obsessed with MySpace's dominance.

MySpace CEO Chris DeWolfe confirms that an acquisition was indeed on his mind when he and a small team visited the Palo Alto office in early 2005. Zuckerberg arrived late because he'd been at the colocation site dealing with a server crisis. DeWolfe was impressed but, after the Accel funding, thought that the value of Facebook was too high. That summer, MySpace was purchased by Rupert Murdoch's NewsCorp for $580 million. With bigger coffers, MySpace was ready to spend more to buy Facebook. Zuckerberg wasn't interested.

In general, Zuckerberg seemed nonchalant about MySpace's huge lead in users. He saw the NewsCorp purchase not as a threat but as a validation of the worth of social-media companies. The Facebook team believed that MySpace wasn't a technology company, and didn't have the rigor that came from focusing on products. Zuckerberg didn't bother to hide his views on this, even to the MySpace founders themselves, much to their annoyance. (DeWolfe disagreed with Zuckerberg: "I think we were both media companies and both technology companies," he says, though he admits Facebook was driven more by engineering.) At a later NewsCorp retreat, Zuckerberg told Rupert Murdoch that the future of media wouldn't be people tuning into Fox News or getting *The Wall Street Journal* on their doorstep, but getting links from their friends online.

For a time, Viacom, via its MTV unit, was pushing hard to buy Facebook. After a number of meetings, Zuckerberg turned them down. He also spurned a Google offer.

But one company could not easily be rebuffed. Yahoo! was then a multibillion-dollar Internet giant with hundreds of millions of users. Facebook's product head, Doug Hirsch, a former Yahoo! exec, had apparently tipped off his former boss about Facebook's value. Yahoo! CEO Terry Semel came to view Facebook as the crowning purchase in a cluster of acquisitions of socially themed websites, including Flickr and Delicio.us. Semel had once passed on buying Google, and this deal would offer redemption.

"Originally, to get us to the table, they were like, we might offer three billion dollars," says Zuckerberg. "I was like, *Well, okay.* When it got down to it, they were talking about a billion dollars."

That was still a jaw-dropping figure, almost otherworldly. A billion dollars for a small company that was still in its infancy. Hundreds of millions of dollars in the pocket of its twenty-year-old founder.

Zuckerberg was no more inclined to sell to Yahoo! than to anyone else. One day a newly hired engineer showing up for work on his first day saw Hirsch packing his things and asked Zuckerberg how he might avoid getting fired like that guy.

"Don't try to sell my company out from under me," said Zuckerberg.

At a board meeting around that time with Thiel and Breyer, Zuckerberg dramatically looked down at his watch. "Eight-thirty seems as good a time as any to turn down a billion dollars," he said. It was a gauntlet thrown to fellow board members who might be wishing for a quick, lucrative payout.

ZUCKERBERG'S HAND WAS weakened by a disturbing development. In mid-2006, Facebook had stopped growing. The dashboards told the story: user numbers had stopped going up. College students were already on board. Facebook wasn't finding the same instant success in high schools. Workplace networks was a failure.

"We had tried something with high schools that didn't work out—it wasn't growing that quickly," Zuckerberg recalls. The News Feed hadn't launched. Soon Facebook would open its gates to all comers—but some people in the company were warning that Open Reg might be the biggest risk of all. Some thought Facebook should double down on colleges and build other services on top of that market. *Own* that market! But Zuckerberg was committed to playing his real-world version of Risk. Colleges were a tiny square on his game board.

"It was very clear from the beginning that this was a utility for everyone on the planet," says Cohler. "He felt, *No, I'm not going to go deep in university, I'm going to go broad to the world.*"

Zuckerberg tried to stall Yahoo! Semel complained in one meeting that the Facebook team wasn't moving fast enough, that they were too inexperienced to make a deal. Indeed, throughout the meeting, Zuckerberg went into his coma routine. Finally, when it was apparent that everyone in the room was desperate for him to say something, Zuckerberg spoke.

"Well," he said, "we think that companies suck."

Yahoo!'s president, Dan Rosensweig, broke the tension with a quip. "At Yahoo!, we like to think we suck less." Everyone had a good laugh. But the stalemate continued. "Terry's Hollywood approach to negotiations definitely wasn't working with Mark," says Facebook's general counsel at the time, Chris Kelly.

Kelly was among the few who supported Zuckerberg's stance. He knew that his boss's resolve might be faltering, and felt it might help to put him in touch with someone with a lot of experience in the Valley who might give some contrarian perspective. Kelly knew a well-known investor named Roger McNamee, and arranged a meeting between the two. Before Zuckerberg said a word, McNamee correctly analyzed the situation. Zuckerberg didn't talk for a long time, and finally blurted out that he didn't want to sell, but wondered if he should. "I don't want to disappoint everybody," he said. McNamee endeared himself to Zuckerberg by telling him that he should feel free to follow his heart.

The pressure was enormous. Owen Van Natta was a strong believer in selling. One night he and Zuckerberg were heatedly arguing the issue in the office. It went on past one in the morning. "If you don't sell the company," Zuckerberg recalls him saying, "you're going to regret the decision for the rest of your life!"

Zuckerberg knew he'd have regrets if he *did* sell it. But he wasn't sure how to deal with the enormity of the offer. Could you really turn down something that big? He had no framework on how to value his company. "It was very hard for him," says Chris Kelly. "He was a very nervous person at that point and was sometimes kind of paralyzed as he ran through things."

Indeed, Zuckerberg himself had trepidations. Ever since Thefacebook had exploded at Harvard, and each step of the way in its journey, he had been opportunistic and ambitious. But he also felt the doubts that anyone in his early twenties might feel when suddenly tossed into the deep waters of high finance and monumental decision-making. Would things really work out? Who was he to do this? "I definitely had this imposter syndrome," he says. "I'd surrounded myself with people who I respected as executives and I felt like they understood some things about building a company. They basically convinced me that I needed to entertain the offer."

Zuckerberg actually did crumble at one point and tentatively agreed to take the money. But Semel overplayed his hand. In the future, when Facebook would make its own huge acquisitions, it employed shock and awe to get founders to sign away their companies before they knew what hit them. That wasn't Semel's style. Instead of locking down Zuckerberg in his lawyer's office and not letting him leave until the deal was done, he reopened the negotiations, believing that he now had an advantage to press. He noted that since Yahoo!'s share price had fallen about 20 percent since negotiations began, the deal should be for the same percentage of Yahoo! as it was before the drop, thus yielding less than a billion for Facebook.

Zuckerberg used this as an excuse to reverse course. "Yahoo! made it easier because they kept reneging on their offer," he says. "And at each step along the way the team was just so spooked and was like, *Look, we should just take it.* I was like, *Can we at least all agree that if they renege on this that we can be done?*" He gathered his resolve and made a final call. He would *not* sell. Cohler now supported him. Thiel, as always, respected a founder's wish. Moskovitz had been with Zuckerberg all along. The rest would have to live with it.

On a late August afternoon, Zuckerberg showed up at the house that the company was renting. Facebookers were hanging out around the pool, beers and conversation, just one more day in the perpetual office party that was Facebook. For weeks it had been a question whether Facebook would continue on its dramatic trajectory or cash out and become part of Yahoo!, which by 2006 was already a company past its prime, with a low probability of empowering its new acquisition to fulfill its destiny. Though negotiations had been handled off-site, and only a few knew exactly what was happening, the possibility of a Yahoo! takeover had loomed over the company. Now Zuckerberg told them it was over: no sale.

On one hand, ending the drama was a relief to Facebookers. After all, they did believe in the Facebook mission. All those all-hands meetings with Zuckerberg's ranting had been like imbibing some potent Kool-Aid. Furthermore, going to Yahoo! would have meant the end of the dream as well as the end of a period of their lives that would never be matched: working like crazy on a project that millions of people loved while being involved in a daily geek spring break of office romances, video games, and gonzo coding binges. No one was excited about becoming part of Yahoo! "It was obvious," says Kate Losse, employee number 51. "Yahoo! was already not cool. And Facebook was very cool at the time."

Still, some of them had been dreaming about a once-in-a-lifetime financial windfall. Their stock options could have bought them houses even nicer than the very nice ones they grew up in. And there would still be enough left over to goof around in luxury for years. All before turning twenty-five!

"We loved what we were doing," says one Facebooker from that time. "But holy shit, three or four million dollars?"

Adam D'Angelo, who had finally graduated Caltech, returned to

Facebook that fall and was struck by how miserable people were. "Not everyone but, like, eighty percent of the people were [experiencing] really low morale," he says. "They were all disappointed that it hadn't sold, and there was no way it was going to live up to that valuation."

Zuckerberg didn't have the experience or the personal touch to rally his troops. Shouting "Domination!" wasn't the same as a few million dollars. "I don't think he had a plan," says D'Angelo. "He didn't know what was expected of someone in his role at that time. It's hard to be a great leader when it's your first time and you're that young. But it was bad leadership."

Zuckerberg later would blame himself for the malaise of the Yahoo! aftermath. In the commencement speech he gave at Harvard in 2017, he said that his failure was not effectively communicating the purpose of Facebook to the employees who had come on board. Without strong internal support, he felt isolated. "That was by far the most stressful time in my life," he told me—*after* Cambridge Analytica.

He did not forget who stood with him and against him. "The relationships were so strained within a year and a half, every single person on the management team was gone," he'd later recall, with satisfaction in his voice. "Some of them I fired."

The commitment to stay independent reflected Zuckerberg's belief that Facebook was now on a mission—to connect the whole world. He had the tools in place: the News Feed and Open Reg. They would take Facebook into another dimension.

All he had to do was ship.

AS NEWS FEED got close to its launch, the team that built it felt they'd done something amazing. The rest of the company began to "dogfood" the product (dogfooding means testing a prototype by using it yourself), and most were already addicted to it. But the Facebook workplace was a social space where everybody knew one another's secrets anyway. To them, the News Feed simply sped and automated the gossip mill that already ran at a high idle. As for privacy, the thinking was that since

Facebook users were already looking at one another's profiles all the time—it was the key activity on the site—it was no big deal to have the news of your friends delivered proactively. It was all information that people had chosen to share, right?

Still, some Facebook workers anticipated trouble. Only relatively late in the process did the customer-support team see the new product. Since they were the ones who dealt with customer complaints—and they knew all too well that many if not most users had no concept of what Facebook did or didn't know about them—they understood right away that people were going freak out.

The warnings were brushed off. "We thought, *Whatever, people are looking at each other's profiles all the time—what's the big deal?*" says Matt Cohler.

There *was* one concern within Facebook that the company did debate, but it was more a commercial concern than one involving privacy. The inefficiency of Facebook's flow had one built-in advantage for the company—all that clicking to different pages to find out what your friends were up to meant that people saw more ads. Some of the new executives that Zuckerberg had hired were concerned the News Feed would reduce impressions and thus diminish even the relatively meager revenues that Facebook was earning those days. But with Zuckerberg's support—not surprising since the product was based on the scrawlings in his own private notebooks—the News Feed team felt that their product would be best for Facebook in the long term.

Another reason to expect user pushback was that News Feed was not only a new disruptive product but a redesign of the entire site. Redesigns always raised a fuss: no matter how great the new look was, people would clamor for "the old Facebook," even when the company was less than a year old. "Just on that front alone, we knew there was going to be a shit show," says Ezra Callahan. "Forget News Feed—the redesign alone was going to cause a catastrophe."

Zuckerberg was untroubled. He had already come to the belief that user objections are a transitory distraction. If you just kept your head down and ignored the noise, people would get over it, and in a couple of weeks it would be like the outcry never happened. "He thought that's what was going to happen again," says Callahan. "And that was wrong in a really bad way."

Facebook typically launched its products late at night, never with advance warning. New features would appear, like Easter eggs. Users would then adopt them, and any design flaws or bugs would be addressed later. In the case of the News Feed, the shift would be particularly abrupt: when users logged in, they would be greeted by a screen informing them that Facebook had changed. To use the product, they'd have to click on a button labeled "Awesome." There was no alternative. When they clicked, their familiar front page of Facebook would be entirely replaced with a flood of information about their friends.

They'd love it, right?

AT 1:06 A.M. Pacific, News Feed went live. A good portion of the entire Facebook workforce, clad in regulation hoodies and jeans, joined the News Feed team at the 156 University Avenue office. No Facebook product had even been close in terms of the time and effort this took—more than six months. Zuckerberg had built the original Facebook in not much more than a *week*. What's more, News Feed was a new direction for the company, a new and perhaps addictive way to share personal information. It was as if the company's raison d'être were embodied in this single product.

Sanghvi had written an item for the company blog called "Facebook Gets a Facelift," explaining why Facebook suddenly looked so different. "We've added two cool features," she wrote; chief among them was News Feed. "[It] highlights what's happening in your social circles on Facebook. It updates a personalized list of news stories throughout the day, so you'll know when Mark adds Britney Spears to his Favorites or when your crush is single again." The other cool feature, the Mini-Feed, would inform you of all the things that everyone was finding out about you.

Users, though, didn't see the blog first, just the "Awesome" button and then an unfamiliar vertical crawl of everything that was happening in their social world. *Angie posted a photo. Ryan is going to the Snoop Dogg concert. Bobbie is no longer in a relationship.*

It was like you had been making out with someone and an interloper suddenly pulled away the blanket that shrouded your canoodling.

Sittig was among the group at 156 University waiting for the responses. And the first one he saw was "fuck mini feed." This hit him hard, since he'd designed it, based on Mark Zuckerberg's meticulous pencil diagrams. Paul Janzer, who headed customer support, saw it as a bad omen. If they hated the Mini-Feed, what might they think of the full News Feed on the front page?

Still, the team figured that these were just the normal bumps for a redesign. "Our thinking was just, *Maybe this will die down in the next few hours*," Sanghvi recalls, and by 3 a.m. they had all drifted home.

When they returned the next morning, it had not died down at all. "It's fair to say there were riots about News Feed," says Sanghvi. People had lined up outside University Avenue, a thoroughfare usually accustomed to those taking strolls to coffeehouses and falafel shops and a smattering of polite homeless folk. TV satellite trucks blocked the street; Matt Cohler was on the phone with his girlfriend telling her how crazy it was when a crane from one of the TV stations swung a camera inches from the second-floor window near his desk. The Palo Alto Police Department, claiming that they didn't have the resources to handle large public protests, called to ask Facebook to turn off the News Feed so the demonstrations might end. For the first time in the company's history, Facebook's leaders felt it necessary to hire a security guard. (No one had a clue that one day it would contract for a small army of people who worked at Facebook every day to secure its property and protect its employees.)

An even larger conflagration was igniting on Facebook itself. When twenty-one-year-old Northwestern University junior Ben Parr had woken up on September 5 and been buried by an avalanche of information about his friends, he was agog. After IMing some of his buddies who shared the same outrage, he quickly started a group, Students Against

Facebook News Feed. When he checked it at lunch, 10,000 people had joined; by the end of the day, it had 100,000, and Parr was being interviewed by *Time* magazine.

What Facebook simply hadn't realized about the News Feed was that pushing information to people was qualitatively different from publishing it on someone's home page. (More accurately, it had shrugged off the early warnings to this effect.) One case in particular stood out as a symbol of the difference: the "relationship status" that Facebook encouraged users to append to their profiles, kind of a mood ring for the state of their romantic life. At any time, it could signify married, single, in a relationship, or the weirdly fraught "it's complicated." When someone changed the status on their profile page, visitors would encounter it as a straightforward self-description of someone's love life. But when instantly broadcast to all of one's friends if changed, it hit your social graph like a stack of tabloid newspapers crashing on the sidewalk. Your girlfriend dumped you, and suddenly your buddy list would explode with lookie-loos demanding the lurid details. All because of Facebook! The corporate inbox overflowed with howls from people whose relationship status and other "news" had become the unwelcome content of a brand-new media channel.

"We heard them," says Matt Cohler. As if they could be ignored. The small team handling customer support got more email complaints in the first day after the News Feed launch than they usually got in three weeks. Janzer estimates that about 30,000 emails came in the first day.

From his hotel on the East Coast, Zuckerberg was considering options with a panicking Van Natta. There was serious talk about a reset—turning back the site to pre–News Feed and then offering people a chance to opt in. Van Natta wanted it shut off, as did one of Facebook's major investors. *Guys, this is simple*, the funder emailed them. *You just turn it off.*

Facebook had recently hired its first full-time PR person, Brandee Barker. She had considered the job somewhat of a step down, as she was fifteen years into her career, but had been swayed by the promise of the company and the energy of its employees. She hadn't worked much with

Zuckerberg before then, and now found herself on transcontinental chat with him deep into the night. "He told me, 'I think we need to write a blog and apologize,'" says Barker. "That was the first of many times where I was like, *Wow, this twenty-three-year-old is going to teach me more about how to do [PR].*"

Meanwhile, Sanghvi and her team were looking at the logs and finding something amazing. Even as hundreds of thousands of users expressed their disapproval of News Feed, their behavior indicated that they felt otherwise. Users were spending more time on Facebook than ever before. It was a validation of the entire concept. She went to Moskovitz and told him that turning off the News Feed would be a bad idea.

The protest's massive traction was actually a vindication of the very product some wanted to smother in its crib. The anger against the News Feed was being fueled by . . . the News Feed. Bosworth, Cox, and others had ginned up an algorithmic amplifier—when a few of your friends took the same action (like joining a specific group), it was ranked highly on your feed. And as more people joined Ben Parr's group, and others like it, a snowball effect kicked in. People's feeds were flooded with invitations to join, and when they did, their own friends would learn about the group. By the end of the week almost one in ten Facebook users had joined Students Against Facebook News Feed, with others signing up for groups called I Hate Facebook and Ruchi Is the Devil.

"It was suddenly giving people a voice when they had no other platform," Sanghvi says. "Not only is it freedom of speech, it's giving people a platform to actually articulate how they feel and what they think and gain support from it and make it known, which you couldn't do unless you were being interviewed on TV or by a reporter for a newspaper prior to this."

At quarter to eleven on the night of September 5, Zuckerberg posted his response, entitled, "Calm down. Breathe. We hear you." Barker and Chris Hughes helped edit it, taking it through many drafts. Its condescending headline was indicative of the tone of the rest of the post, which acknowledged that "many of you are not immediate fans" while insisting that the product was great. Zuckerberg already had the data that told him that no matter what people said, they actually behaved like they liked the News Feed. So he could afford to stand his ground. "We agree, stalking isn't cool; but being able to know what's going on in your friends' lives is," he wrote, while also noting that Ruchi was *not* the devil. So the News Feed was staying, but he promised to add privacy controls to address the complaints.

For the next few days the News Feed team worked all-nighters to gin up the protections that should have been in the product to begin with, including a privacy "mixer" that let users control who would see an item about them. "I don't think anyone ever used it," Jeff Rothschild would later comment. But just knowing the controls were there seemed to quell the rage. In a breathtakingly short period of time, people got used to the idea that the stuff they did on Facebook would wind up spread all around Facebook.

Facebook took a huge lesson from its first public crisis, possibly the wrong one. It had rushed out a product with serious privacy issues—issues its own people had identified—but launched it anyway. "We were pretty cavalier when it came to this stuff—not because we were callous but it was like, in order to build great things, you just got to go for it," says Katie Geminder, looking back at the episode years later. "You can't be afraid."

Yes, a crisis did erupt, but quick action and a dry-eyed apology defused the situation. People wound up loving the product.

"It was a microcosm of Mark and the company," says Cohler. "The intent was good, there were misfires along the way, we acknowledged the misfires, we fixed it, and we moved on. And that's basically the way the company operates."

There was also a lesson about privacy. Though people might complain in the abstract about it, they loved even more to share things with their friends, and especially to see what their friends were up to. What's more, they moved a step closer to Zuckerberg's vision of a new standard of privacy, where people shared more and more with one another.

But Zuckerberg had learned that from Facemash. *People are more voyeuristic than I would have thought.*

FACEBOOK WAS A little more cautious when it rolled out Open Registration soon afterward. Especially since Open Reg reflected such a huge change in Facebook's philosophy, discarding built-in privacy for Zuckerberg's grand ambition of connecting the world. "News Feed and Open Reg had been on the master plan for a long time," says Ezra Callahan. "The ironic thing is that the feeling was always that Open Reg was the minefield." Instead of a sudden launch, Facebook pre-briefed the press for a softer release.

The public response to the idea of a gateless Facebook reflected a new skepticism toward the company. "Facebook could be hurt when users start drawing comparisons to MySpace," said Fred Stutzman, a graduate student at the University of North Carolina at Chapel Hill's School of Information and Library Science, who had become a media go-to person on the subject of how students used Facebook. "There's a backlash with every change at Facebook," he added. "This is now the point of no return."

But Open Reg appeared, and there were no angry groups. Instead, there were millions of new people signing up. "It ended up going better than we thought," says Zuckerberg.

In the last months of 2006, and into 2007, Facebook's flat numbers began to rise. "Within a week of launching we'd gone from probably fewer than ten thousand people joining a day to sixty or eighty thousand people joining a day, and then it grew quickly from there," recalls Zuckerberg.

Open Reg allowed billions of users to flock to Facebook. And the News Feed would keep them there, making the site as totally consuming for everybody as it was for college kids when Thefacebook first appeared. It would also breed bullying, hate, and deadly misinformation. The impact of Mark Zuckerberg's Book of Change, despite its minuscule circulation, would have an impact broader than a blockbuster bestseller.

PART TWO

7

Platform

DAVE MORIN WAS a Montana kid who grew up on computers. He'd paid his way through the University of Colorado by running a web-development firm out of his dorm room. After graduating in 2003, he joined his dream company, Apple, working for its higher-education marketing team. His job was getting college kids to use Apple tools, and he was put in charge of its campus-representative program. At the time, there were about 100 reps across the United States, mostly geeks, and their task was providing technical support to their peers. Morin shifted the program to evangelism, expanding to 900 students who pitched Apple to classmates. Morin believed in communities, and he always urged his reps to join social networks—Friendster, LinkedIn, even AIM. One day in early 2005 the rep from Harvard called him. *You have to see this thing called Thefacebook.*

Morin still had his .edu address from Boulder and he signed on. He was stunned. One of his complaints with AIM, which was the de facto communications system of college students then, was that you couldn't find anyone or know who they were because everyone used oblique screen names. Facebook used your real name—and it also provided your AIM handle in the profile. He was also dazzled at how privacy was elegantly baked into the network structure—you could browse or message anyone on your campus, but not in other schools. *Game over*, he thought. *This is genius.*

He immediately tried to get hold of whoever was running this thing, and soon found himself in the tiny Palo Alto office whose walls looked like they had been tagged by talented and horny vandals. The leader, Zuckerberg, was clearly super intelligent, but he hardly uttered a word. Morin bonded more with Moskovitz and Parker.

At that point, the only other major brand working with Facebook was Paramount Pictures, which ran a big promotion on Facebook for its *SpongeBob SquarePants* movie. Morin wasn't interested in buying ads. He wanted to start a Facebook group to promote Apple—someplace where people could learn about the products, share videos and other content, and exchange tips about using Macs. Apple would lure them there with giveaways of iPods and iTunes cards. They cut a deal where Apple paid Facebook $25,000 a month. The total contract might have been a million dollars. Parker would boast about the contract when negotiating the Accel deal.

By then Morin had broken through Zuckerberg's ice-wall demeanor. Morin and Zuckerberg and Moskovitz would have endless conversations about graph theory, identity theory, signal theory. The latter dealt with how humans signal their identity by things like status indicators. Morin came to understand that Facebook was the ultimate status indicator, as well as a lubricant for a new social order. It was a workshop for the way people would live with one another in the future.

Zuckerberg and Moskovitz urged Morin to join Facebook, but it was hard to leave Apple's beautiful headquarters for a start-up crazytown in downtown Palo Alto. Once Moskovitz and Ezra Callahan visited Morin at Apple's sprawling Infinite Loop campus. "This place is pretty nice," they told him. "But one day we'll be bigger."

Really? Morin thought. *Come on!*

Morin tried to get his bosses at Apple excited about Facebook. His dream was for Apple to make a social operating system. Instead of organizing your system around files, why not around people? Maybe Apple could buy Facebook, as the basis of this new system. The matter came before CEO Steve Jobs. No go. Jobs was open to buying companies, but why join forces with a college-only site of a few million people when MySpace had fifty million?

Morin kept talking to Facebook. One day in the fall of 2006, Moskovitz visited him at Cupertino again. *Wouldn't a social operating system be amazing?* Morin asked him. Moskovitz point-blank stared at him. This was what they were talking about at Facebook all the time! *You need to come to Facebook now and do this,* he told Morin.

Steve Jobs had just given his famous graduation speech at Stanford where he told students to approach each day with the realization that death could come anytime. That gave Morin the courage to cut the cord with the company he'd always wanted to work with and join the company he thought was poised to make an even bigger splash. He approached Jobs at an employee event and told him he looked in the mirror that morning and realized he had to go to the start-up he'd been gushing about. Jobs asked only one question. *Are they offering you a good stock package?*

Yes, they were. Dave Morin is now worth at least a hundred million dollars.

That weekend the newly hired Morin was at Facebook's office talking to Zuckerberg. It was late at night, in the corner room that Zuckerberg often used for one-on-ones. It was all white—white table, white walls, white Eames chairs, and pretty much all the wall space was covered with whiteboards. People called it the Cloud Room. But sometimes it could feel more like an interrogation room.

Zuckerberg told Morin that while Apple was an innovation company, Facebook was a revolution company. Morin felt a jolt of energy. For the first time, he felt he understood Zuckerberg and Facebook. *Facebook creates revolutions.*

And Morin would be part of it, creating the platform that would catapult Facebook into the ranks of the very top tech companies. It was already in the works.

Platform.

. . .

THE FIRST ADHERENT of the platform was an engineer named Dave Fetterman, who had arrived at Facebook in January 2006. Hailing from York, Pennsylvania, he had graduated Harvard the year before Thefacebook appeared, and taken a job at Microsoft. He was among a group of ace engineers in their mid-twenties who left Seattle for Facebook that winter and became known as the Microsoft Five. (Andrew Bosworth was also in the group.) The newcomers took a house together that they dubbed Facebook Frat. The first task Fetterman performed was adding a few more options to the "relationship status" field on profiles. This was the moment when Facebook added "it's complicated" to the choices. (A phrase destined to launch a thousand Facebook headlines in later years.)

As he checked off task after task at the growing company, Fetterman couldn't stop thinking of a casual question that Moskovitz had tossed out at his job interview: What would a Facebook development platform look like? A "development platform" would mean a technological gateway for other software developers to create programs that used the data from Facebook for social applications. The first step would be to create an API, or application programming interface, which was kind of a software socket that people could plug their programs into, allowing access to the data on the platform.

Fetterman asked Moskovitz if he could write that API. Moskovitz said no. The next week he asked again, with the same answer. Finally, Fetterman decided to just do it. He built the gateway, and coded up a prototype application, something software developers might create to make use of the API. It was called Owen Van Natta's Balloon Store. Using the API, the app knew the birthdays of Van Natta's friends, which it could access from Facebook. "It was the ugliest piece of HTML you'd ever seen," he says.

Fetterman demoed this to his colleagues. *Wouldn't it be great,* he asked them, *if, say, you could go to Amazon and find out what your friends were reading? Or you could go to any site and find out what your friends were up to there? It would be like Facebook was everywhere!*

It would also mean that the birthday app would now know the birthdays of Van Natta's friends though they did not give permission or even know about the transfer of information. And that if someone used that Amazon app, the world's biggest bookstore would know the reading habits of that person's friends without their knowing it.

It was a problem that Facebook would grapple with for years.

Fetterman's idea came before what was then the Facebook brain trust. The reaction was almost unanimous—*Why would we give away our network?* As Fetterman recalls, only one person thought it worth pursuing.

"I think we should look into this," said Mark Zuckerberg.

That summer, Facebook released Fetterman's API. It flopped. "We said, *Hey come one, come all, use the Facebook platform to build interesting things*," says Fetterman. "But no one noticed."

It turns out that simply releasing an API wasn't enough. For one thing, there had to be a way for Facebook users to know that there was some other social application using the API, and that their friends were using that app as well. It was a problem of distribution.

As it happened, Facebook itself was in the process of creating one of the most effective ways of distributing data about everybody's friends—the News Feed. Why not employ it to lure users to apps running on the new platform?

Facebook and Zuckerberg were eager to keep pursuing the concept. That had been their reason for hiring Morin: to spearhead developer relations. Facebook's chief technical officer, Adam D'Angelo (finally full-time now that he graduated Caltech), took charge of the platform-engineering team, with Fetterman as the tech lead. After endless whiteboard sessions, Fetterman's original idea of just an API evolved to a much broader undertaking, where the apps would not be on someone else's website but Facebook's, living on pages that were called canvases. Users would learn about them through the News Feed.

"We said here's a canvas inside the trusted blue-and-white borders where you could build whatever your dreams dictate," says Fetterman.

The difference between plan A (Fetterman's original API) and the new plan B was that the latter positioned Facebook as not just a platform but an operating system. This was the pinnacle of Silicon Valley's pyramid of value. If you owned the OS, you had your own little monopoly. The most successful OS of the previous era was Microsoft's Windows, which a judge had determined was in fact a *big* monopoly. While many Silicon Valley leaders still viewed Microsoft as the industry's Darth Vader, Zuckerberg admired Bill Gates's company. The Windows system was unbeatable because a huge majority of PC users had computers that ran it. To reach those customers, software programmers had to write their software in Windows. Zuckerberg came to imagine Facebook as the social equivalent of that. Just as Microsoft owned the desktop world, Facebook would own the social world.

Building a social operating system could be a stunningly complex task. Take a photo app, for example. Each photo came with potential privacy restrictions: in order to keep its pledge to users that they controlled who saw their information, Facebook had to maintain restrictions at every step: was this photo available for everyone to view, or just friends?

But now Facebook was promising outsiders that they could make their own photo apps, or anything else, getting the same information that Facebook itself had for its in-house apps. That was part of the appeal to developers. But could those outsiders be trusted with the information?

Max Levchin, a former PayPal executive who had started a company called Slide, felt that such information sharing would be the essence of the Facebook operating system. He had been lobbying D'Angelo to give developers maximum integration with Facebook. That presented a privacy issue. The very definition of a social app requires that the developer get not only a user's personal information but details about the people they're socializing with. Because users would, in effect, be exporting their social network, by necessity some of that data belonged to other people. These "friends" of the user who actually signed up for the app might not know that their information was transferred. Should they have a chance to vet this exchange? Furthermore, users might have labeled some personal data as restricted. How would Facebook ensure those restrictions would be honored when developers got access to all that information?

Zuckerberg understood that Facebook had to honor the trust of its users. But he also believed that the social apps that would emerge would be worth the risk of leaked information. "There was a lot of deliberation about which data to share," says an executive at the time. "There was a very strong thing coming from Mark that was, *We need to be able to make it so that other developers can build things as good as what Facebook can build*. Facebook was a small company at the time, so they needed to give developers this data to just make the platform desirable."

Facebook did take steps to prevent leakage. Generally, it required developers to keep certain information in temporary memory caches rather than downloading to permanent storage. And developers would pledge to Facebook that they would not sell or release that data to others. That would be the absolute worst case.

Ultimately, the safeguards were built on an optimistic view of what developers might do. Facebook's executives at the time now admit that the protections were relatively weak in part because the data held by Facebook in 2007 wasn't seen to be as critical as it would later be. The stakes were lower and the norms were different. In that time period, the tech community was urging Facebook not to lock down its information but to be more open. Facebook, said its critics, was a "walled garden." This was the term used when the owner of an online destination owned all the services and features that people used when they visited. These digital "company towns" ran counter to the democratic ethos of the Internet. They smothered innovation. Tearing down the walls of your garden meant you were being a supporter of the free Internet.

So the next great project after Open Reg and News Feed would be Platform. It would cement Facebook's status as the dominant company in the social-networking world. It would give Facebook a huge edge over competitors. (And help it surpass MySpace, which was already hosting third-party apps.) It would make millions of dollars for creators of popular apps. By allowing others to make use of the accounts of millions of its users, Facebook would become the de facto global arbiter of people's online identities. And the influx of new users, and the increase of time spent on Facebook, would bring in the revenues it had yet to realize.

Much of that happened, and less. Platform's legacy also would result in frustrated developers, angry users, and, ultimately, the worst catastrophe in the company's history.

THERE WAS MORE than the usual move-fast imperative for Facebook to introduce its platform in a hurry. That January, Apple CEO Steve Jobs, to astonishment and acclaim, had introduced the iPhone. The announcement had created a frenzy, and people marked their calendars for the time in June when they would actually be able to buy one.

In theory, the iPhone would not provide competition for Facebook's platform. Steve Jobs had brushed off criticism that Apple was not allowing software developers to write applications directly to its operating system. In any case, Apple wanted nothing to do with social networks.

But Facebook was wary of Jobs's intent to close the iPhone to software developers. As a student of Jobs, Dave Morin had seen Apple go to market with a strictly focused product that later took on new powers, hitting competitors with a delayed punch. The iPod was out for two years before the iTunes store.

So Facebook set an ambitious May 24 date to unveil and ship. And it rented the San Francisco Design Center, a large venue in the South of Market neighborhood where start-ups abounded, so that it could invite almost a thousand people for what would be its first developers conference. It called the event F8, a reference to its frequent all-night hackathons where engineers would spend eight or more hours blasting away on a blue-sky idea. Maybe it was a coincidence that it also invoked the word "fate," implying an inevitability to Facebook's impending dominance. Maybe not.

In the weeks before the announcement Facebook gave a select group of developers an advance look at the platform so they could prepare their apps to be ready at launch. Some of them had been building widgets—small applications—on MySpace. Others were well-known software companies. Morin learned that Amazon was working on a digital reading device, to be called the Kindle. He tried to convince Amazon to work with Facebook to make it run as a social application, but had no luck. But as a consolation prize Amazon agreed to release an app called Book Reviews, where Facebook users would share their reading experiences. Amazon had no desire to actually write the app, so Fetterman and Sittig whipped it up.

Microsoft and *The Washington Post* were also launch partners. But Facebook's favored app in the first batch was a collaboration between two old friends: Joe Green, who had been friends with Zuckerberg since Kirkland House, and onetime president Sean Parker. The two were building a website to empower activists using social networks. When Morin asked them if they wanted to do a version for the platform, Parker saw an opportunity to weave it so deeply into Facebook that people would think it was part of Facebook. "It should feel like a feature of Facebook," he said. Its code name was Project Agape, but wound up with the name Causes, because Parker wanted it to evoke other official activities on Facebook like Groups or Events. The Causes team scrapped their website, choosing to run directly on Facebook. Zuckerberg liked it so much he offered to buy it for 1 percent of Facebook. "I was like, *Okay!* But Sean didn't want to sell—he already had plenty of Facebook stock," says Green.

All told, seventy developers would be ready with apps when Platform launched. And they would be part of a spectacle that helped change the world's view of Facebook.

Normally Facebook released product in the dead of night, with at most a blog post to mark the occasion. But Platform was to be Facebook's symbolic ascension to the top of the tech food chain, a signal that Mark

Zuckerberg's dorm-room slouchers were graduating from the *Crimson* to the big-boy business pages.

In brainstorming the event, Morin had a single template in mind: Steve Jobs's celebrated Apple keynotes. To produce the graphics for Mark's speech, Facebook used Ryan Spratt, who had worked so much on Jobs's slides that Apple eventually gave him an office. To help conceptualize the message, Morin tapped Stone Yamashita Partners, a consultancy with extensive Apple experience.

All of this would require something that Mark Zuckerberg had never done: keynoting a glitzy public event. Zuckerberg, of course, couldn't be expected to match the glib elegance of Steve Jobs. "He's an amazing communicator now, but in that time period he was still learning," says Morin, perhaps overly kind in his current assessment. For Zuckerberg, the stress of speaking triggered unusual amounts of perspiration. In coming years, he would demand that the backstage area at his speaking events be cooled to well under 60 degrees. Brandee Barker would often wind up blow-drying his armpits before he went on stage.

In the course of brainstorming concepts for his speech, Zuckerberg acquired some of the language that would pepper his explanations of Facebook's mission for years to come. The most important term was what he called the "social graph." Though the concept had been batted around for months in late-night discussions—and was pursued as far back as Adam D'Angelo's Buddy Zoo—that single term seemed to embody what Facebook wanted to unlock for its users.

Social graph refers to the nexus of connections people have in the real world. By expediting connections to those people who were on your friend-and-acquaintance radar, Facebook was unlocking a network you already had, keeping you in close touch with people huddled next to you on this virtual constellation, and drawing lines to those who were one, two, three degrees away.

"We don't own the social graph," Zuckerberg would explain to me later that year, going slow so even a mainstream journalist might understand this dive into network theory. "The social graph is this thing that exists in the world, and it always has and it always will. A lot of people think that maybe Facebook's a community site, and we think we're not a community site at all. We're not defining any communities. All we're doing is taking this real-world social graph that exists with real people and their real connections, and we're trying to get as accurate of a picture as possible of how those connections are modeled out."

Once that picture was captured, Facebook and all the other companies on the platform could exploit the social graph to, as Zuckerberg puts it, "build a set of communication utilities that help people share information with all of the people that they're connected to."

Unsaid was Facebook's ambition to be the only company that captured the full picture of the social graph. It would be like a search company having exclusive access to the World Wide Web.

For weeks, Zuckerberg rehearsed his speech over and over, down to the hand movements and where he would talk on stage. But he'd still be his authentic self, wearing his signature fleece and jeans. That extended to his feet, which would be shod in the unfashionable flip-flops he wore just about everywhere. At the last minute, he discovered the Adidas sandals he favored had been discontinued and an assistant had to scramble to find a pair still on sale. (They bought ten, to stock him for the future.)

As much as he hated public speaking, he knew he had to do it. Zuckerberg would enter the arena, a Cicero of the software world, and explain how Facebook would create the next great platform. His first line would set the tone, and past midnight on the eve of the speech he was still rehearsing, intoning the line over and over.

Today, together, we're starting a movement.

At 3 p.m. on May 24, 2007, Zuckerberg took to the stage. All that practice paid off, as he got through the speech with no long silences or flop sweat. In any case, it was the substance of the speech that wound up impressing people. Though Facebook had been welcoming everyone for months, the tech elite had still regarded it as a college site. F8 forever changed that perception. Zuckerberg recited a list of statistics to prove this. Facebook's user base of 20 million was growing by 100,000 users a day, he reported, with the twenty-five-and-over demographic the fastest growing. It was the sixth most trafficked site in the world. And Facebook was already the most popular photo site on Earth.

Following the event, Facebook hosted a giant "hackathon" where developers could spend the night coding new apps for the platform. Facebook didn't invent those all-night coding parties, but they dovetailed perfectly with the move-fast company ethos. While coders cobbled apps to run on Facebook, Zuckerberg, D'Angelo, and Moskovitz were in the lobby of the nearby W Hotel, making sure that the system would not crash.

The night before the event, Zuckerberg and Morin had sat on the lip of the stage speculating how many developers the new platform might draw. It was hard to judge. After thirty years, Apple had only 25,000 developers. Google had about 5,000 developers making widgets for its user-customized iGoogle home page. "I remember thinking, *If we could grow things at that speed that would be pretty amazing*," Morin says. So his goal was 5,000 developers. He dreamed of it happening in a year.

It took two days.

ILIKE WAS FOUNDED by twin brothers whose parents had come to America as children when their family fled the Iranian revolution. Both Hadi and Ali Partovi had earned computer-science degrees and worked for Microsoft. Then they started a company where people could share their musical preferences with friends and buy things like concert tickets. iLike had existed as a website for a year, and had an app running on MySpace, but was not making much of a mark on the world. When they learned about Facebook Platform, Hadi, the president, urged his brother (and CEO), Ali, to go all in. "In the history of computing, there was the personal computer, there was Windows, there was the Web, and now the Facebook platform," Hadi told his brother.

They had hardly returned from the launch when the gamble seemed to pay off. In the first day, they got 40,000 users, which roughly doubled the number of people who used the site on a given day. The user base rose into the millions. Each new user was not only downloading the app but also uploading huge databases of music they owned. "That has a tremendous impact on your infrastructure," says Nat Brown, who was iLike's CTO at the time.

Ali Partovi called up Morin in desperation and asked if he knew anyone in the Bay Area with extra servers. Morin knew of a company in Oakland, so the iLike guys flew in from Seattle, rented a U-Haul at the airport, and trucked the servers from Oakland to one of the data centers Facebook was using. Morin also arranged cages of new servers for other apps, many of which were taking off like a bottle rocket. Others were also topping a million users. Since Facebook only had around 20 million users at the time, the numbers seemed otherworldly.

So why did the Platform instantly surpass Facebook's most optimistic guesses? The secret turned out to be that the News Feed was a more powerful engine of distribution than even Facebook suspected. Less than a year after its introduction, Facebook was still tinkering with the algorithms that determined the ranking of possible stories on people's feeds. The developers were way ahead of them. In order to popularize their products quickly, they had been experimenting in techniques, sometimes dicey ones, to take advantage of the peccadillos of various platforms. They also understood human nature well enough to know why people click on some things and not others. Some had already mastered the mysterious art of "going viral" on MySpace and other networks and knew just what to do to exploit it for their own gains—and the detriment of Facebook's users.

Two companies with virtual black belts in virality were Slide and RockYou. They had both built huge followings on MySpace. But despite all that engagement—or maybe because those apps had used scorched-earth tactics to grow—MySpace had become unhappy with its developers. It felt that some weren't adding much value, and others it came to view as competition. "MySpace at that time was unfriendly to third-party developers," says Lance Tokuda, CEO of RockYou. "In one meeting,

Chris DeWolfe actually said that they might kick everyone off the platform." So when Dave Morin promised them that Facebook planned to give developers that same access to its system that its own engineers had, RockYou and Slide jumped right in.

Slide and RockYou specialized in activities that wasted time. It was almost as if they had a competition to produce the most mindlessly addictive activities. Their first Platform offerings didn't even bother to build something original, but embellished features that Facebook already offered. Slide's most popular app was called SuperPoke!, which extended the powers of Facebook's dumbest feature. Its CEO, Max Levchin, bought the small company that developed the app and unleashed it in the Facebook ecosystem, like introducing invasive Asian carp into American waterways. The theory was that Facebook users were bored with just poking one another and were hungry for sillier means of nudging a friend. The super poke that caught on was "throwing sheep," which became a symbol of the mindlessness of Facebook apps. (Levchin still defends SuperPoke!, claiming that it added "vibrancy and zest" to Facebook communications. One could also argue that those hurled sheep and such were precursors to the emojis that would rule a decade later.)

RockYou had its own version of Poke called Hug Me. "Hugging was our most popular action," says Tokuda. "Also, you could smile at, you could dance with—any kind of verb that was fun for users." The two companies were always trading accusations of ripping each other off.

RockYou's signature app, though, was Super Wall, which allowed people to replace the Facebook wall on their profile with a gaudier version that let people upload videos and other media. Because Super Wall would work only if your friends were using it as well, RockYou's distribution strategy was to do all it could to clutter people's Walls and News Feeds with invitations. "We would get the list of IDs of their friends," says Tokuda. "And with that list, we could let them invite other friends to invite all their friends to join Super Wall and share content, because for Super Wall we needed everyone connected."

"It did turn into a Wild West," says Levchin of the scramble to corral new users. "Companies would compete with each other on who was loudest and who could incent[ivize] the users to share the most." He admits that Slide was an offender, consciously creating viral loops that would suck in users.

Another company that courted massive engagement was an entertainment app called Flixster. Ostensibly, it was a diversion for movie buffs, creating quizzes they could take to show their cinema knowledge. But that was a subterfuge to draw traffic. "Practically speaking, it was a viral engine to get kids to basically create quizzes and spam their friends," says Brad Selby, senior product manager of Flixster. "And it worked very well."

And then came the games, which were in a class by themselves in terms of disrupting the News Feed.

Mark Pincus smelled the social-game opportunity first. He was the guy who, with Reid Hoffman, had been part of the angel investment that, in his words, was like winning lotto. In late 2006, Matt Cohler tipped him off that Facebook was going to launch a platform and was looking for entrepreneurs to come up with apps. *We don't want any money from you,* he told Pincus. *Just build cool stuff and we'll expose you to our traffic.*

Pincus had already identified that games were the missing piece on Facebook. It was something he had always wanted to do at tribe.net, his own failed social network. He started it now, calling his new company Zynga. "Games," he says, "were the perfect thing to drop in the middle of this cocktail party."

Specifically a poker game. What would be more social? "A poker game is like an always-on bar, like Vegas or something," he says. "You can go online there with your friends, you can meet other people." Pincus had tried to build online poker on the web, but the technology didn't work well. Building on Facebook, though, would solve a number of problems: you would know whom you were playing against (because Facebook used real identities), and you could play with your friends.

Pincus met with Zuckerberg often, a lunch or dinner every other month or so. Just the two of them. They became friends, and Pincus would attend things like the parties Zuckerberg had on his birthday. "I was the non-Harvard, non-Facebook person there," he says. Pincus was in awe of how relentlessly the younger man could soak up knowledge. Zuckerberg was a learning machine. As a poker player, Pincus respected Zuckerberg's skill at not showing his virtual cards. He would always walk away from the table a winner. But at times when his own interest wasn't threatened, he would be generous with advice and aid. "He was always being pitched by a lot of people all the time and he had good judgment to pull things out of what you're telling him that were useful," he says. "When he said, *Okay, I like that idea*, you knew he meant it, and you knew he was probably going to do something with it."

Pincus knew that Zuckerberg personally didn't think that games were the ideal use of the platform. "Their vision was Causes," says Pincus. "They felt that [the Platform] was going to bring out our best selves." But though Causes at first gained a lot of users supporting worthy movements, Causes didn't make money like silly programs or games that displayed ads on their little slices of Facebook. The investors in Causes, including Bill Gates, would eventually lose their money.

Zynga, though, was a booming business. Hold 'Em Poker was an instant hit, the first of many. People got invitations to join games, and notifications when their friends joined a table. When a wildly popular online Scrabble game was quickly closed down when the copyright holder, Hasbro, threatened legal action, Pincus jumped in with his own version, Words with Friends.

Then Zynga came up with a social game called Farmville. Users acquired livestock, crops, and equipment to tend to virtual farms, infesting the News Feed with invitations and status reports with every new chicken and tractor. It was the epitome of a time waster. It was also a giant money maker. Besides advertising, Farmville reaped revenues by selling virtual goods. People became obsessed with developing their farms and sped the process by purchasing make-believe equipment, seed corn, and even trees. Farmville also pushed the envelope in getting its users to become its pushers. The first thing you did on Farmville was give "gifts" to your friends, luring them into the quicksand of virtual agriculture. They'd learn about the gifts, of course, through News Feed. At Farmville's peak, 80 million people would become virtual farmers. *Eighty million.*

With hundreds—thousands—of developers tapping into Facebook's API to distribute content about their apps, the News Feed became jammed with junk posts, a tsunami that engulfed its normal operation. Users also got bombarded with notifications, which developers also could use to distribute "news" of their apps.

So at the same time Facebook was celebrating the takeoff of Platform it was worried that bad behavior might be poisoning the system. "We've got venture capitalists, we've got entrepreneurs, and there's developer events and all these things going on. But this is impacting the users' experience and it's spammy," says Dave Morin. "I believe the word 'spammy' literally became a trending word in the world that year."

At any given time Facebook might show the average user around 1,500 possible stories involving the activities of their friends. Its ranking algorithm would try to narrow that down to a hundred or so. During the average session, users might look at only the very top-ranked half dozen or so. Instead of learning about what their friends were up to, or seeing pictures of cool parties, or learning who was in or out of a relationship, people were scrolling through dozens of posts that someone tossed a sheep, scored high in a dumb quiz, or invited them to play some silly game.

"If you can get a friend to bother ten friends to get one more user on your app, you're very happy because you got one more user," says Josh Elman, who joined the Platform team in 2008. "Facebook, though, has nine other people who've just gotten bothered."

Definitely not the revolution Facebook had in mind.

FACEBOOK BEGAN TO correct course, limiting developers' access to the News Feed and notifications. "The number of developers and the amount of stuff like spam that we had to deal with just grew much faster than we were ready for," says Adam D'Angelo. "So there was this whole crackdown we had to do."

Naturally, the developers hated those new rules. Slide's Max Levchin considered it a bait and switch: Facebook had encouraged them to court user engagement in their tactics. "They said, *Go for it*," he recalls now, noting that Facebook itself used engagement as an internal metric, and all that activity Slide created was bolstering Facebook's business. "One person's spam is another person's entertainment," he says.

But the ones who hated the restrictions most weren't the ones who wantonly flooded the News Feed. It was the law-abiding developers. They felt that they were being punished for someone else's bad behavior.

Joe Green of Causes griped to Morin, "You need to punish the wrong-doers." But that would require Facebook to have its managers make actual judgments on developer behavior. Which was not the Facebook way. The company was operating at a scale where only algorithms or armies could make choices at such scale, and it didn't want to hire armies. "Facebook didn't want to have to do things with human oversight—they want it all to be [automated]," says Green. (It would take a long time for Facebook to understand the limits of algorithms and the necessities of armies.)

The rule changes dampened excitement for many developers who had bought into Facebook's promises that Platform was the next Silicon Valley gold rush. Thousands of entrepreneurs had been starting up companies on the idea that Facebook's Platform would be like the web, supercharged by the juice of social activity—a must-do for any business. Now it wasn't so clear.

iLike was one of the hardest hit by the crackdown. It had been the most popular Facebook app; its CEO told *The New York Times* of dreams that it would be "the next MTV." Though it did use the News Feed for things like telling the user's friends that he or she scored high on a quiz, says Nat Brown, iLike's chief technical officer at the time, it didn't make stuff up to get on the News Feed like other apps did. "We felt like we were at a disadvantage because we were more respectful of the rules than other players," he says. "We were this great place where users are really interested in music, but they're saying all apps are bad because RockYou messages their friends 100 times an hour." With its access to notifications and the News Feed curtailed—Facebook used the term "deprecated"—iLike's growth hit a wall, and it began a slow decline. And then a rapid one.

"There was no way to maintain the business we had built on our Facebook app, and so it became clear to us that whatever business we had in our Facebook app was short-lived," Ali Partovi would later say in a deposition. In 2009, iLike, which had once gathered tens of millions of Facebook users, sold itself to MySpace for the fire-sale price of $20 million.

"Facebook is a rocket ship," says Nat Brown. "It turns out iLike was not strapped to the rocket ship. We were the fuel."

THE NEWS FEED spam wars were only the start of a push-pull between Facebook and its developers. Facebook would change the rules and developers would figure out how to get around those rules. The developers would share techniques with one another. When they tried a particularly sketchy feature, they learned not to show it to anyone they identified as a Facebook employee, or would use geo-tagging to exclude the Bay Area. "We were playing a little bit of a cat-and-mouse game and a lot of times I think we were behind the mice," says Facebook's Josh Elman.

One of the more serious instances of misbehavior came when developers sold space on their pages to low-quality ad networks. Premium advertisers usually weren't interested in the mind-numbing apps that ran on Facebook. The companies buying ads were often bottom-feeders, involved in the dicey practice of lead generation, where they would use deceptive tricks to try to get people's money or data. An example would be ads enticing a user to click, instantly installing a browser that would thereafter stealthily scoop up all of the user's subsequent web behavior. Getting rid of it would almost require a computer-science degree.

In a 2009 exposé of these practices, *TechCrunch* described how the lead-gen advertisers would abuse Facebook customers who fell for offers of low-value items like game currency, introductory offers for services, and other trifles. Facebook's Platform, it joked, should be dubbed

"Scamville." Its author, *TechCrunch* co-founder Michael Arrington, described one scam where users are asked to send their mobile phone numbers to get the results of a quiz. When the text arrives, it gives them a pin code to enter for the score. They didn't know it, but they just signed up for a service that charges them ten dollars a month.

Arrington noted that while Facebook had policies against these abuses, "those rules are routinely ignored by developers, and are rarely enforced by Facebook." (He also cited MySpace as the host of similar abuses.)

Zynga's products were among those hosting deceptive ads, but Pincus says it wasn't his fault, noting that the ads were automated. "We had no control over what the [advertisers] put in," he says. "We would get paid if our users did something." Besides, he said, the same ads would show up on Google. "We were being held to a higher standard."

But Pincus didn't help matters when he spoke to a group of tech founders at a small Berkeley gathering. "I knew that I wanted to control my destiny, so I needed revenues, right fucking now," he told the young engineers. "So I did every horrible thing in the book just to get revenues right away. I mean, we gave our users poker chips if they downloaded this zwinky toolbar . . . I downloaded it once and couldn't get rid of it. We did anything possible just to get revenues so that we could grow and be a real business."

Pincus now says that he was exaggerating, mouthing off to a bunch of aspirational founders over drinks. So why didn't he refute his statements publicly? Because, he says, he didn't want anyone to know how he was *really* making money. "My customers were middle-aged women in Indiana who'd stopped watching soap operas to play Farmville. Some of them were spending huge amounts, thousands of dollars a month with us. But I didn't want to get that story out. So I had to take the arrows, because I was okay with the rest of the world thinking we're making [money from scamming]."

But though Pincus had lit on a perpetual moneymaking machine, he was caught in a strange dance between his company and Facebook. For Zynga, access to Facebook became an existence question when the company cut back on spammy distribution. Facebook's solution: buy ads from us. Pincus ponied up, becoming Facebook's biggest advertiser. Without steady access to the News Feed, its main pipeline became the ads on the "left rail," the screen space alongside the feed. Two-thirds of its traffic came from people clicking on ads.

By then Facebook was pressuring Zynga by other means: in 2010, the company had introduced its own form of in-house currency called Facebook Credits. It urged that developers use this form of payment, which kicked back 30 percent of every transaction to Facebook. "We had real issues with Credits," says Pincus. "Number one, credits sucked. We were testing Credits for them versus PayPal and we had huge losses on anyone we put through Credits." The second reason was the unfairness that Facebook was *forcing* Zynga to use Credits, while the choice was voluntary for other developers.

Pincus came in to talk to Zuckerberg, who brought in Sheryl Sandberg, saying that since she had worked for the Treasury Department and understood economics so well, she could clear things up for him. "They had this whole argument—they felt like we're the biggest user and we were somehow being subsidized where we were using more than our fair share," says Pincus. "She had this whole explanation that it's the tragedy of the commons and all." But Zynga decided not to comply if others didn't have to. "I said, *Fuck that*," recalls Pincus. "When you make it mandatory, I'm in. But until you do, I'm not."

Pincus considered himself a friend to Zuckerberg. He respected Sandberg. But he knew that ultimately it was each side for themselves. "They are amazing people because they're tough as nails but also lovely and nice," he says. "Like soft boxing gloves with brass knuckles. You do not want to get in a scrap with either of them. And I did."

Zynga began exploring alternative venues to Facebook for distributing its games. For a while it was a vicious standoff. In trying to negotiate terms by which Zynga might stay, Facebook offered a contract that Pincus would not sign. Among the provisions were that Zynga could not move its games to other platforms. "We held out and Facebook was getting more and more angry about the user feedback from people who didn't want to see games in the feeds, he says. "We weren't abusing anything, we were doing what Zuck had told us." Pincus even began talking to Google as an alternate partner. Meanwhile, his team was frantically coding up a separate website to host the games if Facebook tossed them off the platform.

It was only Pincus's friendship with Zuckerberg that led both companies to back off. The two had a series of meetings, some going as late as 4 a.m. "He was a night owl who could drink Diet Cokes all night," says Pincus. "He said, Look, there's no one in a position to compete with Facebook. You and Google are it." So each was a threat to the other. They reached a complicated deal where each side got something, but, as Pincus explains, things played out more loosely in practice. "We averted nuclear war," says Pincus. They signed the deal in May 2010. For a few more years things thrived.

"At one point we were eighty percent of their API usage," says Pincus. "At our peak with them, we were sixty percent of their app DAUs [daily average users]. And I heard that by the time they went public, we were something like twenty percent of Facebook's overall revenues." Facebook was so dependent on Zynga at that point (its 2012 IPO) that the prospectus listed it as a business risk.

Nonetheless, tensions still remained, and as smartphones became ubiquitous, Facebook's platform became less valuable to Zynga. "It was clearly all going to be mobile and it didn't matter anymore," Pincus says. In 2012—three years before the five-year timeframe of their contract ended—the two companies renegotiated. Zynga would no longer be a Facebook-first partner. It was a symbolic moment of the dashed dreams of Platform.

"I naïvely thought Facebook would see that their users got the most value," says Pincus now. "They'd stay with Poker, Farmville, and all these things the longest, and Facebook would make more economic value and so they would want to promote our games, and they didn't. They were an ad business."

Pincus, of all people, should have known better.

In a sense, Pincus and the other developers who wrote applications inside Facebook were already fighting the last era's war. Only a year after announcing the original Platform, Facebook in effect created a new way for developers to access information from Facebook—and for Facebook to bring software companies into its ecosystem. Called Facebook Connect, it allowed developers to use Facebook as a log-in on their own services and apps. The applications would live outside of Facebook. It was sort of a resurrection of Fetterman's original API idea, living side by side with the Facebook platform.

Mike Vernal, yet another engineer who had come over from Microsoft, headed the project, which had two purposes. First, it was meant to solve the problem users had of trying to create and remember a login for every online service or site they signed up for. "I feel like I should have one and be able to log in everywhere too," says Vernal. Also, he adds, "We thought that a bunch of apps and industries could be fundamentally better if they were more social."

Facebook Connect was a step toward making Mark Zuckerberg's company the de facto arbiter of identity on the Internet. Your Facebook persona could be used on thousands of other sites. And since you were logging in with Facebook, Zuckerberg's company would be able to monitor your activity.

Facebook already had thousands of developers, but this would raise the number dramatically. And Facebook would also be sharing information it had about users (who intentionally signed up for the apps using Facebook Connect) and friends of users (who had no idea that their information was being passed on to apps that they might never have heard of, let alone signed up for).

What data Facebook gave the developers was supposedly dictated by its rules. But according to some developers and emails that later were exposed due to legal actions, it turned out that in practice those rules were flexible, that a bartering situation emerged when it came to what personal user data was supplied to developers. "There were nominal guidelines, but that was complete nonsense," says Selby of Flixster. "It was absolutely catch as catch can—who can you convince. One week, we'd say, *You know what, we could really do something with a lot of friends-of-friends data about movie likes.* And they would go back in their darkened room and say no. And we'd say, *Let's pitch it to you differently—if you give us this data, we project we'll increase engagement, and it'll flow to you.* And they might say, *Okay, that makes sense*, and they'd flip the switch. Or they might say, *Jump in a lake.*"

In the short term, at least, Facebook was motivated to let the carnival continue. Because if all the developers writing Facebook apps left the Platform, there would be a lot less traffic on Facebook. "What was in it for us was very simple," says Dave Morin. "It was creating more time, and more inventory [for ads]. One of the things about Facebook that's always been very straightforward is, we create experiences that are highly engaging, the business model is ads, and so the more engaging, the more ads, right?"

Some executives at Facebook warned that allowing all the junk posts from developers would begin to alienate actual users. Will Cathcart, an engineer who left Google to join Facebook in 2010, dove into the data and found an alarming trend. "One of my growing fears is that we're routinely erring on the side of avoiding pain for developers and in the process causing user pain," he wrote in a 2011 email. He cited data to indicate how people were tiring of developers' tricks. "Users don't trust apps to do the right thing," he wrote. Furthermore, users didn't trust Facebook to do anything about it. When they reported bad behavior to Facebook, they felt that nothing was done. Cathcart said that people he knew personally had concluded that flagging violations to Facebook was useless, so they stopped doing it. The reply from his boss, Mike Vernal, was that . . . it wasn't so simple. "This is tricky stuff," he wrote. "One week everyone is yelling we're not protecting users enough. The next week everyone is swooping in and saying we're being too aggressive. It's a delicate balance but both sides are right."

His suggestion: be careful about sanctioning developers. "We need to soften the punishment ASAP so we can protect users without screwing developers."

. . .

BY 2010, IT was clear that the Platform needed a fundamental rethink. Hundreds of thousands of developers were using Facebook Connect, but those writing applications on the canvas pages of Facebook were lagging. Facebook made one more adjustment, creating a new API that allowed for even deeper integration into its system.

Zuckerberg had always seen Platform as a way to extend his worldview of sharing to a larger audience, and felt that this would help. Seven years into the Facebook experiment, he now believed more than ever that people would be better off if they knew what their friends, families, and contacts were up to. "Mark has a distinct time in his life when he started talking about sharing information and understanding what your friends were doing," says Don Graham, who besides being CEO of the *Post* was on Facebook's board at the time.

Zuckerberg's new buzzword was "Open Graph." Just as the social graph mapped your personal network, the Open Graph would map the interests and activities of those you knew. Maybe, by noticing your affinities with casual connections on the Open Graph, you'd get closer. Or maybe you'd just learn more about people you knew.

Zuckerberg announced the system in 2010—Graph API V1, the first version. A year later, he was still talking excitedly about it. He explained it all to me one summer day in 2011, not long before the September F8 conference. We were walking in College Terrace, a leafy Palo Alto neighborhood where Facebook had its headquarters at the time.

That year Facebook would be introducing some new tweaks that really showed how user information from apps could be shared with Facebook. The key partners were the music-sharing system Spotify, the video-streaming platform Netflix, and *The Washington Post*, which developed an app called Social Reader. The products introduced were not really stand-alones, but social extensions of their main applications—a means to allow users to circulate to their personal networks what they were listening to, watching, and reading. The idea was that eventually, every application and service would have a companion app on Facebook that enabled people—presumably with approval—to share their exercise routines, media preferences, and purchases. Zuckerberg would soon predict that in five years the top 100 mobile apps would be part of the Open Graph.

That struck me as a potential nightmare of personal transparency. I tried to come up with an example for Zuckerberg. What if one of his employees called in sick—while Facebook was reporting that they were binge-watching *Breaking Bad*?

"I'd ask how they were feeling," he told me.

Originally, it seemed like those key partners were doing wonderfully. But it turned out that it was something of a replay of the algorithmic overkill of the original Platform, where people's News Feeds piled up with reports of what people were doing on those partner apps. "We couldn't believe how many people adopted it, and how much they liked it," says Graham of the *Post* Social Reader. "That was the problem. Everybody's page began filling up with everything everybody read on social media—the Facebook algorithm overweighted them. Then Mark and Chris [Cox] didn't like the outcome and started *under*-weighting it. At that point, it didn't collapse but it sucked."

None of the original applications in that generation performed the way Facebook envisioned, and the tsunami of apps it hoped would follow—apps where people shared fitness data, location data, and other information—did not become part of people's lives the way Facebook had hoped.

By then, it hardly mattered, because developers had found a much better operating system—two of them, in fact. Apple and Android had created their own development platforms for their mobile phones. Developers quickly understood that mobile was the best place to build their businesses.

Facebook's original ambitions for Platform—a thriving operating system where developers would write original apps that ran inside Facebook—were over. "Unfortunately, mobile just completely undermined the entire system and basically relegated the platform to irrelevance," says Facebook's head of partnerships, Dan Rose.

Facebook's scaffolding of its platform still remained, and for various reasons, developers still built apps that could be deemed as social, or at least made use of the socially oriented information shared by users. And Facebook Connect—which worked fine with mobile apps running on Apple or Android—remained wildly popular. The reason was simple: being a Facebook developer gave you the access to Facebook data that could add social juice to whatever you were doing in the first place.

Sam Lessin, Zuckerberg's Harvard classmate who joined the company in 2010, put it to Zuckerberg this way, in a 2012 email exchange about the future of Platform:

> *Right now I believe if you asked an application to implement Facebook Connect but didn't give the friend graph . . . people would have no reason for implementing it at all.*

Facebook wanted to make sure that it got information back from that exchange as well, and in 2012 it drew up a tougher deal. Introducing Platform 3.0, it decided to ask for what it called "full reciprocity" from developers. In exchange for Facebook's data, developers had to share with Facebook the user data that *they* gathered. Mike Vernal put it this way in an internal chat: "When we started Facebook Platform, we were small and wanted to make sure we were an essential part of the Internet. We have done that—we are the biggest service on earth. . . . Now that we are big . . . we need to be thoughtful about what integrations we allow, and we need to make sure we have sustainable long-term exchanges." Translation: we may not ask developers for money to get our information (though Zuckerberg considered that, and Facebook executives discussed it extensively in that period), but we need *something*. Like *your* data.

In an internal email explaining it, Zuckerberg wrote:

> *We're trying to enable people to share everything they want and to do it on Facebook. Sometimes the best way to enable people to share something is to have a developer build a special purpose app or*

> *network for that type of content and make that app social by having Facebook plug into it. However, that may be good for the world but it's not good for us unless people share back to Facebook and that content increases the value of our network. So ultimately, I think the purpose of the platform . . . is to increase sharing back into Facebook.*

Zuckerberg made it clear: by now Platform's key attribute was a means of information exchange between Facebook and developers—the exchange of user data, where the user had little awareness of how his or her personal information was being shared. Reciprocity aside, by far, the biggest movement of data was from Facebook to the developers.

Facebook would not let that stand. As a cache of documents later revealed, Facebook in that time period was blatantly planning to limit or ban developers that it considered potential competitors, and deny information to those that weren't delivering value back to Facebook. Zuckerberg yanked API access to a contact management start-up called Xobni; when Facebook began thinking of creating its own Gifts feature, it pulled support for an Amazon Gifts app it had already approved. And in 2013, Facebook began to consider a more sweeping adjustment that would more generally curtail the widespread giveaway of friend information, smashing the business plans of many companies that had built social apps.

Zuckerberg's original vision had been to give outside developers the same access to tools and the News Feed as Facebook had to develop its own features. Now those software companies who had invested in that dream were being shut out. Facebook had turned its promised "level playing field" on its side.

While most of the executives and product managers followed Zuckerberg's lead, there was some grumbling, in particular by Ilya Sukhar, who had come to Facebook when it bought his company, Parse, which made tools for developers. Standing up for those stakeholders, he was feeling lonely. "I feel like I am the only one with a principled stand here," he wrote in an internal chat thread among officials in October 2013. "I just spent the day talking to many dozens of devs that will be totally fucked by this and it won't even be for the right reason."

The right reason, of course, would have been to close the Friends API because it gave developers the personal information of users who were unaware of the exchange—and Facebook could do little to control that information once it left Facebook's servers. "We were creating user experiences that were just awful from a privacy experience perspective," says one executive on the Platform team. "You would log in to these apps with Facebook and suddenly they knew everything about you and your friends. And they were doing very nefarious things with it."

Now Facebook would stop the practice, not to serve users but because it did not want to give away data to developers for nothing in return. This was not a friendly message to share at a developers conference. So Facebook came up with the idea to announce the change as if it were motivated by concern for user privacy. The move would fit in with a set of privacy features already planned for release. One executive dubbed the PR tactic "the switcheroo." The PR people helped shape the announcement—to be made at the April 30, 2014, F8 conference—to lead with the idea of "giving users more control." And so, despite Facebook's self-serving motivation for what it internally called the "friend deprecation," Zuckerberg spent the first part of his keynote explaining how Facebook was standing up for privacy by, among other things, shutting down Graph API V1 and introducing V2, thus ending friend-of-friend access.

But Facebook allowed some developers to avoid the lockout—if they kicked back data or committed to buy ads. It put those favored developers on a white list that still allowed access to friend data. The white list included big names like Apple and Netflix. The bartering could get imaginative. To solve a trademark dispute with the dating app Tinder, Facebook apparently gave the service "full friend access." At one point, Zuckerberg floated the idea of giving full friend access to game developers who kicked back 30 percent of their revenue to Facebook.

Other developers managed to keep the information flow going by committing to what was becoming a significant part of Facebook's revenues: a program called NECO where developers would pay for "app installation ads" on Facebook. The Royal Bank of Canada, for instance, got access to the extended API by promising to pay for "one of the biggest NECO campaigns ever run in Canada."

Facebook made one concession for developers victimized by the switcheroo. Developers would have a grace period of one year before it would block off access to the friends API. Between April 2014 and April 2015, Facebook, despite Zuckerberg's boasts that the new version of Open Graph was closing a privacy hole, would allow the practice to continue.

"In retrospect, I think we should have given a ninety-day or thirty-day notice, and just moved faster," says Mike Vernal now.

It is ironic that this one sop Facebook gave to its stranded developers would become a critical factor in the biggest scandal in the company's history. One might even call it karma. But Facebook would not learn that for four more years.

8

Pandemic

FACEBOOK WAS ALWAYS supposed to be profitable. Even before Mark Zuckerberg released Thefacebook to Harvard, a business model was in the works, in the hands of classmate and partner Eduardo Saverin. As the site grew to other campuses, Saverin worked at selling ads but became less and less a voice in the company. In part this was because Zuckerberg made it clear that while seeking profits was definitely something he encouraged, it was not the heart of the company. And in part it was because Saverin simply was not around. When Thefacebook moved to California, Saverin decided to remain on the East Coast that summer. Had he lived in "Casa Facebook," maybe he would have picked up the business-model basics of a Silicon Valley start-up. As it was, Sean Parker began angling to replace him, with Zuckerberg's tacit compliance.

Late that fall, Parker tapped his former housemate Ezra Callahan to help with a business plan. Callahan's previous experience in commerce was limited to selling ads for his college paper. No matter. At that point, with Thefacebook looking for investors, all it needed was to concoct a story of how it would make money. Callahan describes it as "establishing a revenue stream in theory for evaluation but not actually pretending to build this thing out." They came up with some vague scheme that sounded like the business model for the website Yelp, which focused on giving small business a web presence. No engineers were assigned to build that vision.

That still left the company in need of a plan. When Matt Cohler joined Facebook, he was impressed that the company was cash-flow positive (though, as is almost always the case with start-ups, the actual bottom line was drenched in red). Mainly the company was living then on the angel funding of Peter Thiel's half million dollars, augmented by the smaller investments of Reid Hoffman and Mark Pincus. The money coming in at that time came from two ad products. The first were banner-type display ads that ran on the side of the page. They were sold by Facebook using the traditional model where an actual human salesperson contracted with an advertiser. This was the non-scalable work that Saverin had tried to do, though not successfully enough for Zuckerberg's tastes.

Soon after the venture round with Accel in 2005 came the first Facebook-specific ad product, which was called Campus Flyers. Flyers was a self-service system that let advertisers use the web to take out banners that they would target to audiences of specified campuses. (This was potential bad news for college newspapers.) "It was very crude," says Matt Cohler. "Thinking about impressions was too sophisticated for our buyers, so it was a time-based pricing model called Cost Per Day that a lot of the Chinese web market used."

By 2006, Facebook was making epochal moves like the News Feed and Open Reg, and needed a business model to match. The company began recruiting for a head of monetization. In mid-2006, Facebook hired Tim Kendall, who had just finished his MBA at Stanford. Facebook had avoided MBAs until then. Kendall was palatable because his undergrad degree (also Stanford) was in engineering.

Kendall recalls the ad business then as stumbling along, taking in maybe $20,000 a week. He knew, and everyone else did, that eventually Facebook would have to create a unique and innovative product, as Google did with AdWords, its wildly successful self-service, auction-driven scheme that put relevant ads next to search results. The guy who had headed that product, Salar Kamangar, was a hero to business schoolers. Kendall dreamed of being Facebook's Salar.

The first deal he was involved with was about Facebook outsourcing much of its ad business. Microsoft had been circling Facebook for months, trying to buy it. After the Yahoo! episode, that just wasn't happening. But Microsoft also was looking for a way to leverage the ad team it built to monetize its struggling search product. It had hoped to place its inventory on MySpace, but the social media leader cut a $900 million-dollar deal with Google. Yahoo! had also tried to get the MySpace ads.

"Owen read about it and said, *Holy shit*," says Tim Kendall. "Let's get Yahoo! and Microsoft, pit them against each other, have them scared shitless, and then one of them will do a crazy deal."

Even though Facebook was a consolation prize compared to MySpace, it indeed was able to cut a deal with a guarantee so large that the word "crazy" applied. "This was their rebound date," says Dan Rose, Facebook's newly hired executive in charge of partnerships, who worked on the arrangement. Microsoft, which still dreamed of one day buying Facebook (no chance) jumped at the opportunity, and within a week the companies formed a partnership where Microsoft would have the exclusive rights to sell Facebook's domestic ads. The deal generated half of Facebook's revenues that next year.

Some of the idealists in the company were upset that Facebook was joining with . . . Microsoft? In 2006, Microsoft had the double whammy of being associated with evil and being stupid enough to squander its viselike grip on the software business. Dave Morin, who had just joined Facebook at the time, stomped into the Cloud Room and griped about it to Zuckerberg. The boss's answer floored him: *We don't want to spend a single resource here working on advertising*, he told Morin. *It's not something we care about.* "Microsoft wants to build an advertising business here," Zuckerberg told him, "and so we're going to give our inventory to them and they're going to pay us. How great is that?"

But Zuckerberg wasn't quite accurate when he said Facebook would spend zero time engineering ad projects. His dream, or delusion, was that Facebook could build ad products that would act as cool social features, and actually be embraced by users as much as the ones that didn't pay the bills.

One day Kendall came into work and found that his desk was moved next to Zuckerberg's. This meant that, for the moment, Zuck cared a lot about what a person or team was doing, and wanted to learn about that subject and eventually weigh in. For the next year, Kendall wasn't sure if he was reporting to Cohler or Zuckerberg, but he went with it.

It didn't take long for Kendall to figure out that if Facebook was going to do anything close to what Google did in monetization, it would have to involve the News Feed. He worked with a small team that included Chris Cox, who had become the soul of the feed, to create "sponsored stories," which worked like display ads (advertisers paid by the impression) but looked like actual posts on the Feed. Cox was normally protective of the News Feed but let this one slide. For the time being.

IT WAS TIME for Facebook to make its big push for revenue growth, if not profitability. In mid-2007, Tim Kendall produced a manifesto about how Facebook would do this. The key was something the paper called social advertising, which boiled down to inserting commerce into the relationship between you and your friends. The idea originally came from Matt Cohler, he recalls, who asked him, *Wouldn't it be a good idea to have real sponsored stories, like Joe Schmoe bought something, and advertisers could then sponsor an ad that would go to their friends with that implicit recommendation?* The idea was further developed by product managers Justin Rosenstein and Leah Pearlman.

"What worked on Facebook was learning about my friends," Kendall says. "And so learning about products and services through the lens of my friends seems like it should work, especially if the ads have pertinent, relevant information about my friends."

That was the theme of what was to be Facebook's big unveiling of its ad business, with the code name Panda, a quasi-portmanteau of the words "Pages and Ads." Later, the code name would morph to something less pleasant: Pandemic. The tagline to advertisers would be that people's most important conversations were between themselves—and now, if you were Pepsi or Walmart or some other big corporation, you could inject yourself into that conversation. It should have been obvious that the reason that Pepsi wasn't in the conversation to begin with was that when people talked with their friends they weren't eager to talk about Pepsi. Certainly not on Pepsi's terms. And if someone bought some Pepsi, why would that be something they'd want to broadcast to their friends?

But that concept was a key distinguishing point of social ads, and a big part of Facebook's strategy.

Another part would prove more significant. Facebook was going to change its current ad system to be less about how many people saw the ads and more about targeting them to the right people. As Google had, Facebook would create an auction-based system where advertisers would bid against one another to place ads in the sidebars alongside the News Feed or—and this was controversial within the company—in the News Feed itself. (The engineers working on the Feed, notably Chris Cox, wanted to keep the stream of stories as pristine as possible.) The metric that people paid for would be engagement-based rather than exposure-based: the advertiser would pay for each click as opposed to how many eyeballs grazed on the ad. "It's very much the Google playbook," says Kendall. "The difference is you're bidding on people instead of search queries."

Indeed, while Google used keywords as the bidding criteria, Facebook would use demographic information, sometimes broad (male college students who like football) and sometimes very narrow (married female foodies in a specific zip code). Facebook already had been using such targeting in its own recruiting, directing ads to engineers whose profiles identified them as working for rival companies.

But that was only part of Pandemic, which became a massive sprint to release an entire package of business-based features that set the tone for the company's ad model that still resonates today. Another component, called Pages, would allow companies and other entities, like rock bands, to have profiles of their own, something previously forbidden by Facebook's policy of allowing only people to have accounts. Pages would act like storefronts, billboards, or even websites within Facebook. They would be like the Yellow Pages, while the Profiles would be like the White Pages.

The Pages product manager was Justin Rosenstein, who had left Google to join Facebook earlier that year. When he'd joined, he emailed his former colleagues to say, "Facebook really is That company . . . the one that's on the cusp of changing the world." He can still tick off the three virtues of Pages. "One, it helps users to discover things that they will find valuable," he says. "Two, it's good for the people running those pages—we can provide value to a small business and help them get more customers, we can help provide value to an artist and help them get more viewers. Third, this would be really good for our own business, because in addition to organic traffic to those pages, we can also have paid traffic."

There was yet another part of Pandemic that Facebook would announce in the package. It made use of non-celebrity endorsement, but in this case one not so directly tied to ad placement. Instead, it was a means of spreading the Facebook sharing ethic to the web in general, and tying commercial clients to Facebook.

It was called Beacon.

It worked this way: Facebook struck deals with forty-four partners to put invisible monitors on their web pages, called beacons. The pitch: *Add three lines of code and reach millions of users.* The beacons flagged activity to Facebook. When a user made a purchase on the site, the good news would be shared on the News Feed of friends.

This broke ground that even some Facebook employees felt should remain unbroken. Previously, the interests of Facebook users were self-reported. Facebook did automate some news about people, like who they friended, or that they posted photos. But at least that was based on activities that took place on Facebook. Beacon would stealthily track people as they bought things on the web and then—by default—circulate the news of their private purchases. "What if someone bought a sex toy or bought some medicine that shows that you have a disease or something like that?" says one executive involved in the discussion. "There's bad stuff that can happen out of this." The only notice that this was happening would be a pop-up warning that gave instructions of the actions needed to disable the feature. If you didn't respond to the warning—maybe you didn't even read it—Facebook interpreted your nonaction as consent. Beacon would let all your friends know what you bought. The entire history of user experience dictated that most users would breeze past that warning.

"There was a big debate whether it should be opt-in or opt-out," says Kendall. The opt-in side of the argument would have Facebook asking first if people wanted to participate in the program, which would take effect only if they expressed interest. The opt-out side believed that the purchase information should be shared by default because, well, that's what Facebook is all about, sharing by default. If Facebook asked people to express that they wanted the feature, Beacon probably would never succeed. But once it was implemented, people might like it. Just like the News Feed. And if they didn't like it, Facebook could always roll it back.

"We were fighting about the controls until two a.m. the night before the event," says Facebook's counsel and privacy chief, Chris Kelly. He was joined by several other executives who warned that bad things could happen if Beacon did not have protections. "Mark basically just overruled everyone," says an executive at the time.

IN THE RUN-UP to the Pandemic announcement, Facebook took care of a potential obstacle. Facebook wanted to be able to sell social advertising without running afoul of Microsoft and its "exclusive" partnership to sell domestic ads.

Fortunately, Facebook once again enjoyed an edge in the negotiations. Now that it was beginning to expand overseas, it could offer the same kind of deal it had originally struck with Microsoft with those international sales. Better yet, Microsoft's archrival, Google, was also eyeing a Facebook ad deal. Microsoft was determined to win the contract.

Even before negotiations got intense, Facebook pressed its advantage to settle an outstanding issue. For months, the two companies had been feuding over the way Facebook was scraping data from Hotmail and MSN Messenger products. Facebook had its own complaint—as retaliation, Hotmail had begun labeling invitations to join Facebook as spam. According to *The Facebook Effect*, Moskovitz said that this caused a 70 percent drop in new users. Knowing that Microsoft was desperate to make the ad deal, Moskovitz, Van Natta, and D'Angelo flew to Redmond to negotiate a truce, and from then on, Facebook was able to scrape and exploit Hotmail with impunity.

Microsoft's famous co-founder, Bill Gates, was no longer CEO, but as executive chair he took an interest in Zuckerberg, who was being called the next version of him. The two would eventually become friends, with Gates offering lessons from his experience. Gates acknowledged their similarities—both were Harvard dropouts forming a paradigm-busting software company. But Bill Gates V.2? Not so fast. "Mark never wrote as much code as I did—that's the most important thing. Put that in your book!" Gates tells me, joking but maybe not joking. Furthermore, "And if Steve Jobs was sitting here he'd say, *Hey, Mark never designed a beautiful-looking goddamned thing, so how can you talk about him as any successor of me?*" (Joke? Probably joke. Bill is a card.)

Zuckerberg also participated in more discussions about Microsoft potentially acquiring Facebook, including one meeting in Seattle. He had no intention, of course, of actually selling. "We threw out some big numbers," says Gates, who now says he never expected Zuckerberg to take the bait.

But Microsoft did want to make the international deal happen, and in October 2007, weeks before the ad launch, it all came to a head. "We told them that if we can't get it done, we're going to re-open the discussions with Google," says Dan Rose. Microsoft's lead negotiator, Chris Daniels, flew down to Palo Alto. (Daniels would join Facebook four years later.) On October 23 at 10 a.m., the two teams sat down at the University Avenue office and tried to finalize the arrangement that day, so they could do a press conference the next morning at 9. Late that night, when everyone was flagging, the teams heard hip-hop pulsing through the offices. It was a regularly scheduled hackathon. "The Microsoft guys say, *What? You do this?*" says Rose. "We had the house music pumping and everybody's eating Chinese food and we got it done. And at six a.m. everybody went to sleep for an hour."

The deal gave both parties what they wanted. Microsoft had snagged a partner that Google coveted, and Facebook got a grab bag of goodies—an inventory for its international ads, clarity to sell its new social ads, and, in a twist that rocked the tech world, $240 million in funding, in exchange for 1.6 percent of the company. That meant Microsoft was investing in Facebook as if it were worth $15 billion, barely a year after people thought Zuckerberg was nuts for spurning Yahoo!'s billion-dollar offering.

A few weeks earlier, tech's most connected pundit, Kara Swisher, then of *All Things Digital*, had commented on speculation that Microsoft might invest at a $10 billion valuation, and reported (presciently) that Facebook might be seeking $15 billion. She scorned the deal. Thinking that Facebook could justify that valuation, she snorted, was "delusional." Compared to Google, Facebook was a "lemonade stand." Microsoft was spending "dumb money" to buy a sliver of Facebook for a "ridiculous price."

By the time Microsoft began cashing out its investment, its 1.6 percent slice of Facebook would be worth more than $8 billion.

FACEBOOK WANTED TO make a statement with its Pandemic announcement. (Wisely, it ditched the code name.) As with the successful Platform launch earlier that year, Facebook hired professional organizers to choreograph a big launch. This time it would be in New York City. "We wanted to do it in the advertisers' backyard," says Brandee Barker. "The sales team wanted it flashy, and we delivered on that."

Again, Zuckerberg painstakingly rehearsed for a big speech. He was way out of his element—the audience was not software developers but people in suits. Though he still had reservations about the ad business, Zuckerberg now posed as its avatar.

"Once every hundred years, media changes," he told the crowd sitting in plastic chairs in a glossy West Side event space on November 6. "The last hundred years have been defined by mass media. In the next hundred years, information won't be just pushed out to people. It will be shared among the millions of connections people have."

The people in the room were either stunned or amused at this pipsqueak boy-CEO declaring himself the conqueror of Madison Avenue. Mostly stunned. After the Platform success, Facebook's young founder was more likely to be courted as a savant than dismissed as a naïve outsider. Few at the event realized that Facebook had just made one of its biggest mistakes.

While the headlines from the Pandemic launch focused on microtargeting and social ads, attention soon shifted to the Beacon component. As Kelly and others had warned, automatically spreading news of purchases made on the designated websites could result in some unfortunate outcomes. If one were to pick an extreme hypothetical example, you might imagine someone buying a diamond engagement ring on the partner site, and having the intended recipient learn about the purchase not by bended knee but via her Facebook News Feed. And that is exactly what happened. People started complaining when their purchases started appearing on people's News Feeds. One of them was a well-known industry analyst, Charlene Li, who blogged that she found it "shocking" that her coffee-table purchase on Overstock.com was reported to her Facebook friends. In a comment to her post, someone named "Will" said his story was worse.

> *I purchased a diamond engagement ring set from Overstock in preparation for a New Year's surprise for my girlfriend . . . Within hours, I received [phone calls of] "congratulations" for getting engaged. . . . I learned that Overstock had published the details of my purchase (including a link to the item and its price) on my public Facebook*

newsfeed, as well as notifications to all of my friends . . . including my girlfriend.

Though no one was able to confirm Will's story, it became the symbol for Beacon's disregard for privacy. Later, it turned out that a verifiable jewelry buyer was outed by Beacon—one Sean Lane, who purchased a 14k white gold 1.5ct diamond "Eternity Flower" ring from Overstock for his wife's Christmas present. That "news" was distributed to her and his hundreds of Facebook friends, along with the fact that he bought it at a 51 percent discount. "Christmas is ruined," he said to *The Washington Post*. As tales like this circulated—including people outraged that their movie rentals on Beacon partner Blockbuster were exposed to friends and family—it was fair to wonder whether the next hundred years of advertising was going to be an improvement.

For days, Zuckerberg did not respond to the growing criticism. He had learned a lesson from News Feed: to let people discover the virtues of a feature they originally hated. But people did not warm to Beacon. Instead, they began thinking for the first time that this fun thing called Facebook was something that maybe they could not trust.

Facebook at that point had called in a crisis PR team. "They were pretty clear," says Tim Kendall. "Look, this is a trust issue. You're going to burn through brand equity."

In an article written for *Fortune*, Josh Quittner wrote, "Facebook has turned all the people who rooted for it into a lynch mob. In the space of a month, it's gone from media darling to devil." The headline on the article read "R.I.P. Facebook?"

"We waited too long in communicating about anything," says Brandee Barker. "There was so much disagreement internally of the direction to take it because, arguably, it was quite innovative as an ad product, despite the fact that how it worked was an incredible violation of privacy. It was, *Should we do opt-out, should we do opt-in, are there variations that we could still keep the product?*"

Facebook finally decided to change the settings to opt-in. The company promised that users had to proactively consent before a story was published on the News Feed, finally restoring the default status to what Facebook execs had begged Zuckerberg to do in the first place. But that failed to quell the objections, especially as experts discovered disturbing aspects of Beacon's operation. Researcher Stefan Berteau of CA Threat Research documented that Beacon was transmitting data even when the user opted out, as well as giving Facebook a lot of other information about what its users did on those outside websites. Beacon even gave Facebook information about people who weren't signed up for Facebook. The researcher reported this on November 29, the same day that Facebook finally made an executive available for an interview. Even as Berteau's report circulated, Facebook's VP of operations Chamath Palihapitiya was falsely assuring *The New York Times* that an opt-out would end the information transfer. Faced with the technical evidence, Facebook confirmed that Berteau was correct. But it did claim that when it got personal data over the objections of its users, it deleted the information.

By then privacy advocates, the press, and users—a petition organized by the political group MoveOn had more than 50,000 signatures—were demanding that Facebook kill Beacon outright. Zuckerberg had learned from News Feed that the best way to defuse criticism was to announce a fix. But he had just announced a fix, and it wasn't placating anyone. It didn't help that he wasn't speaking personally on the matter. Meanwhile, the Beacon partners were getting uncomfortable. Coca-Cola and Overstock suspended their participation, and other partners considered bailing as well.

After another week of pounding, Zuckerberg published a post titled "Thoughts on Beacon." They were not happy thoughts. He conceded that Facebook had made errors in its zeal to help people share information with one another. And he admitted that its paralysis after launch was in some ways even worse. "I'm not proud of the way we've handled this situation and I know we can do better," he wrote. And here was his final fix: "a privacy control to turn off Beacon completely."

The outcry died down, because few took the opportunity to "proactively opt in," and even fewer were aware that the purchase information was still being passed on to Facebook, unless you found the privacy control to stop it. The purchases just weren't showing up in people's News Feeds. Facebook would not pull the plug on Beacon for almost two more years, in an attempt to settle a class-action lawsuit filed by users who felt abused by the product. Lead plaintiff: Overstock ring purchaser Sean Lane.

"I felt bad, I felt like we made a mistake," says Tim Kendall. "But then we moved on, and that's part of the genius of that company, I think. Right? There isn't a lot of self-loathing over failure. It's a huge reason why they're so successful."

Still, Beacon was different from the other crises in the company's short history. People now were asking tough questions about Facebook, and about the privacy trade-off involved in social networking, especially when social networks were funded by advertising. Negotiating this new skepticism required new measures. Zuckerberg began to listen seriously to what was becoming a blaring chorus of voices telling him he needed an experienced leader alongside him at Facebook. This was the familiar call for "adult supervision" that investors often insist should accompany tech-oriented young founders on their corporate journey. So the search began in earnest for a second-in-command, ideally one with the gravitas of a CEO in his own right. Or hers.

9

Sheryl World

"INTIMACY."

That's the first sentence of Sheryl Sandberg's 1991 Harvard thesis. She gave the word its own paragraph to contrast how the warm feelings it evokes are horribly corrupted when violence disrupts the sanctity of a loving relationship.

The thesis is called "Economic Factors & Intimate Violence" and, despite its rifle-shot lede, is a dry-eyed, equation-laden study that painstakingly and convincingly makes an unsurprising case that financial pressure leads women to stay with abusive partners longer than they would otherwise. It's an impressive piece of work for a twenty-one-year-old undergraduate, bearing trademarks that she would be later known for: a quiet passion for women's rights, a deep affinity for hard work, and—here's the twist—a belief that the most personal matters can be wrestled to the ground with logic and data. That's the essence of Sheryl. She draws you in, helps you find the formula to solve your problem, and sends you off in time for her next meeting.

Another aspect of the thesis that would become a Sheryl trademark was the copious thanks to mentors and helpmates. In this case, the biggest bouquet in the acknowledgments fell to her thesis co-adviser, Lawrence Summers, the economics superstar who would later head the entire university.

Sandberg's rise to the chancellor's favor at Harvard is a fable of meritocracy. She was raised in Florida. Her mother taught English and her father was a celebrated ophthalmologist; her brother and her sister both pursued medical careers. Obviously, a high-achieving clan.

In Sandberg's books she describes herself as the type of girl who obsessively organized things, and quotes her sister's wedding toast: "Some of you think we are Sheryl's younger siblings, but really we were Sheryl's first employees," said Michelle Sandberg. "To the best of our knowledge, Sheryl never actually played as a child but really just organized other children's play." Citing that remark was an act of self-deprecation, but one senses an element of hurt in the retelling. Sandberg would later lament that girls who confidently showed leadership skills would be diminished by the "bossy" epithet. She reports that as one reason that she often hid her accomplishments. (Not all: in a newspaper article about thirteen-year-old Sandberg's activities supporting Soviet Jews, it was noted that she had attended her first rally on the issue at the age of one.)

Sandberg's Harvard experience couldn't have been more different from that of her future boss. She arrived in a whirlwind of enthusiasm and activity, clad in leggings, jeans miniskirt, and a Florida Gators sweatshirt. She taught aerobics classes all four years of her time there. But what others saw as a bubbly personality shrouded a need to excel—and a lack of inner confidence that drove her to work twice as hard as anyone else. (She would later write of forcing herself to smile for an hour even during sessions when she wasn't feeling great.) She struggled during her freshman year, finding herself ill prepared for a course on "The Hero in Hellenic Civilization," as she was unfamiliar with the *Iliad* and the *Odyssey*, two works that Mark Zuckerberg bonded with in his prep school. In one political-science class she was assigned a five-page paper, longer than the ones she had written in high school. She labored for days at it and was crushed when the instructor gave her a C, which at grade-inflated Harvard is virtually equivalent to an F. "I buckled down, worked harder and by the end of the semester, I learned how to write five-page papers," she later told her readers. It was a strategy that characterized Sandberg's later approach to the workplace: with sufficient preparation and diligence, one could always bag the A+. Even the most intransigent problem could be defeated with hard work.

After graduating, she took that attitude into the workplace, first doing a stint at the World Bank, then led by Summers, focusing on its efforts to address diseases and other problems of developing nations. (She met Bono there.) Then back to Harvard for an MBA. Following that, for a brief period, McKinsey, because that's what Harvard MBAs *did* in the early '90s. When President Bill Clinton appointed Summers to the Treasury in 1995, Sandberg's mentor asked her to be his chief of staff. "Sheryl always believed that if there were thirty things on her to-do list at the beginning of the day, there would be thirty check marks at the end of the day," Summers would later tell *The New Yorker.* After the Clinton administration ended, Google CEO Eric Schmidt, whom she'd met while studying the concept of Internet taxes, kept calling her. "Sheryl, we're profitable," he'd say, sharing information that few in the Valley knew or understood. "You should join." She wasn't sure what her role would be, but Schmidt said it didn't matter. "All that matters is how quickly the company is growing," he said. "This is a rocket ship—get on it."

"And I went home and I thought to myself, *That's exactly right,*" she says. "And I joined." Later she would offer the same advice to others pondering a break to Silicon Valley: *Get on the rocket ship.* "It's all about fast growth," she'd say.

She eventually wound up in the sales organization at Google, which struck some people as odd. "This is a job for a tractor," said her boss Omid Kordestani. "You're a Porsche." But Sandberg understood that Google was pioneering digital advertising at scale. It was just about to launch its AdWords search advertising product, which would become one of the most successful products in history. "I really believed that was the future of the business," she says. She was fine being a tractor in that effort, building an organization, changing the nature of ad sales from schmoozing to analytics.

Sandberg was never one of the people you would hear talking about "Don't Be Evil" or any of that stuff. She once remarked that, in her observation, a company's beliefs were the *opposite* of its mantras. "My attitude has always been that you'd better keep your head down and do your work," she once told me. "I delivered my numbers and focused on my metrics."

But by the end of 2007, it was time to leave Google. She knew that the CEO-in-waiting was co-founder Larry Page, and she wasn't tapped to replace Kordestani as head of the entire business side. Don Graham, CEO of the Washington Post Company, had tried to hire her after her Treasury stint, and was offering her the top job in his corporation. But Sandberg was now married to an entrepreneur named David Goldberg, whose steady demeanor was the perfect foil to her high-strung approach to business and life. Graham had to admit that Washington did not offer the opportunities for Goldberg that California did. (Goldberg would soon get an offer to head a young company in Palo Alto called Survey-Monkey.) Sandberg then surprised Graham. "Tell me about Mark," she said.

Sandberg was already on the short list of COO candidates when she first met Mark Zuckerberg at a holiday party at former Yahoo! exec Dan Rosensweig's house. (Ironic, since Rosensweig was also on the candidate list.) They made arrangements for a longer conversation at the Flea Street café in Menlo Park, a farm-to-table haven for Bay Area foodies that Sandberg adored. They found plenty to talk about. They were so deep in discussion—Zuckerberg going on about the mission and what he needed to accomplish it—they closed the place down, and moved the conversation to Sandberg's place, until she finally sent him home. "I have kids!" she would later recount to Oprah about the conversation. "My kids were getting up in five hours!"

As the recruitment heated up, Sandberg spoke about the Facebook interest with Roger McNamee, who instinctively felt that Zuckerberg—who grew up in a family dominated by females—could handle advice from a woman older than him. Sandberg, says McNamee, was worried that Zuckerberg was too young to be her boss. (Sandberg now downplays the McNamee connection, saying that he cluelessly advised her to take the *Post* job. She also disputes his claim that she was worried about Zuckerberg's age.)

The discussions continued through January and February. Zuckerberg showed her around the offices, which by then had been spread over multiple locations in downtown Palo Alto. When he asked her what she thought, she intuited that he expected her to remark how cool it was. Instead she told him it was absurd to have to walk around to all those different places. "Get yourselves into one building," she told him. (A year later, Facebook would largely consolidate in a single building in a different part of Palo Alto.)

During the months of conversation with Zuckerberg leading up to her job offer, the two discussed everything about the company, but one negotiation in particular would set Facebook's structure for the next decade and beyond: what parts of the company Sandberg would be responsible for and what parts would be excluded from her reporting chain. Zuckerberg felt that Sandberg should basically take on the things that he wasn't so interested in—sales, policy, communications, lobbying, legal, and anything else with a low geek quotient. His own time would be best spent on product—the stuff that engineers build. That's what defined Facebook. This was basically how he'd dealt with the previous executives like Parker, Van Natta, and Palihapitiya, and even though Sandberg would have higher status than any of them, he hadn't changed his mind on that. Sandberg's role was to explicitly "take a bunch of stuff off Mark's plate," she recalls. "It was very easy—he took product and I took the rest."

Since Sandberg didn't consider herself a "product person" but was obviously experienced in business, that split made sense. And Sandberg believed that through their discussions, Zuckerberg came to understand how important an augmented business model was to Facebook. But the split would result in some odd bifurcation. The engineers creating ad products—like new kinds of advertisements that might live in the News Feed—reported up to Zuckerberg in an entirely different organization from those who sold ads. Sales was Sheryl country. And the people building the News Feed itself would report to Zuckerberg, while those who would be responsible for the policy decisions of what would be appropriate content to appear on people's feeds would be working for Sandberg.

Ultimately, of course, all the responsibility fell to Zuckerberg. "Everything that reports to me reports to him because I report to him," Sandberg says. "So [the division of labor determined] just what I was going to work on."

Still, for the next decade of mega-growth, as the company faced unprecedented issues because of its mind-boggling scale, Facebook essentially had two organizations: Zuckerberg's domain and Sheryl World. And in no way were those equal. Zuckerberg headed engineering, the product side, not only because he was better at it but because he felt it was the heart of the company.

Still, it seemed like a no-brainer at the time. It would take him more than a decade to understand what a mistake it was.

Sandberg's key focus would be taking the company's nascent approach toward monetizing and eventually making Facebook profitable, preferably wildly profitable in the mode of her former employer. But because of Zuckerberg's inexperience, her role was much broader. She would be Facebook's *operating* officer. She made sure that an explicit job was helping Zuckerberg scale Facebook into a major corporation. "A thriving business was part of that, but it wasn't the only thing," she says.

But she did have firm ideas about business from the get-go. On her first day she attended the mandatory new-employee boot camp and listened to the standard inspirational speech delivered by Chris Cox, a tribute to Facebook's lofty mission that was becoming a legendary part of company culture. But then she flouted orientation protocol by making her *own* speech. She explained to the astonished newbies that there was an inverted pyramid of advertising, and to date her former employer, Google, had dominated the bottom by monetizing intent (as people did searches). But Facebook, she said, would have an even bigger business, because it had the potential to create and monetize *demand*. That was the much wider top of the inverted pyramid. People come every day to Facebook to learn what's new and share their interests. So advertisers would be able to sell to Facebook users things that they wanted even before they thought to ask for them.

She and Zuckerberg agreed to meet at the beginning and end of the week, every week. They began to work out what would become a close relationship. In one of the first meetings of the executive team Sandberg attended, she went on about the number scale that recruiters should use while rating candidates (the *only* proper scale is 1 to 5, she insisted). Zuckerberg rolled his eyes. After the meeting he profusely apologized and said he'd never undercut her like that again.

Sandberg set about meeting everyone she could, learning about Facebook's structure, asking how people hired, what the culture was. But she also dove deep into the rabbit hole of her young boss's psyche, quizzing people who knew him well. She had seen how well Eric Schmidt had handled the youthful founders of Google, never once saying publicly they were anything but geniuses. "I gave her a copy of *Ender's Game*, and said, *Read to this to understand Mark*," says Joe Green. (The protagonist is a teenager who saves the world when it turns out the war game he trains on is the real thing.) Though she might occasionally give an eye roll of her own to a friend when talking about this or that Zuckerbergism, in public she spoke rhapsodically of the partnership.

Zuckerberg's previous adjutants—Parker, Van Natta, and Palihapitiya—had all been masters of chaos who let their boss's bro-tastic move-fast ethos run loose not just in releasing products but in the management style of the company itself. Sandberg instantly put a stop to that. She was like Wendy parachuting onto the island of Lost Boys. Early in her tenure, she invited all the key women employees to her house for a cocktail party. (There was no problem fitting them into one room.) She let them know that the bro era was over. As she had at Google, Sandberg would reach out to young women at the company and in the course of a tightly scheduled one-on-one, ask about personal issues, offer advice, and pledge that she'd watch out for them. She joined several women's groups at Facebook. She hosted Gloria Steinem at her house to meet Facebook women.

Dave Morin remembers one instance early in Sandberg's tenure when a problem emerged and instead of having the rank and file cope with rumors about what Zuckerberg and his inner circle was thinking, she gathered the teams together and had people sit on the floor and talk about it. "A lot of us were kids and didn't know how to manage people and didn't have the skill sets to do nuanced communication," says Morin. "She put that stuff in place and brought a level of maturity."

By and large people saw her as the perfect complement to Zuckerberg. "She was everything Mark wasn't," says Ezra Callahan. "She was diplomatic, she was eloquent, she was relatable. She could make all parts of the company feel like they mattered, whereas Mark increasingly was making clear that product engineering runs the show, and the rest of you should shut up and do your jobs. It took us from feeling like this is a billion-dollar company that's going to shoot itself in the foot one too many times to, *Okay this is going to happen now.*"

Not long after Sandberg arrived, Zuckerberg took his first long vacation since he'd started Thefacebook. It was a trip around the globe, lasting more than a month. "Once Sheryl was on board I felt I could do that, and I also wanted to give her time to ramp up," he says. He packed light and went alone, visiting some friends at various locations. He started in Europe and continued east. One destination was a remote Indian ashram. The trip was compelling in part because it was the same ashram that Steve Jobs had been to before he started Apple.

Zuckerberg's travels were rugged, with the same highs and lows that any twenty-four-year-old might experience on a solo trip abroad. He went to Berlin, Helsinki, and Katmandu. He tried to go to Russia but was unable to get a visa. He got sick trekking in Nepal; locals tried to cure him with yak milk, with little success.

The visit to the ashram didn't wind up providing any flashes of enlightenment, though it was extended when a storm hit and what was planned as a single night became several. He spent the time "writing and meditating." His mind was never really off Facebook, though, as the journal he kept brimmed with ideas he might execute on his return. "I remember spending a lot of time thinking about how people communicate and how groups of people can feel and act together as one," he says. "It definitely strengthened my belief in our mission—which focused on making the world open and connected."

That insight would later result in another trip to India. The trappings would be luxurious, an entourage would accompany him, and he would be both greeted as a role model and attacked as a colonialist forcing Facebook on the country's 1 billion people. But that was years away.

WHILE ZUCKERBERG WAS globe-trotting, Sandberg used his absence to gain consensus on Facebook's business model. She didn't have to start from scratch. Despite the total botch of Beacon, the rest of the Pandemic launch—Pages, targeting, and cost-per-click ads—was already working. "We were on a $500 million run rate," says Tim Kendall, who still headed monetization, and was now happily on Sandberg's team. (He would remain at that post for two more years, before he left to join Pinterest and eventually become its president.) Helping Facebook even more were the terms it had negotiated with Microsoft, which guaranteed a certain amount of revenues. "It gave us air cover," says Mark Rabkin, who headed ad engineering. When Facebook didn't have an ad to place through its system, Microsoft would serve one through its inventory, often at a higher rate than Facebook was able to get at that point. On the couple of occasions where Facebook's ad servers crashed, revenues would actually rise. "We would do the full postmortem and find we made an extra $50,000, because when our system was down, we would just hand it all over to the Microsoft system and they were paying a higher guaranteed minimum," says Rabkin. (After several months, Facebook advertisers were paying higher rates, so that circumstance no longer applied, even before the Microsoft deal's sunset in 2009.)

It was obvious to Sandberg that Facebook's business would be advertising and everything else a rounding error. But not everyone at Facebook was comfortable with the idea, especially among some of the younger people who thought that ads sucked and Facebook should do something less . . . smarmy. Even Zuckerberg wasn't all in—after the Pandemic event that fall Rabkin told him that he needed to hire more people. Five engineers were way too little to scale into a billion-dollar system. Google had four hundred! "Well, how many do you think you'll need to build the best system in the world?" asked Zuckerberg. Rabkin mustered up his courage and said . . . twenty. "It seems like a lot—let me think about it," said Zuckerberg. Rabkin says it took a couple more years before he had twenty engineers on his ad product team. (Today there are hundreds, just like Google.)

Sandberg scheduled a series of meetings with a cross-section of key employees on Tuesday or Wednesday nights, and went through the exercise of exploring how Facebook might possibly build a huge business on revenue other than advertising. WHAT BUSINESS ARE WE IN? she'd write on the whiteboard. Charge users? Do research? Everything was examined, and all found wanting—except advertising. Which was exactly where she wanted to go. "I thought at the time it was a pointless exercise," says Kendall. "But in retrospect it was brilliant, in getting buy-in across the board."

The group reached a consensus—Facebook would concentrate on demand, just as Sheryl had revealed on her very first day. Because Facebook knew so much information about its users, it could tell when they were prone to pitches for specific products, or even political candidates. What Facebook would not do, Sandberg vowed, was go for the quick money offered by advertisers to take over the home page with a promotion, or present outsized banner ads. (Zuckerberg had resisted this as well.) Around that time, MySpace turned over its front page to a Batman movie, followed by a promotion where the whole page went green to tout a film about the Incredible Hulk. "The concept of advertising was like the green thing was going to take over your home page," she says. "The first meeting I had at one of the studios, this woman who was head of marketing literally screamed at me because we wouldn't do a Hulk-like takeover for their movie and stormed out." She wanted Facebook to do better. Ads should be good experiences, consistent with the good experience you were having on Facebook. "It wouldn't have to be the Hulk coming out."

Zuckerberg joined the conversation on his return, accepting the conclusions of Sandberg's gatherings. Sandberg began building her team. Because she tapped a vast Rolodex of friends and contacts—whom she had methodically kept in touch with—Facebookers started using the term "FOSS." Friend of Sheryl Sandberg. These included her friend Marne Levine, who would head Facebook's DC office.

Unlike Zuckerberg, whose closest contacts consisted of "the small group"—the subset of Facebook's leaders who were his informal advisory board—Sandberg liked an organized infrastructure of aides and lieutenants. (The larger management group was called "the M team.") She'd have her own chief of staff, just as she'd been COS to Larry Summers.

Her most conspicuous hire was Google's head of communications and policy, Elliot Schrage, who would assume that role at Facebook. Brandee Barker, who had been filling the top communications role, felt rebuked. But working with one of the corporate "coaches" that the consultants were providing Facebook executives, she got through it. "I had to understand that for where Facebook was going, they needed Elliot, that it had grown past what I was capable of doing," she says. Barker found satisfaction working on product communications. Talking about the issue with Zuckerberg was never an option. "Mark doesn't necessarily have a history of talking directly to anybody when he is ready to move on to the next person," she says.

Sandberg's contract with Google dictated that she could not solicit people to Facebook in her first year there. (Schrage had volunteered.) Google was already alarmed that too many of its employees were leaving for Facebook, and a few months after Sandberg started, her former bosses, Jonathan Rosenberg and Omid Kordestani, both called to discuss limiting the exodus. (According to an affidavit submitted in a case involving Google's illegal non-poaching arrangements with other companies, Sandberg says she declined.) But after that, the floodgates opened.

On the day her soliciting embargo ended, she went down a list of former colleagues she wanted with her. The dragnet bagged three executives, including Greg Badros (who had interviewed in Zuckerberg's apartment two years earlier).

That initiative confirmed Google's worst fears—a high-cost talent war bereft of secret no-poaching agreements. For years thereafter both companies spent hundreds of millions of dollars luring each other's workers and retaining those who got offers from rivals.

Of course, Sandberg's net was cast much wider than Google. She poached Microsoft's top advertising executive, Carolyn Everson, to head sales, even though Everson had only recently taken the Microsoft job. (Microsoft CEO Steve Ballmer was so upset that Everson took careful note of the golf club he was swinging when she gave him the news. Fortunately, "He didn't try to hit me," she says.)

In Everson's interview with Zuckerberg, her prospective new boss seemed still to be wrapping his head around why companies needed brand advertising. His mother, he told Everson, always used the same shampoo. Could Facebook ads change that? Everson used the example of BMW versus Mercedes. Maybe younger people weren't thinking of buying one, but if advertising over twenty years created a desire, they might buy one when they reached the age where consumers buy luxury cars.

When Everson took the job—she would be the person who built relationships with the top marketers at the big brands—she found that it would be harder than she expected. "It felt as though money was just flowing into Facebook and they must have everything figured out," she says. "And when I got here we didn't have everything figured out because everything was brand-new and we're still building."

Zuckerberg understood that. As always, he was focused on the long term. If Facebook was indeed going all in on advertising, "Mark knew correctly that the race to get the real product right and to internationalize the product and encourage sharing, and all of that was much more important than early monetization," says Mark Rabkin. As always, *product* was his interest—what new ads should be invented now that

Facebook was set on its monetization path? After the Beacon train wreck, the sponsored stories had been pulled off the News Feed—ads would not return there for years—and Facebook needed to explore other avenues. One step the company took that year was something called "engagement" ads, where an advertiser would put something on a user's home page asking him or her to take some action—RSVP to an event, or visit the advertiser's Facebook page.

Meanwhile, Facebook built on the foundation it had just laid. The Pages feature. Targeting. A pay-per-click auction to place ads. And then an idea emerged that would wind up augmenting the value of it, as well as having an impact on the News Feed, and extending Facebook itself into the world. The Like button.

THE FIRST STIRRINGS of the Like button came in July 2007, when News Feed was less than a year old. Code-named Props, it was an attempt to give users a way to approvingly tag stories on their feeds. Leah Pearlman, a designer on the team, had gotten the idea after a friend of hers suggested a "bomb" button to single out posts. Pearlman suggested that something be installed to indicate enthusiasm. This would mitigate what had been an obligatory necessity to respond to certain posts (new jobs, engagements, cool vacations) with congratulations or other rote kudos responses. "It was just how can we take this vast social network that we're creating and make it as easy as possible to spread little bits of positivity and love and affirmation into the system," says Justin Rosenstein, who conceptualized it with Pearlman. Since you couldn't do that with zero clicks, a single click was the desired minimum. Rosenstein, as the product manager of Pages, also saw this as a way to get users engaged with those commercial efforts. "You could create advertising that was tied to someone liking a particular page, or you could advertise a page to people who liked similar pages," says Rosenstein. So while a low-friction means of approval would obviously be helpful in ranking stories, it might have value for advertising. It could be a subtle way of helping to identify a user's interests without the user conspicuously sharing them with Facebook.

After a flurry of emails, a working group temporarily decided to name it the Awesome button, despite that word's association with the chaotic News Feed launch. Also under consideration: Star? Plus sign? Thumbs-up symbol? At a hackathon that summer, Rosenstein and a small team coded and designed a prototype for the Awesome button, using a star as an icon. But for various reasons progress on the project halted.

Later that year, a start-up company called FriendFeed launched. Its website had its own feed, an aggregation of all the messages and posts from the different social networks the FriendFeed user belonged to. And it had a Like button. According to an informal history of the Like button by Andrew Bosworth, Facebook ignored it. By the time Zuckerberg had decided that "awesome" didn't strike the right tone for a feedback feature, and renamed it "like," Facebook had purchased FriendFeed. (The prize, besides taking a potential threat to Facebook off the market, was co-founder Bret Taylor, a top-notch former Google engineer who would later become Facebook's CTO.)

Paul Buchheit, the other FriendFeed co-founder, was amused to learn Facebook was working on a Like button, but thought it was a good idea. "I can't definitively say that FriendFeed was where that word came from. But it's an interesting example of the difference just a single word makes. It would be weird if it just said 'awesome' everywhere, right? 'Like' is sort of this nice word that's lightweight and sort of meaningless. It isn't a major commitment."

At the same time, there was an effort by the design team to redesign how feedback worked on Facebook. The two teams merged to work on the Awesome button and got some momentum. The decision to make the actual icon a thumbs-up sprang from Facebook tradition—the Poke button was also a hand. Aaron Sittig tweaked it into the style that would become familiar to the world.

Still, it took another year and a half for Facebook to launch the Like button, in part because Zuckerberg was always lukewarm to it in product reviews. Seven Zuck reviews failed to get a thumbs-up from the CEO.

One big reason was a fear that installing a one-click way to comment on a News Feed post would inhibit actual comments, and instead of interesting conversational threads there would only be a mindless accumulation of positive clicks. Bosworth referred to the Like button as a "cursed project."

In late December 2008, product manager Jared Morgenstern inherited the Like-button albatross and tried to figure out how to lift the curse. The big hurdle was proving that a Like button would not cannibalize commenting, a much higher-quality form of sharing. He built in some tricks, like moving the cursor to the comment box after someone hit the Like button. But ultimately Facebook would only know if Likes depressed commenting by trying the button out and measuring the responses. Instead of another Zuck review, Morganstern sent an email to Zuckerberg, casually mentioning that he was going to launch the Like button in the Scandinavian countries. He interpreted Zuckerberg's non-response as an implicit go-ahead. After giving some Nordics the use of a Like button and comparing their behavior to those who didn't have one, Facebook's researchers discovered that a Like button would *increase* commenting.

Zuckerberg ruled that the project go forward. "It's going to be *Like* with a thumbs-up, just build it and ship it," he said. "We're done with this."

The Like button exceeded all expectations. People took to it immediately. As originally intended, it provided a crucial signal to help rank News Feed posts. What could be a clearer indication that people liked a post than an explicit action that expressed that very sentiment? Since the goal of the News Feed was showing people what they wanted to see, Facebook's job became easier.

But a more significant, and somewhat ominous result came when Facebook decided to expand the Like button beyond its own site and spread it to other sites of the Internet. The company essentially made a deal with the World Wide Web: if you put our Like button on your page, whatever you're selling, promoting, or just saying in public could be boosted by implied (though unwitting) approval from millions of users.

It was as if the entire web were posting to News Feed. And it was an unbelievable source of data for Facebook.

So unbelievable that it caused a minor uproar when a privacy expert named Arnold Roosendaal—at the time a doctoral candidate in the Netherlands—analyzed the data extraction and published a paper on what he found. Collecting the information when a Facebook user "liked" something on a website provided valuable information on its own. But Roosendaal found that even when users simply lit on a web page that supported likes, Facebook would plant "cookies" (which are persistent information trackers) on the visitor's browser. What's more, "When a user has no Facebook account, a separate set of data concerning individual browsing behavior can be created," wrote Roosendaal. "When a user creates an account later on, the data can be connected to the newly established profile page." Facebook called the latter issue a "bug," and CTO Bret Taylor told a reporter that the button was not used for tracking. Still, the behavior sounded a lot like the concept of "dark profiles" outlined in Zuckerberg's Book of Change in 2006. "What people don't realize is that every one of these buttons is like one of those dark video cameras," said maker of privacy software Rob Shavell to *The New York Times*. "If you see them, they see you."

Aside from those privacy concerns, the Like button also had other downsides. It created a race for Likes. It was a not very subtle incentive for people to tailor what they posted to court those clicks. People would feel bad when they didn't get Likes for posts that meant something to them. For businesses, the pursuit of Likes became a serious goal. The number of Likes their pages accumulated determined how visible they might be to Facebook's vast audiences. If people expressed interest in their pages, advertisers could write posts that found their way to their News Feed. If a post got a lot of Likes, the News Feed algorithm would distribute it more widely, giving it "organic" traffic that went to the News Feeds of the friends of those people. It was free advertising. Many companies, including some of the world's largest, engaged in an attention war to lure people into "liking" their pages. They sometimes offered goodies to those who gave that thumbs-up. Some pages began patronizing a black market for Likes: for a price, one could buy thousands, sometimes provided by legions of low-paid workers in China or some other nation, sitting in sweatshops, index fingers on mouse buttons, adding Likes to boost a brand.

The Like button would become a symbol of Facebook itself, with the thumbs-up icon displayed at Facebook headquarters. People would take selfies in front of the sign and post the images to social networks. Hoping, of course, that friends would "like" those photos.

So it was that the simplest of features boosted Facebook's business, gave users an easy way to express themselves, and set the company on a disturbing course of overemphasizing trivial or angry content. Not to mention that the Like button was a gateway drug for Facebook's data-gathering to extend beyond its borders. In recent years its founders, Rosenstein, Pearlman, and Morgenstern (none of whom work for Facebook anymore), would express, if not regret, a sober recognition that their work has been a factor in degrading society and empowering their former employer to wantonly gather data on its users. All now feel it was the right thing to do at the time, but all now wish that somehow Facebook could have gotten ahead of the unintended consequences. Which is a sentiment that could largely apply to the entire enterprise of Facebook.

In any case, the Like button's conquest of the web was wildly successful for Facebook, and could be viewed as the revenge of Beacon. While Beacon shared personal data it received from websites with other users on Facebook, the Like button let Facebook use that data for its own purposes, largely to build its profiles of users and power its advertising. Facebook had learned that going beyond its borders to augment its monetization would be transformative. Later, Facebook would take the further step of buying information from data brokers. What was once "blasphemous" to its chief privacy officer was now Facebook's business as usual.

What's more, as Facebook gathered more data, much of it in real time, that data pointed to how the company could exceed even Sheryl Sandberg's revenue expectations. From the day she arrived at Facebook, she thought that the company would be limited to advertising that created demand for a product, a huge market to be sure. But by gathering information about what people were doing on the web—what they were shopping for and fantasizing about—it could also capture the precious information involving people's *intent*.

That was something advertisers would pay even more for. And it put Facebook in a much more powerful position to capture the lion's share of online advertising revenue.

"The idea that we could move up the chain and do more intent rather than demand fulfillment . . . was pretty fundamental," Sandberg says.

And so Facebook finally had its business model. It would require further tweaking, and a deeper dive into personal data, especially as people abandoned desktops and moved their online world to handheld devices. But as the dollars poured in, billions and billions of dollars, now drained from traditional advertising venues and into Facebook's coffers, Mark Zuckerberg's over-the-top introduction to Pandemic seemed less and less hyperbolic. Maybe the next hundred years of marketing history really did start with Facebook.

10

Growth!

EARLY IN HER tenure at Facebook, Sheryl Sandberg had a series of conversations with Chamath Palihapitiya. They knew each other well. He was a family friend, part of a regular poker game with her husband, Dave Goldberg. The two of them would often go on man trips to Las Vegas. Palihapitiya, himself a father, was great with Sandberg's kids. So Sandberg was open to hearing what the thirty-one-year-old executive had to say.

Palihapitiya was at a crossroads at his time at Facebook. He had joined barely a year before, leaving a job as a venture capitalist. Previously, Palihapitiya had been a VP at AOL, the youngest person to hold that position at the company.

That job had been one more step in a heroic rise for Palihapitiya, who had emigrated from his native Sri Lanka at six years old, when his civil-servant father got a post in Canada. When his father lost the job a few years later, the family went through hard times. As Palihapitiya told *The New York Times* in 2017, they moved into a 400-square-foot apartment above a laundromat. His father had trouble finding a job, and "there was drinking and stuff," Palihapitiya recalled. Though his mother was trained as a nurse, she worked as a housekeeper and nurse's aide.

Palihapitiya worked, too, first at a Burger King. He found it more profitable to run a blackjack game in his high school lunchroom, though, raking in $40 or $50 during some breaks. He'd then take that money to a casino—he had a fake ID that gained him entry—and try to augment his winnings. He had a knack for high-stakes poker, and eventually would compete in the World Series of Poker.

After graduating from the University of Waterloo with an electrical engineering degree in 1999, he worked for an investment bank for a year. He later called his job trading derivatives "the most insipid, idiotic use of my time." (No one in Silicon Valley could match Chamath Palihapitiya when it came to erudite invective.) He quit the bank and began applying for jobs in the online sector, ultimately moving to California to work for a music start-up based on the WinAmp music player that had inspired Zuckerberg and D'Angelo back in high school. Though WinAmp was seen as a cool application, Palihapitiya came to view it as a platform; outsiders were free to customize it with decorative "skins" and creative plug-ins that boosted its power. (Synapse was an example.) The platform concept enabled talented engineers and designers to improve their products at no cost, making them more valuable and harder for competitors to beat them. WinAmp was where Palihapitiya met Sean Parker, and naturally the two got along famously.

AOL bought WinAmp in 1999 and Palihapitiya joined what was then the world's most popular online enterprise, a year before its disastrous purchase of Time Warner. And now, in 2005, Parker was calling, saying, "I've just become president of a company called Thefacebook, and I want you to meet me and learn about it and meet the founder."

Parker and Zuckerberg flew to Dulles and met Palihapitiya and another AOL executive, Jim Bankoff. Parker did most of the talking, as Zuckerberg was in his silent mode for most of the meeting. But Palihapitiya was impressed. He later told Bankoff that AOL should consider buying Thefacebook, but AOL was still sifting through the rubble of its train-wreck merger with Time Warner and couldn't consider an offer. Not that Zuckerberg would have gone for it.

AOL wound up doing a small deal with Thefacebook that linked AIM to the start-up's website, so that Facebook friends could find one another on the chat service. Palihapitiya often referred to it as solely to AOL's advantage.

But the biggest outcome of the deal was the connection between Palihapitiya and Zuckerberg. In 2005, Palihapitiya quit AOL—one takeaway from his experience there, he later said, was that "most people at most companies are really shit"—and joined the venture-capital firm Mayfield, in Menlo Park. Every couple of months or so, he and Zuckerberg would get together. Zuckerberg's combination of chutzpah and shyness appealed to Palihapitiya.

Not that Palihapitiya necessarily agreed with the lionization treatment that the young CEO was enjoying in the press and in Silicon Valley's inner circles—he never subscribed to the pervasive meme in the tech world and the business magazines that successful founders were like gods in the Pantheon. To him, they were more like opportunistic beneficiaries of favorable economic and social conditions. Maybe if Zuckerberg had gone to Ohio State instead of Harvard, he thought, none of this would have happened. (The chip on Palihapitiya's shoulder that came from not attending Stanford or an Ivy seemed big enough to leave a permanent depression on his clavicle.)

Still, the boisterous Palihapitiya and the more introspective Zuckerberg had similar views about business and tech. Inevitably the idea arose that Palihapitiya might join Facebook. Nothing formal. But Zuckerberg, in the throes of post-Yahoo! dissatisfaction with Van Natta, wanted someone in his top management rule who was more an ally.

So in early 2007 Palihapitiya went to the University Avenue office—just to check things out, mind you. Zuckerberg wasn't there and he wound up talking to Moskovitz, who asked him why he wanted to work at Facebook. Being regarded as a supplicant offended Palihapitiya. "Hold on a fucking second," he said. "I'm not interviewing with you." Then Palihapitiya proceeded to tell Moskovitz everything that was wrong with Facebook. The list was long, but Palihapitiya wasn't implying that Moskovitz and Zuckerberg were idiots—they were doing the best they could, considering that they had zero experience in business.

But what did impress Palihapitiya—and make him amenable to actually taking a job there—was that Zuckerberg and his coterie seemed open to listening to suggestions that might fix things. To Palihapitiya, it was an opportunity worth taking. Even so, Palihapitiya drew the process out; at one point during his Hamlet-esque indecisiveness, Zuckerberg even called in Roger McNamee to tout the company. "I think [Chamath] already decided he was going to do it, and he was just playing with me," says McNamee.

When Palihapitiya finally joined the company, he was given considerable, if amorphous, duties under the umbrella of "product marketing and operations." These were basically responsibilities yanked away from Van Natta, who, after his lack of loyalty during the Yahoo! episode, was essentially demoted from chief operating officer. (Facebook temporarily retired the title.)

He had an immediate impact. Director of monetization Tim Kendall was impressed with Palihapitiya's talents—and appalled by his demeanor. "He was an incredibly shrewd leader," he says. "I learned quite a bit from him. And I would never want to work for him again."

Palihapitiya thought that the haphazard hiring process over the past two years had brought dead weight into the company, and he introduced a "forced ranking" process to identify and dismiss the laggards. While that made sense, his public ranting about people not pulling their weight was scary even to some high performers. Most of the older, more experienced people who came into Facebook had learned to adjust to the callow culture. Palihapitiya doggedly remained himself—embracing his own worldliness, his tough upbringing, his contempt for the conformity of the striving-for-Harvard crowd.

And Palihapitiya could be a bully. He would humiliate people at meetings, criticizing their appearance. He mocked one barely middle-aged executive for his receding hairline. Another former executive, who would only talk about Palihapitiya if I turned off my tape recorder, was almost in tears as he recounted the verbal abuse he took from Palihapitiya. It was as if he was still afraid that Palihapitiya might emerge from behind the bushes and resume his battering.

When I asked Palihapitiya about this, his reaction was unapologetic.

"Get. Out. Of. My. Way," he says. "They must feel really bad as they lounge around in their chinchilla blankets and their multimillion-dollar mansions right now." He explains that the workplace is not a family, and if people were looking for touchy-feely responses from him, "that was not going to be a good meeting, those people probably felt intellectually bullied."

On the other hand, he could be inspiring. He would stand on a table and tell people how big Facebook would be. At a time when people would be buzzing at the prospect of Facebook being valued at $10 billion, he would proclaim that Facebook would be worth ten times that. (Later, when that no longer sounded crazy, he would talk of it being worth a trillion dollars.) He also had no compunctions about dismissing people he felt were not sufficiently fast-moving and innovative to help Facebook meet those marks. Some of those people had been regarded by coworkers as subpar but lingered in their jobs because Zuckerberg could be timid about firing people.

Yet, for all his talent and bluster, Palihapitiya in his first few months was not a smashing success at Facebook. He had not really added much to the Platform team. What he was most associated with was Beacon, one of the company's most traumatic moments. He felt that his unfortunate quote in *The New York Times*—the one about Beacon not transferring data against a user's wishes—was misinterpreted. ("I learned not to get into technical details with reporters," he now says.) Nonetheless, it eroded what was already a fragile relationship with Zuckerberg.

By the end of 2007, even Palihapitiya acknowledged he had done a miserable job. Maybe he was one of those people whose success was simply due to lucky circumstances. "If I were you, I'd fire me," he told Zuckerberg in a post-Beacon meeting. They agreed that he should figure out a more focused task for himself and take one last shot at things.

NOW, IN SANDBERG'S office, Palihapitiya was about to throw a Hail Mary: he had something in mind that he felt would be instrumental in

Facebook's future success. He didn't anticipate that the same effort would cause its failures as well.

At that moment—early 2008—Facebook's growth had slowed. A similar trough had hit Facebook more than a year earlier, before Open Reg and the News Feed kicked in, but in a way, this was more alarming, because there were no similar groundbreaking products on the horizon. And no one had any idea about the cause. "Everything stopped," says one executive. "And to this day we don't know why."

"Growth had plateaued around ninety million people," Zuckerberg recalls. "I remember people saying it's not clear if it was ever going to get past a hundred million at that time. We basically hit a wall and we needed to focus on that."

Palihapitiya came offering a solution: a high-energy team with a long leash whose focus would be accumulating and keeping users. He felt he had identified the North Star of Facebook, the most elemental aspect of how it defined its business and its financial health. And that was the concept of the Monthly Active User. Other Internet businesses counted how many people were on the site each day, or how many had signed up in total. But monthly was a better indicator, because someone consistently on the service for a full month was likely there to stay. Thus the number took into account the "churn"—how many people were leaving Facebook. Palihapitiya proposed to be utterly obsessive about MAUs—to look at every part of Facebook's business in light of this metric, to learn what can drive MAUs, to fix things that don't increase it, and to build new parts of the company to boost MAUs even higher.

Sandberg was intrigued. "What do you call that?" she asked.

"I don't know," said Palihapitiya. "The job would be to grow things—all to grow MAUs."

"Maybe you should just call it growth," she said. (Sandberg says she doesn't remember that exact meeting—she had a lot of them with Palihapitiya—but confirms that something like that exchange occurred.)

Palihapitiya honed his ideas and presented them at the next board meeting. Palihapitiya claimed he could double or triple the user base by using aggressive techniques, and go far beyond that by making Facebook's whole Platform itself an engine for growth.

The board wasn't terribly excited. *Just give me some time to execute*, said Palihapitiya. If they didn't like what he was doing after a few quarters, then maybe they could all agree on his exit.

Palihapitiya was doubling down on a mania for growth that had begun well before he arrived. Even as Sandberg gave Palihapitiya the green light to lead his team, she felt that Facebook was already a focused growth machine. Facebook's origin story was a fable of growth: in 2004, in the space of a month—February, the shortest month of the year!—Zuckerberg expanded Thefacebook from Harvard to other campuses. In 2005, an early employee named Noah Kagan (who later got fired for leaking product plans to the trade publication *TechCrunch*) once suggested to Zuckerberg that Facebook sell tickets to events listed on the site's pages. Zuckerberg's answer was to walk over to one of the omnipresent whiteboards and scrawl one word with the marker: *Growth*. "He said that if any feature didn't do that, he was not interested," wrote Kagan. "That was the only priority that mattered."

Indeed, even in 2005, the company decided to hire specialists who would dive into the information Facebook gathered to draw more users. That was to be the main role filled by Dan Plummer, the scientist who died in a bicycle accident in January 2006. His replacement came to Facebook by a chance encounter involving Zuckerberg's sister Randi, whom he'd asked to come work for the company. At her going-away party in New York, Zuckerberg recognized an engineer named Jeff Hammerbacher as someone he'd been in a seminar with at Harvard. (Hammerbacher's girlfriend had been pals with Randi and had dragged him along.) On Zuckerberg's suggestion, he applied for work at Facebook. His plan was to establish California residency for a year and go to grad school there paying in-state tuition. But he was impressed with the Facebook engineers who interviewed him. And he was intrigued when he saw that Adam

D'Angelo's business card read "data mining." Hammerbacher was all about that. And so was this little Palo Alto start-up, even in 2005.

Hammerbacher, usually collaborating with others, analyzed behavior on Facebook, with an eye to see if the data they unearthed could help growth. More important, he built a system where Facebook would collect huge amounts of information and analyze it to come up with conclusions to help Facebook work better. He set up a system to log every single click made by Facebook's users. He would later refer to it as an "Information Platform." On the first day, he had 400 gigabytes of information.

Working with Naomi Gleit and Matt Cohler, he looked at schools where growth took off and compared them to ones where growth had stagnated, seeking to understand the ingredients for success and failure. When Facebook unveiled Open Reg, he drilled down on the data to find how users came to Facebook, and discovered that the biggest source came from a program an engineer had built that imported people's contacts from Microsoft's Hotmail.

The program was called "Find Friends," written by Facebook engineer Jared Morgenstern. It worked with Hotmail, Gmail, and Yahoo! mail. Users would provide their login and password, and Facebook would scan all the people who were contacts or connections and import those into its database. Those who matched with people already on Facebook would be sent friend invitations. Others would be presented to the user and a simple check alongside their names would trigger an email inviting them to join Facebook. Upon completion, Facebook would delete the login information.

Microsoft had objected to this. "They were blatantly stealing, trying to build their social network on the backs of others," says a Microsoft executive involved in the discussions. Zuckerberg shrugged it off when confronted with the charge. "He was like, *Yeah, I know it's kind of annoying, if it bothers you we'll stop*," says the executive. "But he didn't stop." In Microsoft's view, it not only violated their terms of service but was arguably an unethical data grab. Just because you were someone's contact on Hotmail, that didn't mean that you would be okay with winding up in a Facebook database. Find Friends created a tension with Microsoft that was resolved only on the eve of the larger company's investment in Facebook in 2007.

Such techniques were as vital as oxygen to Facebook, because people often did not bother to fill out their social graph. "A huge percent of Facebook users barely even sends friend requests," Zuckerberg told me in 2011. "They won't even set up the most basic relationship. They only accept incoming connections."

To keep that growth going, Facebook needed to scrape not only Hotmail but numerous other services. The process had to be done separately for each email provider, a time-consuming project that could not possibly be executed by the single engineer Facebook had assigned to the task. An early Facebooker named Jed Stremel, who had been an ace dealmaker at Yahoo!, took care of the problem. He discovered that the global wizards of contact scraping were found in a two-person company in Malaysia called Octazen. Stremel quickly made a deal for them to write their little data gobblers for Facebook. He recalls that he paid about $400 for the deal. "It was in keeping with the spirit of 'Move Fast and Break Things'—just getting something done quickly is what mattered," he recalls. (Facebook would buy Octazen in 2010.)

Some early Facebookers contend that email scraping was the single most valuable factor in helping Facebook grow when for the first time it found itself competing not in closed networks but in the world at large.

By the time that Palihapitiya moved to Growth, Hammerbacher had grown doubtful about the mission he was fulfilling at Facebook. He left the company in September 2008. "It was turning from a place to explore to a place to exploit," he once told an interviewer. He became a co-founder of Cloudera, a company that stored data in the Internet cloud, and later became involved in trying to solve cancer with data analysis. Though he felt no animosity toward his former employer, at times he has expressed its motivations in phrases that speak volumes. In 2011, assessing the jobs of data scientists at Facebook and its peers, he made a remark to a

BusinessWeek journalist that would reverberate for years: "The best minds of my generation are thinking about how to make people click ads," he said. "That sucks."

IN THE OLD television show *Mission: Impossible*, each episode began when the leader, Jim, flipped through the dossiers of spies, strongmen, and honeypots, putting rejects in one pile and tossing the photos of the talented ones who were perfect for the mission into a stack on his coffee table. Palihapitiya did something similar in handpicking his team, plucking his diamonds from the ranks of other groups as well as outside prospects. He had a great eye for talent. Among them were Naomi Gleit, one of the first few Facebook employees; Javier Olivan, an engineer from Spain; Alex Schultz, a UK-born marketer; Danny Ferrante, a data wizard; and Blake Ross, a star hacker who had been the co-creator of the open source Firefox browser. They were a diverse bunch. A majority had either been born overseas or at least had a parent from outside the United States. Two were gay. One of the leaders, Gleit, was a woman. They were a team of outsiders, a data-driven Dirty Dozen, armed with spreadsheets instead of combat rifles. The choices proved brilliant, especially Oliver, Gleit, and Schultz; more than a decade later, that threesome would still be at Facebook in the most powerful leadership cadre, "the small group."

"Chamath had the rare ability to do the crazy-speech thing and get everybody fired [up] but those were like the three musketeers," Mark Slee, an early Facebook engineer, says of the trio. "At the end of the day it was the three of them working with a boatload of engineers and product managers to do a lot of really unglamorous work. Everybody wanted to work on the cool product stuff, but they were doing the stuff that actually gets more people on Facebook."

As one of the first employees of Facebook, Gleit had a quiet moral authority. In a sense she had helped own Facebook's growth since the start, working, as everyone in the company did, to map Mark Zuckerberg's bottomless ambitions onto products and initiative. "We always had growth projects," she says, noting that from the very start Zuckerberg began spreading Facebook to other colleges. When Gleit joined in 2005, Facebook was moving to high schools and already preparing for Open Reg. Some people on the team would say that it was she as much as Palihapitiya who provided the vision for this new initiative.

Palihapitiya wanted Growth to be a power center in the company with special status and a distinctive subculture. To distance it from any other group at Facebook, its members would not refer to themselves as a "team," but a "circle." The Growth Circle. They had their own little grotto in one of the Palo Alto buildings, and in 2009 when Facebook moved from downtown Palo Alto to a new headquarters, Palihapitiya claimed space apart from everyone else, in a dark corner of the ground floor.

"Even the name reflected how special it was," says Gleit. "We were a growth circle, we were tight-knit, we were the A-plus team. It was just magical, almost."

It also could be brutal. Palihapitiya saw himself as more worldly and pragmatic than the idealistic and largely homogenous workers at Facebook. He had a sort of contempt for the charmed lives they led. The way that he saw it, the Facebook hiring process rewarded the kind of people who spent their lives making checkmarks on life-achievement boxes—astronomical test scores, Harvard or Stanford, enviable internships. You could see it in their Facebook feeds, as each post seemed to unlock one more level on a video game. But the problem with those box-checkers was that they were locked into following predetermined courses.

The idea of going off course, abandoning the neatly arranged checkboxes entirely, didn't fit their mental model. They couldn't handle it. So Palihapitiya undertook to mess with their minds, doing psychological tricks to rewire their brains so they wouldn't be so fixated on validation and affirmation. Calling his team a circle was part of that—an attempt to jar them out of the idea of a hierarchy where the most important person sat at the head of the table. The idea was to empower anyone on the team.

While the circle was supposed to equalize the power dynamic in the team, Palihapitiya's personality still dominated. His style was geared to provoke reaction. Every third or fourth word out of his mouth seemed to be an expletive, usually some variation of the synonym for fornication. When he felt someone did or said something stupid, a reliably frequent occurrence, he would unleash a fusillade of insults. Such assaults weren't the kind of thing that you would take to HR, though. At Facebook that behavior was regarded as Chamath being Chamath.

And what happened in Growth Circle stayed in Growth Circle.

The core members of Growth still speak highly—reverently—about Palihapitiya. But even within this group, the feeling could be mixed. "I didn't love working for him, and he knows that," says Alex Schultz. "But he was generally right. And so I really thought as a leader he was pointing us in the right direction."

Ultimately, Facebookers shrugged off Palihapitiya's behavior because he clearly had the company's interests at heart. "I had a good relationship with Chamath," says Chris Kelly, whose job included reining in the Growth Circle's excesses. "We would have conversations about where the boundaries were, usually ahead of time. Transgression wasn't their goal, growth was their goal. Maybe there was a little bit too much comfort with transgression, but growth was the goal."

EARLY ON, GROWTH identified a lot of low-hanging fruit to help Facebook's numbers. One of them was search engine optimization (SEO), the practice of raising the visibility of content in Google search rankings. If a Facebook virgin using Google came across a friend's profile, he or she might be inclined to join the service. But Facebook's SEO was so weak that when you typed the word "Facebook" into Google, you had to scroll far down the rankings until you saw the actual site.

In the previous year (2007) Facebook for the first time allowed its users' profiles—or an abbreviated version of them—to appear in search results. But they weren't appearing high in rankings, in part because they weren't easily found within Facebook, and Google's web crawlers would have to burrow deep into Facebook to find them. Schultz and Gleit put together a directory for Facebook that interlinked people's profiles in a manner that was catnip for Google. It resulted in the profiles being ranked higher, and when people came across them, they could ask to friend those people right there in the Google search engine. It got Facebook some new users.

"It wasn't like doubling [new users] but was it five percent, maybe ten," says Schultz. "But the thing is, each of those things you try, even if it gets one percent, and you get one percent and one percent and one percent [from other tricks], it really adds up. Doing a lot better in terms of growing as a company. And so we were willing to fight for each of those individual wins."

"We were very open about trying new things," says Palihapitiya.

But the masterpiece of Growth—its *Mona Lisa*, its "Like a Rolling Stone," its *Godfather* 1 and 2—is a feature that became almost as much a part of News Feed as weddings, vacations, and political outrage. It's called People You May Know, referred to internally by the acronym PYMK. Officially launched in August 2008, People You May Know is a feature that identifies personally selected prospects for one's friend list. PYMK proved to be one of Growth Circle's most effective tools, and also one of its most controversial ones, a symbol of how the dark art of growth hacking can lead to unexpected consequences.

It wasn't a Facebook invention—LinkedIn, a growth-crazy company in its own right, did it first. (Reid Hoffman would later put a ribbon on the growth-at-all-costs phenomenon and dub it "blitz-scaling.") But Facebook took the idea of presenting current members to new and current users to dizzying heights.

On its face PYMK seems innocuous enough: a carousel of profile pictures on Facebook presumably connected to you, but somehow not your Facebook friends. Its impetus was to address an imperative that the Growth team's researchers had unearthed: a new Facebook user is likely to abandon the service if he or she doesn't connect with seven new friends—fast. For someone without a core list of friends, Facebook is like playing soccer solo.

Facebook did have some tricks to use for newcomers who had yet to connect with friends. "At one point we created these fake stories called fluff stories," says Palihapitiya. "We were like, *Why don't we create stories about people's birthdays, or suck in an interesting article or whatever?* It's not like [friends] generated a story I'm seeing, it's system-generated. The point is, it's fluff." The idea was to stave off entropy until people made enough friends to have real stories.

But that's no substitute for actually having Facebook friends. The Growth team's data scientists did a study that emphasized how critical it was for Facebook to find friends for newbies—especially active ones. As the study called "Feed Me: Motivating Contribution in Social Network Sites" stated,

> *It is vital for developers of social networking sites to encourage users to contribute content, as each individual's experience is dependent on the contributions of that person's particular set of connections. It is particularly important, if rather difficult, to encourage continuing contributions from newcomers.*

Thus PYMK was essential for Facebook. Exposing potential friends is a way to improve a member's experience; it increases the chances they will share more, and, most of all, it makes people less likely to bail on Facebook.

For many people, PYMK is a welcome feature: a helpful prompt to get in touch with connections who would help them get value from their Facebook experience. But sometimes PYMK can be unsettling, raising questions of what caused those cameo appearances on your News Feed by people whose connection to you was obscure, and sometimes downright unwelcome. A sex worker found Facebook recommending her clients, who did not know her true identity. A sperm donor got a suggestion for the biological child he never met. A psychiatrist learned that Facebook was recommending that some of her patients friend each other on the service. And millions of people went *Ew!* as Facebook suggested they develop relationships with friends of their children, spouses of their casual acquaintances, or disastrous blind dates of a decade ago.

Journalists who studied the feature—notably *Gizmodo*'s Kashmir Hill, who spent part of a year trying to get to the bottom of the mystery—were never able to get Facebook to divulge exactly how the product works. Hill was the one who unearthed the story of the woman who got a Facebook suggestion that she friend her long-absent father's mistress. And Hill herself was stunned to find that someone on her own PYMK suggestions turned out to be a great-aunt she'd never met. Facebook did not provide her the information she requested on how it made this connection.

Later, Hill would also write about the psychiatrist who discovered that PYMK was suggesting that her patients make friend connections with each other—even though the psychiatrist did not friend her patients on Facebook. Hill conjectured that it might have something to do with the fact that the psychiatrist at one point shared her phone number with Facebook, and the company scraped her contacts, as well as the contacts of her patients. Once again, Facebook would not provide an explanation.

Neither would Facebook respond to Hill's queries about whether PYMK's instant suggestions for new users means that it was storing data on people not signed up on Facebook, and making use of "shadow profiles" when someone joins. Years later, Mark Zuckerberg would testify in Congress that the company does not engage in that practice. It does keep some information on nonusers, he said, but only for security purposes, to fight fake accounts. (Zuckerberg did not mention his early cogitations in the Book of Change about dark profiles.) In a more elaborate explanation provided later, Facebook said, "We do not create profiles for non-Facebook users," though it also says it keeps certain data, like what device and operating system version a nonuser has, for things like "optimizing registration flow for the specific device" should someone decide to join.

But Palihapitiya now indicates that dark profiles did exist, and the Growth team took advantage of them. He says that Facebook would take out search ads on Google using the names of Facebook holdouts as keywords. The ads would link, he says, to those dark profiles of nonusers that supposedly do not exist. "You would search for your own name on the Internet and you'd land on a dark profile on Facebook," he says. "And then you'd be like well, fuck it, you'd fill it in and then PYMK would kick in and we would show you a bunch of your friends."

Some of the mysteries of PYMK were addressed in a 2010 talk by Facebook data scientist and engineer Lars Backstrom. Reporting that the feature "accounts for a significant chunk of all friending on Facebook," Backstrom went through the technical process of how Facebook chooses its suggestions. The most important hunting ground is the "friends of friends" region, according to the presentation. But that is a very large set.

Users have an average of 130 friends, he said, each of whom has their own 130 friends. (This is close to the so-called Dunbar number, named after the sociologist Robin Dunbar, who discovered that most people can reasonably maintain relationships with no more than 150 people.) So the typical user has 40,000 friends of friends (FoFs), and a power user with thousands of friends might have 800,000 FoFs. That's where the other data comes in—to find signals like the number or closeness of mutual friends and mutual interests, along with "cheaply available data" to identify which ones are likely to cause someone to click when spotted in a PYMK list. As the data gets refined, Facebook uses machine learning to make the final suggestions.

Backstrom also revealed that one's behavior on PYMK helped determine which suggestions Facebook would offer—and how often it would show you the list. Once Facebook determined you fell for the feature, it would keep coming back, stuffing your friend list with weak ties.

The Backstrom presentation omits any specific information about what data sources besides FoF analysis Facebook uses in the feature. To be sure, those sources have evolved steadily since Facebook introduced PYMK in 2008. It's almost certain that Facebook watches your email and sees whom you are contacting. Probably your calendar as well, to see whom you're meeting with. Other sources have indicated that if someone glanced at your profile, that act might increase the odds that the person might appear on your PYMK list. It's doubtful that simply *thinking* of someone is enough to put that person on your PYMK lineup. It just seems that way.

Alex Schultz says that a lot of guesses about what data goes into PYMK are conspiracy theories. He says that people often misremember that they permitted Facebook to use their contact lists or email (maybe they got insufficient notice they had given permission?). Anyway, he says, the biggest reason for someone showing up is that they are friends of friends deemed likely to know you.

Or, as Cameron Marlow, who once led Facebook's Data Science team, puts it, "The goal is to try and find relationships that you have that already exist on Facebook, but that you were not aware of."

As troubling as PYMK is, the scary thing is that it could have been worse. Facebook's chief of privacy, Chris Kelly, says that he blocked the use of some questionable techniques that the Growth team had suggested. "There had to be some rules," he says, declining to share the ideas he snuffed.

Other problems with PYMK are subtle but no less troubling. (Warning: we're going down a rabbit hole here.) The early Facebook executive Dave Morin came to view PYMK as an insidious means of boosting retention numbers at the expense of a good user experience. Since a key goal of PYMK was to boost the value of Facebook for new users—making sure that they had enough friends to fill up their News Feed—the suggestions were tilted to help those newbies more than the people they friended. Particularly valuable to Facebook would be suggestions of users who posted promiscuously, because (as the "Feed Me" study proved) early exposure to super-active users will influence newcomers to share more throughout their Facebook life.

As Morin puts it, "When Facebook shows you people you should connect to, it can make a choice as to how that algorithm works. It can either show you people you'll become closer to and who will make you happier if you add them to your world. Or it can show you people that are advantageous for Facebook, the system, to show you, because it increases Facebook's value and wealth and it makes my system better." He says that Facebook takes the latter course, benefiting itself at the expense of its users.

This might give the experienced user a *worse* experience. The News Feed is zero-sum—people view only a limited number of stories. Facebook would prioritize stories from your newer, weaker ties that it wanted to keep on the service. And you would see fewer things from people you did care about. "The system knew that if I said yes to you, you would become more engaged," says Morin. "You'd be effectively stalking me because I'm like a person distant in your social graph who you want to know. It's almost like watching a tabloid." Morin says this semi-stalking factor "became the primary variable in PYMK."

Some people pushed back on Palihapitiya on this issue, arguing that such behavior was not Facebook-ish. Palihapitiya said since the ultimate goal is getting everyone on Facebook, it doesn't matter in the long run. Though he said it more colorfully than that. "He was basically like, *Go fuck yourself*, and he'd walk out of the meeting." says Morin.

Eventually Morin would quit Facebook and start his own social network, called Path. The idea behind it was that it would limit one's social network to only meaningful connections; following the dictates of Dunbar, one could have 150 friends and no more. (He later increased the number.) Despite winning plaudits from critics, Path ultimately failed, unable to compete with Facebook.

Zuckerberg defends PYMK, and the way he does it illuminates his thought process and product acumen. When I bring up the above conundrum to him, he gets very serious. "This gets to a really deep philosophical thing about how we run the product," he says. He concedes that if users take the hint from PYMK and friend their weak ties, their experience might be somewhat degraded. But there is a more important issue at stake, he argues—the health of the network in general. "We don't view your experience with the product as a single-player game," he says. Yes, in the short run, some users might benefit more than others from PYMK friending. But, he contends, all users will benefit if everyone they know winds up on Facebook. We should think of PYMK as kind of a "community tax policy," he says. Or a redistribution of wealth. "If you're ramped up and having a good life, then you're going to pay a little bit more in order to make sure that everyone else in the community can get ramped up. I actually think that that approach to building a community is part of why [we have] succeeded and is modeled in a lot of aspects of our society."

Furthermore, Zuckerberg believes that by friending your weak ties—which includes people you hardly know—you become closer to them. Facebook might even violate the physics of social interaction by stretching the number of meaningful contacts that people can handle. "There's this famous Dunbar's number—humans have the capacity to maintain empathetic relationships with about 150 people," he says. "I think Facebook extends that."

In a social-science sense, that would be like surpassing the speed of light. But if anyone could do it, it would be Facebook's Growth team.

SOON INTO THE Growth effort, Palihapitiya met with Sandberg and told her he was going to focus on the most fertile territory for growth: the international market. While Facebook had captured a significant portion of the North American market, it had yet to make a huge impact on the rest of the world. He wanted to expand the team to aid the push to other countries. She suggested he look at what other companies, like Yahoo! and eBay, did when they drew users internationally. Palihapitiya came back and told her what he wouldn't do—he wasn't going to hire white Ivy League grads. He wanted street fighters who knew the local language.

He tapped Javier Olivan to lead the effort. Olivan was a Spanish engineer who'd arrived in the United States in 2005 for a Stanford MBA. His dream was to start his own company. He was struck immediately with the way that Facebook had captivated campus life, and decided, basically, to clone the idea. Working with friends, he started a college social network in Spain while he was still at Stanford. One day Zuckerberg himself spoke to one of his classes, and Olivan chatted with him after the session about expanding to other countries. Soon after, he was at Facebook, on Palihapitiya's team. "The assignment I had was international growth," he said. "The most obvious thing to be done was to get the site in as many languages as quick as possible."

Facebook had already embarked on the process some months earlier. A small team working on internationalization had created an app called Translate Facebook. It enabled people in various countries to translate words from the original English into their home language. Of course, Facebook wasn't waiting around for people to volunteer. It checked its logs to identify people who were using the English version of Facebook in some foreign country. Those chosen—by algorithm of course—would see a message on top of their News Feed asking if they'd like to help to translate Facebook. These unpaid helpers (another benefit of crowdsourcing) would do a first draft, setting up a scaffolding that identified the pitfalls of moving Facebook to the individual language.

After that first step, Facebook might open up the process of translating the terminology to everyone. Sometimes it would prompt people for hints: a native speaker might get a pop-up saying, "Hey do you speak this language, can you help us with these ads?" Sometimes the crowdsourcing would be used to refine a machine translation.

The hurdle was getting good translations. In order to be put into production, a translated version of Facebook needed verification that the translation was accurate and nuanced, not to mention avoiding any ugly American faux pas. And often even translating simple words could be difficult in certain languages. Some languages couldn't easily make the metaphorical transition of a "wall" from a structural element in a building to a virtual bulletin board on a profile page. And forget about getting an easy translation to "pokes," which even English speakers had difficulty grasping. One engineer on the team who had been a huge fan of the Reddit website suggested that Facebook adopt the Up and Down buttons that Redditors used to sort content.

The crowdsourcing approach ran counter to accepted wisdom. The tried-and-true method was known as the 80/20 rule, where resources were concentrated, and professional help sought, on the key languages known as FIGSCJK—French, Italian, German, Spanish, Chinese, Japanese, and Korean. That would get you the vast bulk of the online audience. "As a company we've always had the mission of connecting the world," says Olivan. "And the 80/20 rule is not going to cut it. We really wanted to make sure it was available for everyone." To do that Facebook needed to improve its crowdsourcing tools so those professionals weren't needed. The idea was to have Facebook everywhere. Even in many languages that no one at Facebook was familiar with.

Crowdsourcing would only go so far, though, and Facebook had to accept that it needed professional translators for what were deemed the most important languages, which they calculated by a country's gross domestic product. Someone from the Internationalization team would set up a meeting with native speakers who taught in English in a given country, and then spend days verifying the translations of certain words used on the site for a given language. Still, the priority was shipping rather than perfection. "We had lists of the words that were used most often on the site and we would be like, *Well, let's just only ask them about those*," says Kate Losse, an early employee on the team. "And then there's hundreds of words that only appear once and we don't care if that one's wrong, because you just have to triage. Essentially it was proofing that the translations that were coming into the app were going to be good enough to launch."

There were limits to this low-cost, high-velocity approach. When Facebook officially announced its entry into Japan, for instance, it hit a wall of criticism that seemed to stun the high-level executives who traveled to Tokyo for the event. Zuckerberg himself was there, on the way back home from his 2008 personal journey to India.

Facebook Japan's announcement uncovered a culture clash that would play out in many countries. On one hand, Facebook was proud of its process. Over three weeks, 1,500 "amateur and professional translators" took a stab at translating the English text on Facebook—pages, dialogue boxes, and messages—into Japanese. Then they debated and revised. But local sentiment was mixed. Some Japanese users found it offensive that Facebook thought a superficial translation, and not a thoughtful product redesign, could make its American product meaningful to the unique Japanese audience. "It did not set up a Japan office or select a local representative to lead the effort," complained the *Japan Times*. "In fact there isn't a single unique feature about Facebook Japan."

When the *Japan Times* asked Javier Olivan, who attended the launch, about this, he confirmed their fears that the company was exporting American values along with their product. "Facebook is exactly the same all over the world," he said. (Facebook Japan would eventually vanquish its competitors, reaching three-quarters of the social-media market in Japan by 2016. By mid-2019, however, that share would plummet by two-thirds.)

Under Olivan's direction, the process worked astonishingly well. When it deemed the translation ready, the Internationalization team would push it in a given country and watch the usership explode. Then they would plant another flag in a corner of their office where they had a small cluster that marked the national symbols of the countries where Facebook spoke their language. In a few months Facebook went from no languages to hundreds of languages.

In 2012, Facebook hired a woman named Iris Orriss, who helped further refine the process. She had begun her career in software testing, then moved to engineering teams, and realized that her most satisfying tasks were working to extend technology to other countries. She was passionate about language and cultures and globalization. It was almost a calling for her. When Facebook recruited her from a job at Microsoft, scaling speech software to other languages, she researched the company. She concluded that something about internationalization at Facebook was different from the previous places she worked. There was a mission driving the entire enterprise.

And the key to it was growth. Other companies view international expansion as an operational exercise. "If it's operations, it's more a cost-center mentality: 'Get me as many translations as cheaply as you can,'" she says. "If it's in growth, it's all about the opportunity: 'How can we actually integrate this and make connections to everybody and open up the world and make it smaller at the same time?'"

Orriss understood how crowdsourcing would help Facebook pursue that mission, particularly in tongues spoken by small populations. Take the African language of Fula. It has two flavors, one in northwestern sub-Saharan Africa and another spoken around Nigeria. They are sufficiently distinctive that it's almost two different sets of languages. There is no way Facebook could spend resources finding linguists who knew those dialects and understood technology. So Facebook unleashed its translation tools.

It was all worth it. Even in countries where many languages were spoken but official business was conducted in one language or dialect, the local vernacular made for better, and more profitable, Facebook users. People engaged a lot more when the language was theirs. "Local is what's relevant to them," says Orriss. "Who wants to see something translated of what's going on in the other half of the world? They want to know what's going on right here."

But one aspect of this expansion presented a problem that Facebook largely ignored: by crowdsourcing the process of internationalization Facebook started operating in regions where none of its employees, or almost none, actually spoke the language. This meant that Facebook could not provide suitable customer support, or oversight on what people posted. Posts that violated Facebook's standards could be dangerous or lethal. Even when local users reported violations, Facebook was hamstrung in responding, because its people didn't speak the language.

IN 2013, THOUGH, it wasn't those consequences that worried Facebook about its international growth. It was the concern that even with its best efforts, the company might soon hit the ceiling of its possible member base. Yes, it was amazing that Facebook could enlist a billion or two people into its community. But that wasn't the whole world—just the easy part of it, the ones who had Internet connections and enough money to pay for data. Many among the not-yet-Facebooked billions were not in that cohort. As Facebook saw it, they were either too poor or did not have a way to connect to the Internet. Or both.

The solution? Make a version of Facebook that was cheaper to use. And take the initiative to get them connected, even if it meant creating infrastructure.

Mark Zuckerberg embraced this idea with a passion. Though the impetus came from his Growth team as part of its drive to get everyone in the world connected to Facebook, he claimed to view the venture as sort of a philanthropic effort that would transform the lives of those billions. He would come to call the project Internet.org, invoking the top-level domain suffix usually reserved for nonprofits and foundations.

In 2013, Zuckerberg would write a ten-page white paper laying out the vision, entitled, "Is Connectivity a Human Right?" The answer was a resounding yes. "Everyone will benefit from the increased knowledge, experience, and progress we make from having everyone connected to the Internet," he wrote. "Connecting the world will be one of the most important things we all do in our lifetimes."

When he briefed me about it that August, I asked why we shouldn't view the initiative as just a way for Facebook to get more customers.

"We do theoretically in some way benefit from this," said Zuckerberg. But that wasn't why he was doing it, he insisted. "I think the thing that's kind of crazy [about people thinking Facebook is doing this for profit] is that the billion people who are already on Facebook have way, way more money than the next 6 billion people combined. So if we really wanted to focus on just making money, the right strategy for us would be to focus solely on the developed countries and the people already on Facebook and increasing their engagement rather than having these folks join." Facebook, he said, may never turn a profit from its efforts. "But I'm willing to make that investment because I think it's really good for the world."

Facebook devised a number of initiatives to spread the Internet—and itself—to what was called "the next few billion." First, it addressed the problem of many underserved regions: people could not afford to pay the data charges that came from using the mobile Internet. Facebook had two ways to deal with that. The first was to create versions of Facebook that consumed less data. Second, it would partner with telecom carriers to deliver a selected slice of the Internet with *no* data charges. Facebook, of course, would be included in that slice.

The most ambitious part of its plans, though, was to actually create the means of delivering Internet to the next few billion users. For several years at his F8 keynotes he would sketch out his plans with pride.

One scheme was the satellite that would beam connectivity to sub-Saharan Africa. That was the one that blew up on Elon Musk's launchpad when we were in Nigeria.

Zuckerberg had another means of broadband access in mind, though, an even sexier one: beaming down the Internet from solar-powered, super-lightweight drones. By mid-decade it had become almost a fetish for him. The Facebook air force! The drones would be capable of circling at high altitudes while directing broadband signals Earthward. (Google had its own high-altitude Internet-delivery scheme involving modified weather balloons.)

This was a geeky dream that had gotten some serious attention: a company called Ascenta was actually building such an aircraft. Its CEO had formerly created a thrill ride for the Jurassic World theme park. Facebook bought the company for a reported $20 million and began building a protoype drone, dubbed Aquila. Its wingspan, covered with solar paneling, was the same as an Airbus A320 weighing nearly 100,000 pounds, yet the exotic materials of its frame kept the weight down to under a thousand pounds, less than a standard sedan. Aquila became Facebook's unofficial mascot. There was a period when Zuckerberg would lead visitors to a piece of an Aquila wing he just so happened to have hanging around, standing taller than he was, and he would lift it like a kite.

After a few years of relentless fanfare, Aquila was finally ready for a private test flight in 2016. Zuckerberg flew to the Yuma, Arizona, test site and watched as it took to the air. Upon his return, Facebook summoned a reporter from *The Verge* to write about how brilliantly the plane performed. Later, it emerged that the plane had a "structural failure" that led to damage on the landing and triggered a National Transportation Safety Board investigation. Zuckerberg had failed to mention it in his narrative. A second test flight reportedly went better but Facebook gave up on the idea in 2018.

The immolation of the satellite and the crash of Aquila were physical representations of wreckage in Internet.org's policy initiatives, particularly in the company's biggest target, India. This country, with a population of more than a billion, was a shortcut to the Growth team's goal of worldwide ubiquity. Zuckerberg made it a personal crusade. He visited India in 2014, schmoozing with the prime minister and visiting a rural classroom. *Time* magazine's Lev Grossman accompanied Zuckerberg, and while acknowledging the value of connecting the underserved, he noted that there was another way of looking at the program. "However much the company spins it as altruistic, this campaign is really an act of self-serving techno-colonialism. . . . Facebook, like Soylent Green, is made of people, and it always needs more of them."

While the drones and satellites were meant to provide access to the 15 percent of the world without a signal, the real foundation of the initiative was the Internet.org program that partnered with local telecoms to deliver services with no data charges. Facebook launched it in India in 2014, as well as in other countries. The program drew criticism because its inclusion of a group of selected apps—including Facebook—put competitors at a disadvantage. This seemed to violate the principle of net neutrality, which dictated that all developers should have equal access to the Internet. In April 2015, some publishers abandoned the program.

Facebook announced that it would open Internet.org to all developers, but the criticism persisted. Zuckerberg recorded a late-night video from Facebook headquarters and posted a plea that begged critics to back off. Universal connectivity can coexist with net neutrality, he insisted—no one was blocking any sites. Of course, the critics weren't complaining about that but about Facebook giving itself an advantage by including itself in the "free" tier. Zuckerberg's answer for that? "If someone can't afford to pay for connectivity, it is always better to have some access than none at all."

The critics were not quelled. Zuckerberg renamed the program Free Basics to eliminate the complaint that Internet.org falsely promised the entire Internet. But in February 2016, India banned the service. It ran into trouble in other countries as well.

The Internet.org debacle had elements that would later reappear in the Facebook narrative: officials and the general public rejecting a seemingly benevolent initiative with collateral benefits to Facebook. Before challenging Facebook was cool, they did not trust that the company's motivations were pure.

Facebook now claims that the program is successful, with around a hundred million people using it. But even the successes would turn out to be woes, as regions unprepared for sudden access to a huge unmoderated speech platform were exposed to those who used Facebook to manipulate, spread misinformation, and promote violence.

By then Mark Zuckerberg would be keynoting much less on Internet .org. He would have other products to tout—and new mistakes to defend.

PALIHAPITIYA LEFT THE Growth team in 2011, and Javi Olivan took his place. As Chamath's protégé, he was just as single-mindedly obsessed with boosting Facebook's numbers. But he was also an engineer, and under his leadership, he expanded Growth to include coders outside the circle who previously implemented initiatives while reporting to other organizations.

His personality was different. When asked what the differences were, Alex Schultz replies, "Javi is very kind, very considerate, just . . . ethical and a great guy."

The other big change was something more profound. To date, Growth had focused on adding users and engagement for the Blue app, the core Facebook product. With Zuckerberg's blessing, Olivan expanded that domain. Critical portions of the company came under the Growth umbrella.

Growth became the lens through which Facebook viewed almost everything it did. "The way that the Growth team thinks about the product is unlike the way any entrepreneur or any businessperson in the world ever thinks about a product," says Rob Goldman, who joined Facebook in 2012. "I've never seen it before, which was a deep bedrock assumption that everyone in the world should probably be using the product every day." While Zuckerberg extolled the mission of connecting everyone in the world, the actual operations of doing so were owned by the Growth organization—so one could make a credible argument that the company's real mission wasn't connectivity but growth. Facebook's PR person, Brandee Barker, recalls a meeting she had with Zuckerberg around 2009 where he told her that Facebook needed more publicity to increase growth and engagement. "That was the only reason we were supposed to be doing it!" she says.

"In the beginning, things were a lot smaller," says Naomi Gleit of the Growth team. "We were only focused really on growth—we owned the registration experience, we owned the invitation experience, we owned that new-user experience. And then, over time, our scope grew. Mark was like, *Okay, can you take on this? Can you take on growth and Messenger? And then can you take on . . .*" She laughs, knowing there are a number of things she could list to fill in that blank.

Growth was in charge of internationalization. Growth would own mobile. And in recent years, when it became clear that Facebook was going to have to spend vast resources to win back trust—trust that in part the practices of the Growth team had eroded—Growth was put in charge of that too. Some people found it ironic that the Growth organization, notorious inside Facebook for flirting with the algorithmic dark side, was now charged with integrity. Even the area of what Facebook calls "social good" was turned over to what was Facebook's increasingly powerful Growth organization.

Gleit explains the reason is that her team's characteristic approach to tackling problems scales well to other initiatives. "What do these things really all have in common? It's about taking this data-driven, product-driven approach to problems. It really is all just taking the same approach of understand, identify, execute."

The data-driven, take-no-prisoners DNA of Growth, the Palihapitiya way, is baked into all those initiatives. Palihapitiya left the company in 2011 to start a venture capital fund. (Investors included Facebook, and several of its current and former employees, though not Zuckerberg.) In his farewell memo, he said that the journey was all about winning—everything else comes in second, he said—and warned Facebook folk to be alert to spot "the company you don't know," whose big ideas might displace you.

All good, but the memo was remembered at Facebook for one final Chamathism as he slipped out the door.

Don't be a douchebag.

11

Move Fast and Break Things

IN 2008, AN Austin-based poster maker named Ben Barry saw an ad on Facebook. It was well targeted. Facebook, the ad said, was looking for designers. People like him.

Barry hadn't been on Facebook much since leaving college a few years earlier, but he decided to apply anyway, and after multiple interviews, he got the job. He moved to Palo Alto in September 2008. Facebook paired him with another new design hire, Everett Katigbak.

One of the first things Barry did was ask if Facebook was doing anything with the upcoming presidential election. His bosses suggested he get in touch with the only Facebooker working in DC at the time, who was working out of his home. They decided to create a virtual button that said "I Voted." Users would click on it after they cast their ballots. On Election Day, Barry was astonished to check the dashboard and find that each time he looked, the number of people clicking jumped by thousands. Eventually 6 million or so people used the feature.

Hmm. *Facebook could influence voters?*

Barry's biggest contribution to Facebook was something else. During his interview, the idea came up that Barry might create some posters for its offices. Facebook had wisely decided not to continue along the lines of the misogynistic graffiti that David Choe had so lucratively produced. In the age of Sheryl Sandberg, the company was looking to adorn its walls with something more sophisticated but still reflective of its culture—maybe even defining it.

In Austin, Barry had done a lot of silk-screening for concert posters and art prints. He felt that even at a place where design was digital and the canvas was a creation of coding—*especially* at a place like that—it was essential to produce physical art. He envisioned enlisting the workforce to help produce a profusion of posters on the walls, providing a font of persistent inspiration. But there was no space for an art room in any of the cramped offices that Facebook rented in Palo Alto.

In early 2009, Facebook moved to its new 150,000-square-foot headquarters at 1601 California Avenue, in a Palo Alto neighborhood south of El Camino Road, known as College Terrace. When Barry heard Facebook was also taking other space nearby, including a building with warehouse space that wasn't needed, his team claimed it for a poster-production facility. He and Katigbak went to Home Depot and got wood for tables, a small letterpress machine, and other paraphernalia for silk-screening. There was a janitor's room with running water, and Barry set up the tables near it. Since there was no drain, the runoff water was to be collected in buckets. Then Barry saw that someone in San Francisco was selling a guillotine paper cutter, a crucial component for churning out posters en masse. It was an unwieldy 700-pound device that looked like it was still rooted in the Gutenberg era. Moving it required a forklift. He bought it himself and got permission to move it into the warehouse.

Barry called the room the Analog Research Lab. He modeled his vision on early twentieth-century propaganda posters, specifically ones done for a 1920s antiwar group, the National Council for Reduction of Armaments. He had seen them on the Library of Congress website. The posters featured bold uppercase letters, printed in Speedball Fire Red. It was a conscious rejection of Facebook's tidy online look and feel. "I wanted it to be old and I wanted it to be not blue," says Barry. So while the website was modern and anodyne, Barry designed his posters to evoke the trumpet calls in old movies when a spinning newspaper stops to reveal a shocking headline in 36-point type.

The silk-screens inevitably evoked the Big Brother poster from the original movie of George Orwell's *Nineteen Eighty-Four.* The texts would be taken from things Barry heard at Facebook, some consisting of phrases that embodied the company culture as he saw it, and others arising from the company's attempts at self-definition.

These efforts had been under way for about two years, while Zuckerberg was still smarting from his failure to convey the importance of his mission during the Yahoo! crisis. In 2007, on a walk with Microsoft CEO Steve Ballmer, Zuckerberg had asked how his company communicated the qualities its employees should exemplify. Ballmer told him that Microsoft had lists of self-defining qualities. Zuckerberg had gone home and written a bunch of those out and pinned it to the office refrigerator. The list wasn't all that popular—someone took offense to the specification of "high IQ" and scratched it off the list.

In 2009, Zuckerberg felt that a serious definition of company values was in order. A new head of recruiting, Lori Goler, had asked Zuckerberg what recruiters should tell people about what it's like to work at Facebook, and he thought the question was worth exploring with a larger-scale search for broader self-definition. "What do we want to be when we grow up?" Zuckerberg said at one meeting. (Goler was a Sheryl hire who formerly headed eBay's marketing; she would soon replace Cox as head of HR.) Under Cox's supervision, she began to answer that question.

Goler began working with a relatively new hire named Molly Graham, a former Googler who joined in part because of her close relationship with Sandberg and Elliot Schrage (she'd been Schrage's chief of staff). She was the daughter of Don Graham.

Also on the project was Chamath Palihapitiya, Facebook's truth-teller. "Mark wanted him there because he would be so opinionated about it, and not try to make everybody feel good," says Graham. At around that time Netflix had created a well-circulated slide deck that inculcated its values ("*real* values," it specified, not those vague and phony ones that other companies boast about), and everyone in the Valley was envious. Palihapitiya in particular felt that Facebook shouldn't be out-valued.

Graham quickly realized that the answers lay with Zuckerberg himself. *He* was Facebook culture. "Companies are built in the image of their founders," she says. "For a while Facebook felt like a nineteen-year-old's dorm room, but ultimately it's a try-it-and-iterate place, which is how Mark exists as a human. He is a learn-by-doing person, and that is the DNA of the company. Facebook doesn't believe in perfection."

In one sense, Facebook's self-definition could be summed up in six letters. "We gathered people in a room, in small groups, and said, *Tell us how you describe [Facebook] to candidates or to friends, to your mom, to your brother? What did you tell the last three candidates? What language did you use?*" says Graham. One word kept coming up again and again: "hacker."

To the public, the term evoked destructive, high-tech nihilists or crooks vandalizing remote systems or stealing credit-card information. In start-up culture, though, the term was used in its original sense: highly skilled and totally righteous coders who believed that their efforts made a broken world work better. "This word is not associated with goodness in the broader world," says Graham. "But we meant it in a very positive way."

Understanding how Facebook was a hacker company helped shape the four values that they ultimately presented to Zuckerberg. It was like showing him a mirror.

Focus on Impact.

Be Bold.

Move Fast and Break Things.

Be Open.

Zuckerberg liked those but insisted on a fifth: Build Social Value. While the first four were internal guidelines, this fifth value emphasized Facebook's impact on the outside world, which Zuckerberg believed was overwhelmingly positive. (He still does.)

Of those values, one stood out as uniquely Facebook, uniquely Zuckerberg. "Move Fast and Break Things" was, in a sense, already synonymous with the company. No one is sure where those exact words first appeared, but it may well have been in an all-hands meeting at the Hamilton Street office in Palo Alto, at a time when Facebook had hired its first wave of managers, after becoming too big for all the individual contributors to report to D'Angelo or another executive. The idea of having people farther down the food chain with the ability to say no was a concern of Zuckerberg's. So he told everyone that Facebook could not afford to be afraid of moving fast and breaking things.

BEN BARRY WAS able to draw from this well of values, and also to extract other slogans from things he heard. In the Facebook spirit, he did not clear any of this—the style, the language, or even the plan itself—with his superiors. One day his posters just appeared, as if a mad hacker propagandist escaped from a rogue AI lab had been set loose in Facebook HQ.

DONE IS BETTER THAN PERFECT

IS THIS A TECHNOLOGY COMPANY?

PROCEED AND BE BOLD

EVERY DAY FEELS LIKE A WEEK

And finally, the one that would become Facebook's unofficial motto:

MOVE FAST AND BREAK THINGS

Since this run of posters were simply text, they did not visually depict Facebook's supreme leader. But whether or not those words issued from his lips, they were understood to be a means of channeling Zuckerberg's innermost thoughts.

At first some of the Facebook employees objected to what seemed like scary edicts. But when they learned that these slogans came from their beloved leader and other makers of their culture, the objections melted away.

Soon thereafter came a poster with another slogan:

WHAT WOULD YOU DO IF YOU WEREN'T AFRAID?

While "Move Fast and Break Things" seemed to emerge from Facebook's cortex—speed was the tactical advantage that set the company apart—this challenging query mined its heart. By adopting it as one of its unofficial mottos, Facebook was expressing not just an approach to business, but a path to self-realization. In work and in life, fear was the enemy. *Do it*, the slogan urged. *What's the worst that could happen?* Later, bold-colored posters from the lab would feature portraits of fearless people, underdogs who risked it all for selfless causes: Dolores Huerta, Shirley Chisholm, Cesar Chavez. The multimillionaires who got early stock options and the engineers making six-figure salaries straight out of college somehow identified with these heroes of the downtrodden.

Some saw a darker meaning behind the posters. "All of these things had a common theme, which is basically we only care about growth," says Sandy Parakilas, a former employee who spoke to me while at the Center for Humane Technology. "*That* is the singular thing that we are focused on: we don't really measure or focus on any of the other problems that we might be creating, and we will do everything we can to drive growth. And, frankly, we also don't care that much about polish or perfection of product. We want to move fast, ship the thing that will get the impact that we want, and then keep going."

Indeed, the slogans, particularly "Move Fast and Break Things," were prone to misinterpretation. "It meant to iterate and try things and not be afraid of failure—but not to be sloppy," says Graham. "It didn't mean to duct tape the server and run away." But just as Google's motto—"Don't Be Evil"—would be used against it, the "break things" part of Facebook's motto would be used by critics as a cudgel, when people were accusing Facebook of actually breaking things—breaking social order, breaking democracy, breaking civilization itself—like some digital version of a bull in a Restoration Hardware store.

A few years later, Zuckerberg would amend the slogan at the 2014 F8, to "Move Fast with Stable Infrastructure." It didn't have the same crackle. But the spirit of "Move Fast and Break Things" remained pervasive at Facebook. From Zuckerberg on down, everyone believed that Facebook's edge came from its speed and risk-taking. Going slow would mean death.

THE POSTERS WERE only the most noticeable aspect of Facebook's evolving culture, a morphogenesis from its college-dropout DNA. Everyone who describes Facebook's character in its early days—and often up to the present—winds up using "dorm room" as an adjective. Its rival Google, in contrast, sported the traits of a graduate student. Google's elders were professors who wrote the textbooks that its leaders learned from; Facebook hired Mark Zuckerberg's Harvard TA. True, even in 2005 there was a smattering of thirtysomethings on staff—a few of them married, with kids. But while Zuckerberg understood the value of veterans like Jeff Rothschild, at his core he believed that younger people were . . . smarter. He said exactly that in a Y Combinator start-up school in 2007, telling 650 would-be founders to hire people who were young and technical. "Why are most chess masters under thirty?" he asked.

His later apology for that remark (which, if it truly reflected Facebook's hiring policy, would put the company in violation of federal labor laws) didn't cover up the fact that his original statement seemed totally in sync with his worldview.

Of course, Facebook culture was more complicated than that, and as the company grew it became more professional, hiring people with a more cautious outlook. Sandberg had a role in that, but mostly it was a consequence of getting bigger. Those mature newcomers would struggle with Facebook's reckless velocity and wonder how much of their job was to succumb to it and how much to temper it. Since the message from

Zuckerberg, at least for the first few years, was to keep moving fast, the leaders learned to embrace it, though undoubtedly the sheer weight of a larger company acted somewhat as a brake.

This was, to Zuckerberg, not necessarily desirable. The constant theme of his talks at all-hands meetings was that the company was on a grand mission. In the spring of 2009, he gathered the company at the Palo Alto Sheraton; it was just before Facebook moved to California Avenue. As later recounted by *The Wall Street Journal*, Zuckerberg, inspiring recruiters in the ongoing talent war with Google, reverted to his favorite trope, invoking the ancients. This time he didn't cite his hero, Homer, but a recent movie, *Troy*, where a messenger confessed to Achilles that he feared taking on the Thessalonians. "That's why no one will remember your name!" said Achilles. Likewise, he said, recruiters should use this compelling comeback when potential hires asked why they should take a job with Facebook: "Tell them: because people will remember your name!"

By then, Facebook itself had lost some of the most important names in its own history. Among the departures were co-founders Chris Hughes and Dustin Moskovitz, and Zuckerberg's compatriot since high school Adam D'Angelo. Hughes left in 2007, to apply Facebook's lessons to the Obama campaign. Cohler departed in 2008, taking a job at Benchmark. Moskovitz started a software company called Asana. D'Angelo moved on in May 2008.

On the surface, these partings were amiable, with the exiles remarking on the cool adventure ahead rather than attributing the departure to something, or someone, they'd tired of. A former top Facebooker explains, "The thing employees at Facebook value most amongst every value in the world is freedom—because they don't get any of it when they work there. They make money and they get none of the freedom. A lot of these people just want freedom." The understood implication was *freedom from Zuckerberg.*

"Working with Mark is very challenging," Hughes told author David

Kirkpatrick after his departure. "It's much better to be friends with Mark than to work with him."

For Zuckerberg, perhaps the most painful departure was Moskovitz, the workhorse instrumental in spreading Thefacebook beyond Harvard and the leader most supportive of Zuckerberg's refusal to sell to Yahoo! In 2007 Moskovitz gave up his role in management to build software tools that he felt Facebook needed as it grew. That went so well that in 2008, he decided to form a company to develop similar tools. His co-founder was Justin Rosenstein, making his departure two years after declaring Facebook as That Company.

Years later, the Moskovitz exit still appears to have been a friendly one. But D'Angelo's move had a sour aftermath. In June 2009, D'Angelo began his own company, Quora, co-founding it with another Facebook former engineer, Charlie Cheever. The Quora idea was that anyone could post questions, which would voluntarily be answered by the people best able to answer them.

Though clearly Quora was not a competitor to Facebook—and from the start it had actually participated in the Facebook platform—Zuckerberg responded to it with hostility. It was unclear whether he wanted to send a message to others at Facebook who might be considering something similar, or legitimately considered Quora something of a future threat. In any case one of Zuckerberg's top engineers, Blake Ross, began to work on a feature called Questions. It did exactly what Quora proposed to do. The engineers at Quora found Ross creating so many accounts that they banned him as a spammer. When Questions launched in July 2010, many thought that Quora was doomed. How could it compete with Facebook's 500 million users?

But Questions did fade away, and eventually Facebook dropped the feature. Ultimately, it could not match the passion of two founders who had built a start-up around the concept.

Some at Facebook believed that Zuckerberg's motivation was to warn current employees to not start their own social products. In any case, the incident did prove that when Zuckerberg sniffed a threat, he would go full throttle to try to snuff it out.

SOMETHING ELSE WAS beginning to happen at Facebook around then: a realization that as its user population grew, there were more problems with the content they were sharing. And that outsiders would hold Facebook responsible for dealing with the problem.

When Zuckerberg launched Thefacebook in 2004, no one thought that whatever was or was not permitted on the site would ever equate to a de facto global ruling on the nature of free speech. Yet from the very beginning there were intimations that things might appear on this new service that required pruning, or banning. At the very least, there needed to be some way that people could report things they saw on Facebook that might be harmful to users.

The job originally fell to people on the customer-support team. Their job was to sit at a desk and respond to emails on all sorts of things. Most of them were straightforward requests like password recovery. But there were also a number of complaints about things that had been posted. Despite Facebook's organic protections—like people using their true names in a bounded community, and an ability to lock down your profile so people outside your network couldn't see it—there were still instances of harassment, offensive speech, and inappropriate photos. These were "nascent bubblings" of a bigger problem, according Kate Losse, one of the first customer-support people.

In the fall of 2005, the support team was growing quickly to handle the increase in complaints as Facebook's population grew, and by the end of 2006, it might have been a third of all employees. (The pay for these people was much lower than the engineers were getting, but for early Facebookers, the options were such that they wound up getting paid more than Major League Baseball players.) That wasn't ideal, but necessary. "There was recognition that when your motto is 'Move Fast and Break Things,' you can't alienate your users too much," says Ezra Callahan.

The Customer Support team leader was a recent Stanford grad named

Paul Janzer. He hadn't envisioned a career in the tech world for himself—his plan had been law school, and in fact he had been accepted by NYU. But at the last minute he decided to defer, and stay in the Bay Area for kind of a gap year. He figured he'd work as a paralegal to pay his bills, but one day, he was using Facebook and saw an ad in the right-hand column soliciting someone to start a customer-support team. It was August 2005. He submitted his résumé, interviewed, and within a week was the first full-time hire working on customer support and content moderation. Previously, the job had been handled by a freelancer in Berkeley, whose backlog of requests and complaints in his inbox was 75,000 and counting. Janzer was twenty-two years old.

His training was a fifteen- or twenty-minute conversation with an engineer who'd been doing some of the work part-time. The engineer told Janzer about the kinds of emails he might see. There were no set rules; the guy had generally free-styled it. He expected Janzer to do the same.

At first that's what Janzer did, too, as well as what his small team did, bolstered by the new recruits he added. But he quickly realized that as Facebook kept growing, this ad hoc approach was less and less effective. There were a lot of gray areas and a lot of nuance. If there was some picture or comment the team weren't certain about, they would ask a nearby worker, usually one of the veterans. A "veteran," in this case meant a person who might have been there for a month or two before you arrived. Or Janzer himself, who still found there were plenty of cases that had no clear resolutions. He found himself asking Facebook's general counsel and chief privacy officer, Chris Kelly, about them more and more.

Informally, the team began to work out some form of online justice. There was the "three strikes and you're out" system, inspired by baseball. There was the "thong rule," which determined that if a thong was visible, it was too risqué. Same with a skimpy bikini. "What little we had of sales and marketing then were the main advocates [of the prudery]," says Ezra Callahan. "Since this was a college site, they were hypersensitive." The pressure got more intense after Sean Parker's cocaine arrest.

Janzer understood that a more methodical approach was necessary.

Even the process by which his team took down content was an awkward kluge. Using the superpowers available to employees via Facebook's internal tools, you first had to log in using the account of the person who flagged the objectionable post, just to see it. If the post failed the probity test, then you'd use the tools to log in via the account of the offender so you could remove it. Thus violating the privacy of *two* people.

Certainly the issue seemed beneath the attention of Mark Zuckerberg, who was in that period doing CEO kinds of things that the low-level employees were not privy to. The support workers were not engineers, and thus a lower life form at Facebook as far as he was concerned. But since Facebook was a small group of young people who wound up socializing almost exclusively with one another, in another sense they were peers.

In September 2005, Losse brought up a situation that had been troubling her—a group that called itself Dead Bodies Against Gay People. "It was this gruesome group that had just dead bodies and all these slurs against gay people," she says.

She felt, with pretty good reason, this was something that had no place on a college site. But why? What was the line? When did expression bleed over into bullying or hate speech? If someone made an explicit threat against a real person, the decision was a no-brainer. Bam, it's out. When the threat was not explicit but a matter of perception, matters were fuzzier. What about when a whole group was possibly objectionable? And what counted as hate? One group was about hating people who wear Crocs. Clearly no one thought that a facetious attack on a fashion choice should be banned. (Plus, Crocs are awful.) Between the poles of death threat and satire, though, was the slipperiest of slopes.

Near where the support team sat at 156 University was a common area where people would eat lunch, or sometimes just sit around and talk or play video games. One afternoon, a bunch of these liberal-arts people handling complaints sat on the couches there and had the kind of dorm-room bull session they'd enjoyed in their college years, which for many of them had ended only a few weeks earlier. Chris Kelly, who was Facebook's general counsel, sat in too.

It wasn't the kind of discussion that they would bring to Zuckerberg or Moskovitz. "Those thorny interpersonal and sociological and speech issues among users were not the focus of the technical side, which in the hierarchy was the important side," says Losse, who always approached sexual politics at Facebook with a gimlet eye, especially after she left as a millionaire and wrote a critical book about it.

But there *was* a form of guidance from Zuckerberg, and that informed the discussion. Zuckerberg would often talk about how he wanted Facebook to be a place where people could exercise the First Amendment, even if it meant offending people. He wanted the site to be a safe place as well, but he saw censoring the self-expression of his users as a last resort, used in extreme cases.

Out of that discussion came what might be considered the first formal content policy, something to identify the kinds of things that even Mark Zuckerberg would agree did not belong on Facebook. They agreed to build up a kind of internal wiki—a crowdsourced document—that eventually evolved into a set of rules. "Our first bias was toward allowing open communications," says Janzer, following Zuckerberg's lead. "But we knew there had to be certain lines. Anything that made people unsafe using Facebook was something we wanted to protect against."

One built-in tool for regulating speech was Facebook's insistence that people represent themselves with their true identities. Early on, the company drew a line against anonymous or false user names. On Facebook, *you* were supposed to be *you*. "We don't have to wait for you to do something bad," says Janzer. If you started by using a fake identity, you were more likely to be a troublemaker anyway.

By the end of 2005, there were between fifteen and twenty people working on the Support team, a fairly high percentage of the company. Newcomers to the team would be given a short Word document listing the taboos. "It was almost more like a common-law record. Like, *Hitler?*

We're against it. Pants, you need to wear them," says Dave Willner, who joined Facebook in 2008 and would replace Janzer several years later.

After 2006, however, the two big changes Facebook had rolled out that year threw the team into turmoil.

First was the News Feed, which shifted the entire focus of Facebook onto a stream of content thrust directly in people's faces. Then, Open Reg threw wide the gates and let everyone join Facebook. In small networks of colleges or high schools, misbehavior was limited by the probability that if you used, say, misogynistic insults, the people in your real-life community would know who you were and shun you. There were consequences. But those organic protections disappeared when people you hardly knew, or didn't even know at all, could in theory put things in your stream. Or when people you did know, and didn't like, now had a way of bothering you. It was flipping a teenage-only early hours club into Studio 54, the kids suddenly joined by denizens of the demimonde. The privacy officer Chris Kelly "sounded the alarm bells" to Zuckerberg and the management team.

Everyone agreed that people should be safe on Facebook, but actually doing something about it was another matter. Kelly was used to this. "It would often take a blowup to actually kind of get the company to refocus," he says.

The blow came quickly. In mid-2007 a group of state attorneys general had become concerned about child predation and pornography on MySpace and won a settlement from that company. Suddenly, the safety of social networks became a burning concern, and the crusading AGs included Andrew Cuomo of New York (later to become governor), Richard Blumenthal of Connecticut (later to become senator), and Roy Cooper of North Carolina (later to become governor). After MySpace, their attention shifted to Facebook.

That July, *The New York Times* reported that "a concerned parent" had created a fake profile of a fifteen-year-old girl, ostensibly to see how dangerous Facebook was. The fictional teenager announced she was *looking for trouble* on her profile. She proclaimed she was seeking "random play" and "whatever I can get." And she promptly signed up for Groups like "Facebook Swingers" and "I'm Curious about Incest." Not surprisingly, this activity led her to inappropriate friends. (Kelly later claimed to me that Facebook traced the fake accounts of the "concerned parent" to a law firm representing NewsCorp, the owner of MySpace.) Still . . . what were those groups even doing on Facebook?

Kelly began talks with Blumenthal and Cooper, trying to convince them that Facebook was handling the situation. "I would always say to regulators, *Look, bad things happen in human society, therefore bad things happen on Facebook*," he says. He hired a couple of consultants to help promote Facebook's case, one a former Indiana AG and another who had recently been on the Federal Trade Commission. They set up a meeting with Cooper and Blumenthal. It didn't go well, as the AGs presented Facebook with volumes of pornography they had found on the site.

Then came the results of a sting operation run by Andrew Cuomo. The New York AG's team planted fake accounts that purported to be underage teenagers. It didn't take long for predators to begin grooming these fictional innocents.

Blumenthal was particularly unhappy with Facebook. His own children used it. He blamed Open Registration. "I have observed [Facebook's] mutation into a somewhat different kind of site," he told *The New York Times*. "There are now some troubling aspects to its features and culture that were absent before." (Blumenthal would be no less troubled a decade later.)

Cuomo put pressure on Facebook to agree to increase its scrutiny of such misbehavior. After an intense three-week negotiation, Facebook settled with New York. The settlement dictated that all reports of unwanted harassment or pornography had to be handled within twenty-four hours. That triggered a number of changes for the Support team. For the first time Facebook had to monitor content seven days a week.

Though the Facebookers felt that the AGs were mainly motivated to grab headlines, following the agreement helped organize Facebook's moderation efforts. "For all the challenges [the settlement] presented a very immature company at that time, it was very necessary because when else would we have prioritized porn reports?" says Charlotte Willner, who joined Facebook in 2007. (She later got her boyfriend and future husband, Dave, to join her.) "It was taken very, very seriously that we did not bust the twenty-four-hour limit," says Dave Willner. "Not really because of the penalties in the contract or in the settlement actually. It became this matter of honor."

As Facebook's number of users rose, so did the need for more moderators, particularly around the globe. In 2009, it expanded its operations team to Dublin and began using an outside company to help with hiring. It also began hiring more people who spoke non-English languages. The next year, it opened an office in Hyderabad, India. Still, until 2012, most of the people vetting content were full-time Facebook employees, and most of what they did was spot pornography and nudity. That year, though, Facebook decided it would be more efficient—and cheaper—to use contract employees. It hired the outsourcing firm Accenture to set up a huge moderation center in Manila, in the Philippines. Over the next few years, the number of mods ballooned. There would be more moderators in more languages, and the nature of their work expanded beyond just finding nudity, to things like bullying, hate speech, and even cannibalism. (It is a violation of Facebook's terms of service to eat someone.)

By the time Janzer left his job in 2015, there were 250 people spread across four offices: Palo Alto, Austin, Dublin, and Hyderabad.

By then the crowdsourced Word doc was falling short as a guide. Even something seemingly cut-and-dry, like nudity, proved to have its vexing aspects. "We needed to *define* what nudity was," says Janzer. "One of the clearest guidelines was 'no nipples.'"

But enforcing that policy meant that Facebook's support team wound up removing photos of women breastfeeding their babies. This was not unanticipated. Janzer says that his team consciously considered whether breastfeeding should be an exception but decided that the simplicity of the no-nipple policy should remain. No exception for latched nipples.

Some women are quite militant about their right to breastfeed in public, however, and those whose photos were taken down were outraged. To them, Facebook was regarding the apex of nurturing love as something sexual. Their complaints grew more vociferous, and the situation blew up in 2009 when a group called the Lactivists held a protest outside Facebook's offices. Online, 11,000 mothers staged a virtual "nurse-in." Though Facebook originally tried to defend its actions, it ultimately amended its policy, and made an exception for nipples with newborns attached.

There were different policies about harassment for public figures. It was all right to say "Fuck Aaron Rodgers," but what about a well-known college quarterback? What if it were a high school quarterback? Did it matter if the kid was a celebrity in the state? *How good a quarterback do you need to be in order to be trashed without us taking it down?* And when it came to other countries, how the hell could you know where to draw the line on who was a public figure or what constituted harassment?

"The Turkish constitution bans the defamation of Ataturk, the founder of the Republic," says Dave Willner. "They widely consider the idea that an Armenian genocide occurred to be defamatory of Ataturk because, of course, he oversaw it. And so talking about the Armenian genocide is viewed as a cultural insult by many people in Turkey. On the flip side, because of the historical tensions between Greece and Turkey, you will get folks from Greece who enjoy doing things like photoshopping Ataturk to have rouge on him, because they know it bothers people in Turkey."

Dave Willner volunteered for the job of refining the standards. "It was an inductive process from the things that we had documented and all the things that we had seen. Remember, I saw fifteen thousand photos a day. Like, we're not super fans of naked people everywhere, and threatening people is bad and all this other stuff. We actually ended up drawing from John Stuart Mill's harm principle as a way of thinking about how we would justify [our actions]."

But beyond the common framework, Willner says, Facebook needed a high-level set of principles. Why were various things forbidden and others not? What it came down to, he says, was one thing: the mission. Give people the power to share and make the world more open and connected.

It was very much a free-speech-driven philosophy, and totally in sync with the views of Mark Zuckerberg. "Mark's involvement was in setting the intellectual frame of the mission and the milieu of the company," says Willner. "He wasn't like, *Write a set of free speech rules.* In fact, nobody had written a set of rules. That's why I ended up writing them, and we needed them."

The document wound up being about 1,500 words long. But the actual task of applying the rules remained challenging. Inevitably, people made the wrong calls. "It's insanely complicated. And when you have insanely complicated processes involving humans making millions of pieces of content, you're going to have a lot of errors," says Willner. "There are certain distinctions people want you to make for values or moral reasons that are hard to actually make real in description." Even breastfeeding wasn't really settled, because a woman feeding her baby might have some other kind of nudity. "If someone is breastfeeding, but not wearing pants, what is that?" says Willner, answering the question right away. "No pants wins. Take it down."

He sighs. Willner left Facebook and now heads content standards for Airbnb, where people too often defy propriety in apartment and home listings. His wife, Charlotte, is head of trust and safety for Pinterest. "Like, I cannot express to you how completely bananas Facebook would be if it were not for the moderation," Dave adds. "It is basically a miracle that it's as calm as it is."

FACEBOOK'S MOVE TO 1601 California Avenue was a chance to physically express what Zuckerberg saw Facebook to be. In ultra-egalitarian fashion, everyone worked in front of giant display monitors—at least two—on a long trestle table, with executives sprinkled among the hoi polloi. That included Zuckerberg, who used his placement in the digital mosh pit to favor teams working on high-priority issues, moving them close to him for a spell. It's fairly common to hear Facebookers talk of the time when they worked a few feet away from the CEO.

Zuckerberg took meetings in "the fishbowl"—a glass-walled area in the center of the vast workspace on the first floor. When Facebook later moved to the former Sun campus in Menlo Park in 2011, he opted for an even less private situation, in a ground-floor office with a large window fronting the courtyard, where a steady stream of Facebookers and visitors passed. Inevitably, some of the gawkers treated his window as if it were an actual aquarium, with a nonaquatic resident as bizarre as anything the ocean had to offer. (Which should not have been surprising to the Facemash creator, who found that "people are more voyeuristic than I thought.") Facebook put up a sign discouraging lingering.

Zuckerberg liked to take walks, and often, before a visitor crossed the threshold of the Aquarium, he'd ask if the meeting could proceed on foot. He'd accompany the guest through the workspace, out the tiny lobby, and onto the bucolic streets of College Terrace. (Zuckerberg, by now living with Priscilla Chan, had finally acquired a house, with actual furniture, only a few hundred feet away from the office.) When Facebook later moved to a remote area of Menlo Park, near salt marshes, he'd continue the practice; during one walk, an executive noticed a large snake close to Zuckerberg's path. Zuckerberg just kept talking.

As the company grew, Zuckerberg stuck to the arrangement he made with Sandberg where he would keep himself focused on both Facebook's product and the long-term plan for the company. The huge parts of the company he had assigned to Sandberg—sales, policy issues, investor relations, and the care and feeding of the press—he largely ignored. Still, he was conscious that he was a CEO of what was becoming a major corporation, and took pains to improve himself in areas where he knew he was weak, like public speaking, or dealing with politicians or the press. Chris Kelly, who'd shepherded him in early meetings with politicians and other officials, would push him to engage. "He will sit back and stare at people," Kelly says, though this tendency faded with time. One of the first meetings he had with a politician was with Michael Bloomberg, then mayor of New York City. After an uncomfortable silence, Zuckerberg asked him, "Why are you *doing* this?" It opened a good conversation.

In other ways, he remained the stubborn kid his parents had learned was intractable. He had his own enigmatic habits and private rituals. A more public example was his annual resolution. In January 2009, as the global recession was just under way, he gave an all-hands address when everyone returned from winter break, and wore a tie to emphasize the seriousness of the situation. "When we entered the year, everyone thought that the world was falling apart, and all these other companies were stopping hiring, focusing exclusively on revenue and their finances," Zuckerberg told me at the end of that year. "I was like, *That's not what we're gonna do.* We're not going to chicken out and say because we're not cash-flow positive, we're going to divert all our resources towards revenue. We're going to stay on course, continue focusing on growth." One of his employees remarked that if it was such a make-or-break year for Facebook, he should keep wearing the tie. He agreed to do it. Facebook did indeed grow in 2009; its revenues almost doubled, and it showed its first profit. Zuckerberg didn't warm to my suggestion that he might want to keep the tie on to maintain momentum. "Maybe the tie is a charm," he said. "I think it mostly just chokes me."

He decided to continue his yearly challenges. At first they were relatively low-key, but as people learned about them, they took on an aspect of marketing, as he would announce the year's resolution with great fanfare, and close the year out with a report. In 2010, he decided to learn Mandarin. (Cynics postulated that it was a ploy to curry favor with the Chinese leaders, who had banned Facebook from their country.) Another year, he vowed to read a book every two weeks, skipping from Steven Pinker to William James to Henry Kissinger.

He took a more private approach to 2011's resolution, a vow to go vegetarian, eating only meat that he killed himself. It rose out of his genuine curiosity about what it meant to consume living things. "I think many people forget that a living being has to die for you to eat meat, so my goal revolves around not letting myself forget that and being thankful for what I have," he wrote to a reporter when the news leaked.

A neighbor at the time was Jesse Cool, a well-known restaurateur who owned Flea Street. (It was the place where Zuckerberg and Sandberg had first discussed her coming to Facebook.) Cool had chickens in her backyard, and under her tutelage, he killed one in her kitchen, and took it home to cook. He later graduated to pigs and goats, performing the slaughter at a farm or certified facility, having the animal butchered and then freezing the meat, and cooking for friends. Before he took a life, he would take a moment of silence and put his hand on the animal. It was, thought Cool, the epitome of respect. "It was his journey into really understanding the food he ate," she says. She found it painful when people heard of the challenge and either mocked it or inveighed against it. PETA sent Zuckerberg a basket of "delicious vegan goodies." Even years later, Twitter CEO Jack Dorsey would make headlines when he recalled a dinner at Zuckerberg's house during this period when the main course was goat, undercooked.

Dorsey obviously felt that, unlike revenge, goats should not be served cold. When it came to Twitter, Zuckerberg would prove better at revenge.

ZUCKERBERG'S TWITTER ADVENTURE might have been the first time that he used what would become a familiar strategy toward rivals: Identify a company that poses a present or future threat. Try to buy it. And if they don't sell, emulate it.

In 2008, Twitter's growth and influence were exploding. Like Facebook, it was a social product built around a stream of user-provided content. But it was distinguished from the News Feed in a number of ways. The order of the posts, called tweets, was strictly reverse-chronological. It didn't rely on one's personal social network, in that the tweets users saw were those of people they chose to "follow." There was no "friending" ritual; you needed no permission to follow someone. And it operated in real time.

Twitter's leadership team was plagued by internal conflict. The founder of its technology, and CEO at the time, Jack Dorsey, had come up with the idea while working for Odeo, a company led by entrepreneur Evan Williams. Unhappy with Dorsey's work as Twitter's CEO, Williams and another co-founder, Biz Stone, were in the process of pushing Dorsey out.

Dorsey, meanwhile, had opened discussions with Facebook. He had been taking exploratory meetings with Chris Cox at a coffee shop in San Francisco. The two respective kings of their company's stream jammed on the concepts. "Effectively, what I told him was that we have two different models," says Dorsey.

Facebook, though, wanted what Twitter had. In fact, it wanted Twitter. After Dorsey was pushed out in favor of Williams, Zuckerberg gave the new CEO a call and invited him and Biz Stone to Facebook. They hopped into Williams's Porsche and drove to downtown Palo Alto. Williams correctly predicted that Zuckerberg would try to buy them and they decided to throw out the highest number they could think of. At the Facebook office, Chamath Palihapitiya delivered them to Zuckerberg, who was sitting in a tiny space that seemed more of a phone booth than a conference room. Williams and Stone were crammed into a love seat and Zuckerberg was in the only other chair.

"Should we leave the door open or closed?" asked Williams.

"Yes," said Zuckerberg, totally confusing Williams, who decided to close the door.

Zuckerberg got to the point. He didn't like to begin an acquisition discussion with a number—but if they did have a price in mind, what would it be?

"Five hundred million dollars," said Williams, citing a figure at least twice as high as Twitter's valuation at the time.

"That's a big number," said Zuckerberg. He didn't give a lower counteroffer. He did do something disturbing. While Zuckerberg didn't say it in so many words, what he did say had the Twitter executives convinced that if they didn't sell, he would copy their features into Facebook. They had guessed he would do this anyway, but it was chilling to hear it.

Even at the inflated valuation they asked for, Williams was not inclined to sell. He felt that Twitter was destined to be worth much more. (Indeed, several years later the company eventually went public at a valuation of $14 billion.) Also, he was dubious about the Facebook stock Twitter would receive instead of cash. (On this he was wrong—$500 million of Facebook stock in 2008 would one day be worth many *billions*.) Ultimately, he simply didn't trust Facebook or Mark Zuckerberg. Something rubbed him the wrong way about this young CEO. Performing his fiduciary duty, though, he brought the matter to his board, and advised against it. The board agreed.

Since Zuckerberg could not own Twitter, he decided to neutralize its effect by Twitterizing the News Feed.

In a sense, Facebook had been borrowing from Twitter since 2006, beginning with giving the user the ability to add status updates to the News Feed. "Status was a very late addition and just a straight rip-off of Twitter," says Ezra Callahan. "No way around it—Twitter got popular real fast, *let's do that here*. That was the first time we just straight ripped off somebody."

Now that Zuckerberg moved to implement some of Twitter's core principles into Facebook, this meant completing what was already a shift in the product's focus from the Wall to the Feed. Facebook was on a path to essentially tear down the wall of comments on a user's profile page and make the News Feed itself the repository for public interaction on the service. Twitter envy seemed to speed the process.

Mark Slee was the product manager for this 2008 redesign. "I would say that Zuck was the real product manager," he says, describing himself as translating and implementing his boss's ideas. Those ideas did come from a lot of internal discussion, because it changed the nature of content on Facebook. The News Feed would change from the point of view of third-party reporting activity on Facebook (*Mark has posted a photo*) to people actually doing the posts themselves (*Hey, I'm posting a photo!*). Facebook began to encourage users to share more kinds of content besides text and personal photos. The News Feed welcomed more outside media—links to articles and videos.

While this was done in the spirit of improving Facebook and acknowledging the power of a stream of content, lurking behind all that decision-making was Zuckerberg's determination to thwart any rival. "I couldn't tell you if Mark thought that we should have Twitter-like features," says Slee. "My read is that Mark is extremely competitive and absolutely would not want to give someone the opportunity to overtake us. Facebook should be Facebook, but we need to play defense against this potential threat."

The redesign would have a profound effect on the News Feed, on Facebook, and, it isn't a stretch to claim, on humanity itself. Till that point, the stream of stories in the feed had been determined by signals that indicated their importance to you in regard to your network of friends. The News Feed algorithm was called EdgeRank. It depended on three main factors: Affinity, Weight, and Time Decay. Affinity was measured by how close you were to the person making a post; something by your brother or your best friend would get a high score. Weight was determined by a formula that predicted how likely you were to engage with a post, based on your interests and previous behavior. Time Decay dealt with how recent the post was—newer ones were prioritized. There was a lot of computer science involved in assigned scores according to those criteria. Where a post would show up on your feed—or whether you would see it at all—depended on how each of those factors was weighted. It was largely a question of Facebook's turning the knobs that measure how much influence each of those three factors would have in determining the score for each possible post. At any given time, this algorithm might change, reprioritizing the importance of one factor over the others.

Becoming more like Twitter meant that the News Feed put more of an emphasis on engagement and timeliness, to replicate Twitter's ability to reflect what was happening in the world at a given moment, and capture Twitter's dynamism.

In a 2009 redesign, the News Feed became even *more* Twitter-like. A project code-named "The Nile" gave it more of a real-time flow. One of the fundamental differences between Facebook and Twitter had been the way that the social graphs worked. Twitter was more of a nano-broadcast medium than a pure social network. Tweets were distributed to people who "followed" whoever posted them. And unless someone designated their tweets as private (very few did), any one or every one of Twitter's millions of users could follow a given person. Celebrities and influencers could have hundreds of thousands, even millions of followers, and their tweets could act as a news service, a giant comedy club, or a 140-character performance space. On Facebook, friendship was reciprocal—privacy dictated that your posts would be distributed to a limited universe, under your control. But now Facebook was encouraging you to use its News Feed as you used Twitter—to keep up with celebrities and experts in various fields. If you engaged with them, their posts would probably find you.

Facebook did make some adjustments to ensure that the influx of distractions didn't outright kill social value on the site. A famous story at Facebook is that one of Zuckerberg's relatives had a baby around that time and he was furious that the post never reached the top of his News Feed. "You're not going to want to have to go through a hundred stories to see that one of your friends had a baby," Zuckerberg told me at the time. "That better be at the top, or you're going to be angry, and we're not doing our job." So Facebook made sure that it picked up on signals like the wording of birth announcements, weddings, and death notices. Also, when someone replied "congratulations" to a post, that was a serious signal that it was a life event, and the post was ranked highly.

To use Slee's words, Facebook was still Facebook—but it now expanded its role from an information and entertainment source about your social network to something that aspired to be your source of *all* your information and entertainment, whether it was news of someone you knew or a friend of yours sharing news of Beyoncé. If you were interested in a subject or a person, Facebook would be interested in serving you stories about it. And the main way it detected interest would be whether you engaged with something similar. Those seeking to circulate stories more widely on Facebook learned the rewards would come when people reacted to their stories—clicking on them, liking them, or even just having their eyeballs linger on them.

In a sense, News Feed was cultivating a replay of the spam attack that certain developers launched in the early days of Platform. But in this case, the "spam" was not annoying notices that someone had downloaded a game or threw a sheep at you, but something that provided a bit of distraction: a heartwarming news story, a cat picture, or an opportunity to learn what *Star Wars* character you are most like. The techniques of virality were much the same; the difference was that this time around, Facebook was now not so subtly encouraging such posts. The highest-ranked stories were the ones most likely to deliver the equivalent of a sugar rush. In Facebook's view, it was giving people what they wanted.

Early in the News Feed's history there had been a discussion about how to handle what Facebook's algorithms determined were the most interesting stories. Some people thought that they shouldn't be the first ones delivered, on the premise that people would keep scrolling until they got to them. Zuckerberg decided that even at the risk of satisfying users too quickly, the cream should rise to the top. "If you only see three stories, they should be three pretty good ones." In practice, three good stories would be just as likely to keep the person scrolling for more.

Facebook was thinking not like a social director but like a publisher. If Facebook knew that a story would appeal to you, it would stretch to the edges of your social network—your weakest ties—to put it at the top of your feed. Or maybe no one in your network posted it, but someone made a comment on an article posted by one of their friends. In that case, your friend would be spreading something that was less likely to have a personal angle but might be something that could amuse you or make you angry. And maybe get you to comment on it too.

All of these factors led to the transformation of the News Feed into a viral engine. Those who understood this early, like the political action group MoveOn or the meme-generation factory BuzzFeed, realized they could now build movements or businesses on Facebook's capacity for rapid distribution of stories that evoked outrage or tugged at heartstrings.

Facebook made this shift with its eyes open. Its Data Science team—part of the Growth organization of Facebook—studied it, not as a menace but as a phenomenon to be understood in order to best exploit it. A paper out of the Core Data division titled "Gesundheit! Modeling Contagion through Facebook News Feed" studied a data set of all Facebook Pages created between February and August 2008, including 262,985 Pages where "diffusion events" occurred. (This term seems to be a fancy way of saying "going viral.") The idea was to conduct "an empirical investigation of diffusion through a large social media network." It found that the mechanics of the News Feed helped foment huge spikes in people "fanning" a Page. (The study predated the Like button.) The paper uses the lexicon of epidemiology throughout. If conditions are proper, the News Feed can trigger "global cascades" where a comment gets spectacular readership.

Though it was not strictly part of their research, the scientists could not help but remark on who could benefit most from the viralized News Feed. "These models have significant practical implications for marketers," the researchers noted, "particularly those who are interested in advertising through social media."

Meanwhile, now that it had appropriated some of Twitter's innovations, Facebook shut off Twitter's access to the News Feed. People had long been able to "cross-post" their tweets to Facebook—but in 2011, Zuckerberg called the Twitter CEO at the time, Dick Costolo, and told him Facebook was cutting off Twitter from the API, so cross-posting would end. He didn't offer an explanation and didn't need to. "We always knew it was going to happen," says Costolo. "It was sort of known that [if] you're becoming big, they would shut you off. And if we got stronger, they were going to cut off our air supply more and more and more. If Facebook's got you in their sights, you're in trouble."

THERE WAS YET another consequence of Facebook adopting aspects of Twitter, this one involving what was becoming a treacherous field for Facebook: privacy.

Posting on Twitter had always been a public act. The company offers an option to make one's tweets visible only to those who are granted permission, but the vast majority of people on the service use the default, making their content available to anyone who uses the service. Tweets were also visible on search engines and even collected by the Library of Congress.

Now Facebook was going to follow on that public path. Specifically, Zuckerberg wanted to change the Terms of Service contract it had with its users. The key difference would be flipping the default setting from "friends-only" to "everyone." Unless users took specific action to limit exposure, their posts, Likes, friend lists, and certain profile information not only would be public within the service but would be visible to Google and other search engines. (Previously, Facebook had made public only the users' names and which network they used.)

Twitter was only a partial inspiration for the move, though Chris Cox conceded in press calls explaining the shift that it was indeed a factor. The real impetus came from Growth. Having Facebook's information more prominent on Google would lead people to find more friends there, and perhaps nudge holdouts to sign up.

This was an egregious break from the original bargain between Facebook and its users. The essence of Thefacebook and the service in its early years was that all personal information would be kept inside the community. "We understand you may not want everyone in the world to have the information you share," Facebook stated in its 2006 privacy policy. Even when Facebook rolled out Open Reg that year, allowing anyone to join the service, it promised that the change would not mean your profile would become public. "Your profile is just as closed off as it ever was," said the blog item announcing the change. "Our network structure is not going away. College and work networks still require an authenticated email address to join. Only people in your networks and confirmed friends can see your profile."

The idea that the default audience would be changed to "everybody," meaning everybody in the world, was an unimaginable concept in 2006. Now it was an imperative.

And Facebook would be implementing this in a period when its post-Beacon behavior had done little to restore trust. Earlier in 2009, Facebook had released a new Terms of Service agreement that, on its face, seemed to give the company total freedom to do what it wanted with all the personal details that people shared with it—even if their account was closed. A writer for *Consumerist* summed it up in a headline: "Facebook's New Terms of Service: We Can Do Anything We Want with Your Content Forever." The outcry was instant and wide-ranging. Seventy thousand people joined the "Facebook Users Against the New Terms of Service" group. The Electronic Privacy Information Center, in concert with eight other organizations, made a formal complaint to the Federal Trade Commission, which had been said to be interested in investigating Facebook. So intense was the pressure, the company rolled back to the previous set of terms within a week, with Zuckerberg conceding "mistakes." Soon after, Zuckerberg came up with an innovative idea to forestall criticism: from that point on, Facebook would allow its users to vote on policy changes. Since Facebook's user base would make it the sixth largest country on Earth, he reasoned, the citizenry should have some say. These decisions would be binding.

"We took last week as a strong signal of how much people cared about Facebook and how much they want to govern it," Zuckerberg said at a contentious press conference in February announcing the concept, which even privacy advocates regarded as bold and intriguing.

But Facebook's proposal to allow users to determine policy proved to be somewhat of a con. There was a huge loophole: Facebook considered an election binding only if 30 percent of all users cast a ballot. Such a turnout was unlikely, considering the huge population on the service and the relatively few who followed seemingly arcane issues like privacy regulations. Facebook held three elections in this experiment, and none garnered even 1 percent of the base. Facebook would quietly discard the idea of a user-centered democracy.

The late 2009 change in privacy settings was never put to a vote. But it is not clear that it would have won an election even within Facebook.

"The company was split in half on this decision," says Dave Morin, who was among those against it.

By then the original privacy chief, Chris Kelly, had left the company to run (unsuccessfully) for attorney general of California. His replacement as Facebook's internal privacy watchdog was an Internet-savvy lawyer named Tim Sparapani. Sparapani was a privacy expert and an advocate for users. He had worked for the ACLU.

Sparapani also had duties as Facebook's first DC-based policy official. He was the second Facebook employee in the nation's capital, and opened its first office there, in a space that had previously been used to shoot on location for the television show *The West Wing*. Though it was never confirmed, the new tenants believed that the conference table they used to discuss privacy and policy issues was the same one where members of fictional president Jed Bartlet's staff tangled over weighty issues. At Facebook, discussions would sometimes be more bare-knuckled.

The 2009 privacy terms would bloody those knuckles.

On one hand, Sparapani and others at Facebook concerned with privacy were pleased Facebook would be announcing the change in tandem with a host of options that actually increased privacy options. For the first time, people could designate individual posts to be visible only to a selected group of friends, or just friends of friends. It created a "transition tool" that helped users set their privacy levels under the new rules. It would be a definite improvement over the then-current controls, which had become increasingly complicated, and harder to even find in the first place. "You'd have to have a PhD in Facebook to understand what it was you needed to do to set the settings, and you'd have to take lots of time," says Sparapani.

But Facebook knew that most people would never bother to use even those more transparent controls, as it is an industry truism that the vast majority of users simply stick to original settings. (Indeed, Facebook later noted that 80 to 85 percent of its users did not change the defaults.)

On one hand, Facebook had good reason to rethink the privacy settings—with hundreds of millions of people communicating with friends and contacts in multiple networks, the original idea of restricting information to your college class was a shaky foundation for a privacy mode. "That model broke," says Colin Stretch, a DC attorney hired at Facebook just as the new policy was rolling out. "Once [Facebook] was opened up to everybody, it didn't really make a ton of sense."

Still, even inside Facebook, some felt that moving the default settings for the personal data of Facebook's then 350 million users seemed like a massive corporate betrayal. What's more, it might not even be legal.

Sparapani argued that, as the shift was proposed, Facebook would not be meeting either the spirit or letter of privacy law, which required clear notice about any changes and not implementing them without getting clear consent. His opponents in the argument were Palihapitiya and his Growth team, whose objective was to draw and retain more users, and to get those users to share more.

The final call, as always, was Zuckerberg's. He sided with Growth.

To be sure, those pushing back against Zuckerberg didn't feel he was being unethical, or harboring a conscious thought that he was violating the trust of the users. "He just had more of an ends-justifying-the-means than I did," says an inner-circle leader at the time. "I would have quit immediately if I thought he didn't care."

Zuckerberg explained his reasoning not long after in an onstage interview. "A lot of companies would have been trapped by convention and their legacies of what they've built. Doing a privacy change for 350 million users is not the kind of thing that a lot of companies would do. But we viewed that as a really important thing, to always keep a beginner's mind and what we would do if we were starting the company now, and we decided these would be the social norms now, and we just went for it."

ANOTHER CHANGE THAT Facebook made in that period was an alteration to the APIs that developers used to tap into Facebook's warehouse of user information. This was what would be known as the Open Graph, or Graph API V1. It was one more effort by Zuckerberg to spread Facebook beyond its borders. This meant continuing the dubious practice whereby developers imported not only the information of those who signed up for their services or logged in with Facebook Connect, but data on the users' friends. There was nothing those "friends of friends" could do to protect that information, which could include things like birthdays, email addresses, Likes, and relationship status. Arguably, these were necessary for the apps to work properly—a dating app, for instance, might need to know if someone was married or not. But the gravest fear was that a developer might suck up a huge database of Facebook information and use it for itself, or, heaven forbid, sell it.

Sparapani knew that he couldn't stop this practice. But before signing off on the new APIs, he demanded the company publicly declare that it would do audits of the developers making the data culls that gave them personal information. Facebook assured him it would take affirmative steps and build products to track what information it had given away, and to verify that developers did not retain it.

But according to multiple executives at the time, Facebook *didn't* build them. It's unclear whether engineers who might have gotten the job were drawn to other tasks, or whether someone actually said not to follow through. But clearly it was not a priority.

It's not like Facebook didn't have frequent warnings about the trouble this issue might cause. In October 2010, for instance, *The Wall Street Journal* found that Facebook was handing over not only data like friend lists, interests, and gender to developers but even the secret user IDs that Facebook used to identify its users. The leak of user IDs was particularly egregious because outsiders could use those IDs to bypass privacy protections. If a developer had an ID, it could access information that people had specified not be shared even with close friends. The ID could also be used to link Facebook identities with people's real-world information, like their address and financial information.

Facebook said its release of user IDs was unintentional, and the developers *The Wall Street Journal* contacted said that they hadn't asked for the numbers and weren't using them. But it should not have taken a news article for Facebook to address the issue. It now appears that Facebook heard objections from at least one developer who didn't want the data. "We would tell Facebook in our face-to-face meeting, *You guys are giving us user IDs of friends and the user IDs of the friends of the friends from our 30 million users!*" says Nat Brown of iLike. "*We can see 300 million people and we know all this information about them—and we don't want that!*" Worse, those valuable IDs were falling into the hands of data brokers who *were* using them, to market to or track Facebook users.

The data broker that *The Wall Street Journal* highlighted was called RapLeaf.

RapLeaf's very existence pointed to a silent privacy crisis in the way outsiders received information about Facebook users. It was the tip of the iceberg in an underground economy of user information, of which Facebook data was just one part. Public knowledge of this widespread exchange of personal data was sketchy, despite investigations like *The Wall Street Journal*'s. Facebook seemed fine with that. But if the company was paying attention, it might have realized that one day this problem would blow up in its face.

RapLeaf was co-founded in 2006 by a savvy entrepreneur named Auren Hoffman, who became its CEO and public face. Part of its seed funding came from Peter Thiel, who of course funded Facebook and was on its board. Hoffman was a data broker, gathering personal information that was exposed by people who used the Internet and selling it to marketers. "We crawled Facebook, we crawled LinkedIn, we crawled MySpace, we crawled blogs, everything," says Hoffman. The goal, he says, was to sell marketing information to companies. Facebook data was particularly useful because it would give specific data on interests and status, such as whether someone liked the Beatles or if they were single, and where they lived.

Hoffman now says he was simply taking advantage of the opportunities Facebook offered. He claims Facebook (along with other companies he crawled, he adds) knew all about his activities. Facebook *helped* him. "All the highest-level executives knew. They're smart people, they knew everything that was going on about their site. We basically told them the methods we were using. They actually gave us advice on how to make it more [efficient], so we didn't tax their servers. They were watching everything. They would also tell us there were forty other companies doing this. I think they're still doing this today."

Sometimes Facebook felt that RapLeaf went *too* far in sucking up people's personal details. "They would be like, *Hey, this is too aggressive*, or *Don't collect that*," says Hoffman. "We definitely did do a lot of stupid stuff. We're like hackers, so we're out there just like doing stuff."

Hoffman's company was apparently among a number of those buying the leaked user IDs from developers and selling them to marketers. The brokers would be able to take the names on the lists they had, and instantly add all the information in everyone's Facebook profile, making it much more valuable. For instance, if the user was a gun lover or a women's healthcare advocate, that would be of great interest to a political campaign. That was not a hypothetical. "They [RapLeaf] were selling that to political ad campaigns," says Sandy Parakilas, who worked in ad compliance in Facebook (and would later become one of the most vocal apostate critics of the company).

In response to *The Wall Street Journal* article, Facebook closed the loophole that leaked user IDs. It also "reached an agreement" with RapLeaf, which would delete the Facebook user IDs it had collected and leave the Facebook platform.

Hoffman considered this part of the game. "Facebook guys are basically going to let lots of things happen," he says. "They're going to be very tolerant. And then if there's a [critical] article or if something happens, then they'll crack down on it. That's certainly true in our case."

Facebook felt that it quelled the controversy when it cut off RapLeaf. But the episode would be one more unheeded warning.

IN FACT, WITH 2010's Open Graph, Facebook was seeking *more* ways to pass information to developers in order to extend sharing beyond its own site. At its F8 conference that year, Zuckerberg was excited about an initiative called Instant Personalization. The scheme would allow website developers to Facebook-ize themselves by implementing a piece of code that essentially would use the personal data from Facebook—like friend lists, gender, and pretty much everything shared with "everyone"—and make use of it as soon as a Facebook user visited their site. From then on the site would be "personalized." It was supposed to be a service akin to a hotel's knowledge of a customer's preferences, so upon arrival the room would be supplied with the right pillows, a favorite beverage, and the customer's favorite music on the stereo. Visitors to Pandora, one of three launch partners, would be greeted by the music they had already indicated that they liked on Facebook. (The other launch partners were Microsoft and Yelp.)

The privacy issues were clear and disturbing. Instant Personalization would kick in automatically. (To stop this from happening, a user would have to know of the existence of the feature, then find the setting to turn it off.) Again, there was dissent among Facebook's leaders. The problem was not just Instant Personalization but the Graph API itself, which gave developers deeper access to user profiles, even information that users had specified should be shared only with friends. Nonetheless, Zuckerberg rolled out the API and Instant Personalization at the 2010 F8.

Instant Personalization was, in the words of one commentator, a "privacy hairball." Multiple tech sites ran stories on how to negotiate the tricky process of opting out. And Zuckerberg was grilled on it during an interview at a tech conference called "D: All Things Digital," run by tech journalists Kara Swisher and Walt Mossberg. Sitting in the signature red chair that the conference provided interview subjects, Zuckerberg wasn't handling direct questions about privacy well, and by the time Instant Personalization came up, he was sweating so much that he had to remove his hoodie. Why, Mossberg asked several times, did Facebook design the feature to work without asking for the user's consent? Zuckerberg explained that asking people to click even once on a permission button can create too much friction, discouraging people from discovering sharing habits they would ultimately appreciate. That's what happened with News Feed! One day, Zuckerberg said, people will look back and marvel that there was a time when websites *didn't* use features like that. "The world is going in a direction where it's going to be designed around people, and I think that's a powerful direction," he said. His performance was not convincing.

And there was another consequence of Instant Personalization: "It was right after Instant Personalization was rolled out that the FTC inquiry really got legs," says Colin Stretch, who later became Facebook's general counsel.

COMPLAINTS HAD BEEN rolling into the Federal Trade Commission about Facebook's behavior. This aligned with a growing concern in the agency that young tech companies were crossing legal boundaries. "We had begun to focus on some of the tech companies, because we found that they just didn't have their act together. They were new, they were growing fast, and a lot of times they made commitments they didn't keep," says a former FTC commissioner. It initiated an investigation of Mark Zuckerberg's company. "Facebook was involved, potentially, in a violation of law, and we take it really seriously."

Zuckerberg himself did not help matters with some of his public statements. Appearing onstage at an award ceremony in 2010, he opined that social norms about privacy had changed. "People have really gotten comfortable not only sharing more information and different kinds, but more openly and with more people," he said. "That social norm is just something that has evolved over time." As an observation it was defensible, but it failed to reflect his own role in shifting those social norms. He believed that the world would flourish if people adjusted their view of privacy to one that embraced more sharing.

Even as Facebook's lawyers and policy people were dealing with the investigation, Zuckerberg still seemed to believe that privacy could be handled in a trial-and-error fashion. "I think, related or unrelated to privacy, whenever we make a change of this magnitude, I think we should expect that some people will like it, some people won't, and then we'll have a rollout where we give people a chance to try it out if they want," he told me in mid-2011. "And then we'll take some period of time to adjust [to] all the feedback and we'll kind of go from there."

Though the commissioners all agreed that Facebook had misled its users and violated their privacy in multiple ways, they disagreed on sanctions. Some thought that Zuckerberg should be personally cited; this would have had serious repercussions. If he were named, and Facebook continued to misbehave, Zuckerberg could be liable for civil or even criminal penalties. In an early draft of the agreement, Zuckerberg indeed was personally cited. But after ongoing negotiations between Facebook's lawyers and the agency, the Zuckerberg citation did not make it into the final version.

In November 2011, Facebook and the FTC finally settled. While Facebook admitted no wrongdoing, it signed off without disputing charges of misrepresentation. It also agreed to a twenty-year period of oversight, to be conducted by outside auditors, paid for by Facebook. The FTC cited seven specific misdeeds, including several that were implemented at Facebook despite warnings or objections from some of its executives. Here's how the agency described them:

- In December 2009, Facebook changed its website so certain information that users may have designated as private—such as their Friends List—was made public. They didn't warn users that this change was coming, or get their approval in advance.
- Facebook represented that third-party apps that users installed would have access only to user information that they needed to operate. In fact, the apps could access nearly all of users' personal data—data the apps didn't need.
- Facebook told users they could restrict sharing of data to limited audiences—for example with "Friends Only." In fact, selecting "Friends Only" did not prevent their information from being shared with third-party applications their friends used.

- Facebook had a "Verified Apps" program and claimed it certified the security of participating apps. It didn't.
- Facebook promised users that it would not share their personal information with advertisers. It did.
- Facebook claimed that when users deactivated or deleted their accounts, their photos and videos would be inaccessible. But Facebook allowed access to the content, even after users had deactivated or deleted their accounts.
- Facebook claimed that it complied with the US–EU Safe Harbor Framework that governs data transfer between the US and the European Union. It didn't.

The "Verified Application" label gave users the impression that apps designated with a checkmark under this program had been vetted for trustworthiness. In actuality, developers attained that status by paying Facebook a fee.

When writing his Facebook post on the settlement Zuckerberg must have felt on familiar turf. It bore similarities to previous episodes when he'd been caught in errors or misdeeds dating from Facemash, and continuing to the News Feed, Beacon, the 2009 Terms of Service, and the privacy settings. "I'm the first to admit that we've made a bunch of mistakes," he wrote. "In particular, I think that a small number of high-profile mistakes, like Beacon four years ago and poor execution as we transitioned our privacy model two years ago, have often overshadowed much of the good work we've done."

Facebook might well have added a new poster to its collection:

MOVE FAST AND BREAK THINGS. APOLOGIZE LATER.

Because that, too, was part of Facebook's culture.

12

Paradigm Shift

AS 2012 APPROACHED, Facebook was on track to have 1 billion users before the year was out. Advertisers were embracing the service, raising Facebook's revenue to almost $4 billion. A billion of that was profit.

But all that Facebook had built was in peril. Mark Zuckerberg was insufficiently prepared for a huge change in his industry, one that presented an existential challenge to Facebook.

The world was moving to smartphones, and Facebook was bungling the transition.

Facebook wasn't exactly a stranger to the world of mobile technology, but for years, its approach was oddly ad hoc. In 2005, the person making deals with mobile carriers to put a few of Facebook's features on the primitive "feature phones" people used at the time was the former Yahoo! business development exec Jed Stremel, who had begged for a job at Facebook after someone showed him a slide deck with its amazing metrics. Just as he got the job, he attended the party Peter Thiel threw in December to celebrate a million Facebook users. Stremel, then thirty, felt out of place wearing a sport jacket.

For the next two years, he was virtually on his own in making deals with carriers. "The mobile team from 2005 to 2007 really didn't exist," he says. "It was me." In 2006, his deals with Cingular, Verizon, and Sprint resulted in Facebook's first mobile product, allowing users to send text messages on those services. (Facebook got a cut of the message charge.) But because the primitive phones of the time couldn't support the most popular activity on Facebook—photos—the company didn't devote much attention or resources to the mobile world. Stremel quietly made deals for Facebook texting to carriers across the world, but almost no one in the company agreed with him that mobile was the future of technology.

So when Apple began that future with the iPhone in 2007, Facebook was ill poised to take advantage. Help came from an unexpected source. Joe Hewitt had come to Facebook in July 2007, when it bought his two-person start-up, Parakey. Hewitt's partner, Blake Ross, had been excited about the sale; Hewitt was not so thrilled to join what he thought was a stupid college site. "I personally had zero interest in working there," he says. "I wasn't going to work there for more than a couple months, just to see how it was."

He wound up remaining for the full four years of his vesting term, earning every penny of the millions he would reap from stock options.

Two weeks before he began, Hewitt got a just-released iPhone. The apps on it looked spectacular. But it would be very hard for outsiders to match their slickness and power. Those original apps were "native"—specially written to access the hardware of the phone. Since Apple didn't give outsiders direct access to the hardware, software developers couldn't write native. Their iPhone efforts would have to be web pages designed to work with the Internet browser on the phone. Hewitt began experimenting with developing web pages that emulated the look and feel of native apps. When he arrived at Facebook, he asked if he could keep working on that track. He was invited to join the small team making Facebook applications for handsets like BlackBerry, but he considered those junk phones and not worth his time. After a couple of meetings he decided to work totally on his own. "I didn't particularly care about integrating myself into the company. I just did what I wanted and they tolerated me doing what I wanted and it kind of worked out for a while."

What he wanted was to create a great Facebook iPhone application.

He briefly worked out of the University Avenue office, but he loathed the open office plan and wasn't charmed by the graffiti mural. One day he decided to work from home and from then on people seldom saw him. When his group moved to another building, he never even secured a space for himself.

Stremel, Facebook's mobile leader, was a bystander, but cheered from the sidelines. "We had no engineers on our project, and didn't have the ability to do it. Joe's a brilliant engineer and had no responsibilities and no one to report to. He just went and did it."

By August, Hewitt was finished. He'd written the app in two months. Though arguably it represented the future of the company, he released it sans fanfare. "I didn't really have to ask anyone permission because it was kind of the Wild West," he says. He doesn't even recall running it by Zuckerberg. "He probably saw it before we went live. But I didn't have to meet with him and do any design consultations." Hewitt didn't get around to even posting a blog item about it until a day later.

The press was rhapsodic, with some calling it the best app for the iPhone yet.

A year later, Apple lifted its restrictions and allowed developers to create native apps. Hewitt, who was still the one-person team in charge of Facebook's iPhone app, was thrilled. Steve Jobs himself came to the Hamilton Avenue office to discuss it with Hewitt and Zuckerberg. "We met for several hours," says Hewitt. Especially fascinating was the dynamic between Zuckerberg and Jobs. "He was treating Zuckerberg like sort of an apprentice, trying to transfer a lot of knowledge to him, telling him stories and just sharing random, irrelevant stories about Silicon Valley," says Hewitt. "Mark definitely respected Steve and was open to learning from him, but he had a lot of confidence in himself and it wasn't like he was begging Steve Jobs for advice or anything."

By 2009, the iPhone had taken off, and Facebook was its most popular app. Amazingly, Hewitt was still basically doing Facebook's app by himself. But inside the company he was alienating people with his aggressive independence. Facebook's communications team in particular had issues with him. "I had this habit of not going through PR for new releases," he says. "I would just announce them on Twitter or respond to some random, small-time journalist." He also was continually annoyed with Apple. He disagreed with its strict curation of its App Store, considering it somewhat of a bully. At one point, peeved that Apple wasn't moving fast enough to approve version 3.0 of the Facebook iPhone app, he wrote a choleric blog item about it.

Hewitt had an annual tradition of announcing his resignation from the company, and 2009 was no different. "During the first two years, I didn't think the stock was going to be worth that much and I wanted to get back to my own start-up," he says. "The last two years I realized, *Oh crap, it* is *going to be worth that much.* But I still tried to quit." Chamath Palihapitiya convinced him to stay, promising he could do whatever he wanted.

Hewitt had an idea to develop a programming language for the Apple mobile operating system. But it ran into trouble when Apple released a new developer's agreement in April 2009. It would have prevented him from using the new language he was creating to make mobile apps. "It had some very restrictive stuff in there that I strongly disagreed with," he says. "I was pissed." Hewitt dashed off an angry email to Steve Jobs and Apple's software head, Scott Forstall.

Jobs called Zuckerberg directly to complain. Zuckerberg called in Hewitt. The incident amused Zuckerberg—he had been trying to reach Jobs on some other matter, and it took Hewitt's outburst to get the Apple CEO to finally contact him. Zuckerberg told Hewitt that he was on his side, but Apple was important to Facebook. "Steve's being a little crazy," he told Hewitt. "But if you antagonize Apple one more time, we're going to have to fire you."

In a way, all of that was moot, because a disgusted Hewitt had quit working on Facebook's iPhone app. In typical fashion, he had announced his move on Twitter, and later released a statement. "I respect [Apple's] right to manage the platform however they want, however I am philosophically opposed to the existence of their review process," he wrote. "I am very concerned that they are setting a horrible precedent for other software platforms and soon gatekeepers will start infesting the lives of every software developer."

Facebook assigned a group of engineers to do Hewitt's work. It hadn't been ideal for one person to be so instrumental to such an important effort anyway. "They put a team on it, and the team grew, and now it's a whole company," says Hewitt, who now grows organic vegetables in Hawaii.

But not before Mark Zuckerberg made a product decision that the CEO would later describe as the biggest mistake he ever made.

Hewitt's amazing work had allowed Facebook to postpone dealing with a critical problem: it was on the wrong side of a paradigm shift. From the very start of Thefacebook, Zuckerberg had chosen to use the web-based computer language PHP. It was a choice that in 2004 older computer scientists might have rejected—but Zuckerberg had grown up creating quick-and-dirty online projects, and using the younger PHP system was like breathing to him.

PHP's great strength was that it had a built-in safety net. Programs written in traditional languages were released in discrete versions. If the programmers wanted to add a feature, or even fix a bug, they would include it in the next version, which wouldn't take effect until the user downloaded an update. A popular program could have multiple versions of different vintages floating around, and old bugs would keep resurfacing. But PHP was always up to date. You could quickly push out changes or new features, send it to the web server, and the server would shoot out the markup code that generated web pages. If you screwed up, fixing things was easy—you just wrote new code and the next time the user refreshed the browser, the new version would be running. The user always ran the new, presumably less buggy, version.

In effect PHP enabled the speed that had been the secret jet fuel powering Facebook's growth.

Now a new era had come where people would use the desktop less and less . . . soon hardly at all! And mobile was different. Apps did not reach the user directly but they came from a curated store run by the hardware designers like Apple or Google. Every single version had to meet certain standards and pass muster with those gatekeepers. And Facebook was suddenly a legacy company trying to keep up.

Worse, Facebook did not have the proper troops to fight this new war. Of four hundred Facebook engineers, Vernal estimates that only five were skilled at iOS, and maybe three knew their Android. "We didn't have enough people who were actually good at this stuff, and it was slowing down mobile product development in a giant way," he says.

"You had a company that didn't know how to build native applications," says another executive who looked into the problem. Facebook's hiring process was actually *filtering out* people talented in coding mobile apps. Starting around 2009 and 2010, the best young engineers were writing native apps for iPhone or Android. But when Facebook interviewed them, it asked questions geared toward the desktop world. The best candidates would tell them that they didn't know the answer and didn't care, because they just wanted to build cool mobile apps. And when the hiring committees met, they would determine that these coders—the kind Facebook really needed—were poor prospects because they didn't know about desktop development and they had bad attitudes, besides. "We just wouldn't hire these people," says the executive.

But suddenly, it looked like a miracle technology solution would fix everything.

The technology was HTML5. It referred to a new version of the "markup language" that was the lingua franca of the web. It was supposed to be a panacea for a knotty problem: how could software companies like Facebook deal with the issue of getting their product on multiple mobile systems? People increasingly wanted to use Facebook on their smartphones. But different people used iPhones, Android phones, BlackBerries, Palm, Windows, and other systems. Each had its own operating system and unique hardware. While the best apps were native, optimized to the specific hardware, that path required writing a separate product for every single OS.

"People didn't want to have to re-implement the same thing on iOS and Android," says Mike Vernal. "And so the technical question was, could we build a framework that let us write the mobile app experience once and have it be on both iOS and Android. And there was a little bit of a maybe it will work on Windows phones."

HTML5 promised a solution: write once, run on many systems. It especially appealed to some newly hired engineers on the mobile team who had come from Google, which was a hotbed of believers in the open-web philosophy that HTML5 embodied.

The Growth team loved the HTML5 approach. Its interest was spreading Facebook to places where the service had yet to dominate. Many of those regions were developing nations where people got Internet only through low-cost phones. A situation where Facebook's programmers could write a single program that went to all those phones was the Growth team's dream.

What Growth wanted, Growth usually got. Especially in mobile: on the reasoning that new customers, especially overseas, would be using cell phones, the Growth team "owned" the entire mobile effort at Facebook. As with the other areas that Growth controlled, its own mission of chasing and retaining users was setting the priorities for Facebook's mobile efforts.

"There were a lot of people who wanted to believe that HTML5 could be made good enough," says Vernal. So Facebook began to implement its version of HTML5, dubbing it Faceweb.

Soon Faceweb was the official strategy for mobile apps, which were quickly becoming the dominant means that Facebook's hundreds of millions of customers would use to access the product.

It was a disaster.

CORY ONDREJKA HAD come from a start-up Facebook had bought in 2010, just one of a number of "acqui-hires." These were companies bought for the talent and not the products they made, which were usually discarded. Ondrejka, who had once been a key employee of an online simulation called Second Life, was put on the games team. Facebook's then VP of engineering, known as "Schrep," asked him to fix mobile. (Schroepfer became CTO in 2013.)

It certainly needed fixing, because the Faceweb-created apps were abysmal. The highly touted ability of HTML5 to work as smoothly as a native app was just wrong. Each conversion of a page view from Faceweb to the actual device slowed down performance. The pages stuttered as you scrolled down. And the flagship Facebook feature, the News Feed, couldn't work at all.

"In 2011, you could not read a story about Facebook that did not include the words *terrible mobile app*," says Ondrejka.

He took the job, with massive reservations. "Owning mobile at Facebook was like being the drummer for Spinal Tap," he says.

The first thing he did was along the lines of the old adage *When you're in a hole, stop digging.* He told the mobile team—about twenty people—to stop work. Go home and sleep, he told them, and they would meet next week to come up with a strategy. They gathered in a conference room, along with a few engineers and executives around the company who had strong opinions about the mobile problem. Some people were still devoted to Faceweb, others wanted a different web approach, and still others thought that Facebook should develop native apps for each device.

At the end of the meeting, Ondrejka decided that the best course was to start from scratch and write native apps for each system—as Joe Hewitt had done for iPhone until he stormed off because of his unhappiness with Apple's curation for the App Store. The next step was convincing Zuckerberg. "I went to Schrep and said we need to grab Mark," he says. "We went into a conference room and said, *We're screwed. We need to build a native app.*"

Zuckerberg agreed, and the engineering teams went into lockdown trying to create native mobile apps. Fortunately, the company by then had brought on some mobile-savvy engineers. Facebook had recently acquired a few companies with engineering talent in iOS and Android: a small iOS start-up called Push Pop Press and another group creating a messaging system called Beluga.

Ondrejka made sure the team began recruiting people versed in writing mobile apps. Facebook also began schooling its legacy engineers, setting up a three-week course in mobile engineering. Hundreds of its employees would take the training.

A trickier problem was Zuckerberg himself. He hadn't yet reached thirty, but the technology he grew up on was no longer ascendant, and he had to understand the dynamics of a new one. After all, he would make the ultimate call on the new apps. "I went to him and said, *One of the problems is that* you *don't understand native development. You make a thousand decisions a day, and they're wrong for native,*" says Ondrejka. So the new mobile team started training Zuckerberg, showing him what was different in design, in product development, and in the economics of the mobile ecosystem. One lesson Zuckerberg had to relearn was the cost of a mistake. "Done is better than perfect" doesn't work so well when your version 1 keeps crashing and you have to wait for Apple's approval process to push out your bug fix.

Zuckerberg is a legendarily fast learner, and he soon was asking smart questions that stumped the mobile-savvy engineers. "We're like, *Great, your brain is now in the right place,*" says Ondrejka.

Over the next few months, the Apple and Android teams began crafting native apps. They didn't have to throw Faceweb out entirely: there were some functions where it worked fine, like managing friend lists, and those could be written once and used on different mobile systems. But it was painfully obvious that others could only really be done well on native. Prime among those was the News Feed. It was Facebook's technological tour de force: every time you opened it you got a freshly sorted, uniquely personalized, up-to-the-minute stream of stories. A browser-based technology like Faceweb couldn't handle News Feed on a phone, with its relatively weak processing power and unreliable connections.

Meanwhile, Zuckerberg was making his own statement. One day a team came to the fishbowl for a Zuck review of some designs. *Where are the mobile specs?* he asked them. There weren't any. So Zuckerberg tossed the team out of the office. New rule: *Nobody comes into my office without mobile designs.* No one made that mistake again.

In fact, the whole company went self-consciously mobile. "A lot of us really stopped using our laptops entirely," says CTO Bret Taylor.

The ambitious, and unrealistic, goal was to have native apps ready in February 2012. So it was a victory that by March Zuckerberg had a prototype where News Feed was running smoothly.

Ondrejka's initiative was working. Paradigm shifts in technology like the mobile migration had killed great tech companies: Facebook was making the right moves to survive this one. Then he learned about a secret internal project that could put Facebook on an entirely different track.

Facebook was planning to take on Apple and Google with its *own* phone and operating system.

Chamath Palihapitiya had gotten bored. Growth, in his view, had been solved. Leading the team was no longer the intellectual challenge he thrived on. He was also happiest as the underdog, the dissident, the unpredictable person in the back of the room launching spitballs.

He had been following Facebook's struggles with mobile. But it wasn't the lack of great apps that worried him. He saw the evolving mobile ecosystem itself as an existential threat to Facebook. In order to gain primacy in the digital world, he believed, you had to control your own mobile operating system, or you would be a pawn of those who did. Only Apple and Google had significant operating systems.

Palihapitiya felt there was only one solution: Facebook should build its own smartphone. It would not be easy for Facebook to break into that exclusive club. But the number one thing people wanted to do with their phones was . . . go to Facebook. So why not build a mobile operating system built around people? Specifically a mobile device where Facebook was the center of all activities.

Palihapitiya was a master of persuasion. He got Zuckerberg's approval and began recruiting his team. He would take someone to lunch and tell them they were wasting their time at their current post—or, at the least, that it wasn't as important as what he was working on. Then he would tell them about the phone. One of his targets recalls reacting with puzzlement. *Why would we do that? It seems like a terrible idea! We're not good at hardware. We've NEVER been good at hardware.*

Despite those reservations—which turned out to be prophetic—those engineers would join the team, after Palihapitiya launched a verbal denial-of-service attack on their objections.

Molly Graham agreed to be product manager, and one of the company's best designers, Matt Kale, also signed up. But the big score in Palihapitiya's recruiting effort was Joe Hewitt. When Palihapitiya pitched him, he was skeptical about the prospects of success but thought it would be a good way to run the clock out before he vested. Besides, he enjoyed Palihapitiya, who had always spoken up for him. "I liked Chamath's bombast and his boldness," says Hewitt.

The original code name—it would change multiple times—was GFK, named after a kung fu film villain called Ghost Face Killer whose name was adapted by a Wu-Tang Clan member known as Ghostface Killah. Palihapitiya insisted on total secrecy for the project, taking his inspiration from Amazon's top secret Skunk Works team that built the Kindle. He moved his squad out of 1601 California and into the second floor of an unmarked building down the street. It even had a separate badging system from Facebook's. When Facebookers asked about the rumors, the company denied the truth. "It was the first time I recall Facebook lying internally," says Ezra Callahan.

According to a member of the team, Palihapitiya had an obsession with Steve Jobs and wanted to surpass him—destroy him—by building an even more beautiful phone. Palihapitiya's equivalent of Jony Ive, Jobs's design ace, was Yves Béhar, a much-admired Silicon Valley designer, who contracted to create the look of the hardware. Béhar sketched out a sleek device with an unusual groove in the curved surface, where one could scroll using a thumb.

To provide the microprocessor, Facebook hooked up with a logical partner: Intel. The chip giant had made one of the greatest blunders in its history by missing the first generation of smartphones—both Apple and Android used a rival chip—and apparently saw the Facebook phone as a way to mitigate, if not reverse, that failure.

Intel also had a lot of interesting technology that it was willing to share with Facebook, including an innovative touch sensor that would both unlock the phone and scroll in a single movement. It worked almost like a trigger in a game controller. The way it was configured, though, only right-handed people could use it. "We decided we didn't care about left-handed people," says one person on the team.

The software, written in Hewitt's artisanal programming language that Apple had rejected, was designed around communication with your Facebook contacts. The idea was that the Facebook phone would be so tied to one's social graph and interests that it would be inseparable from the person themselves. As soon as you turned it on, it would present a list of potential activities based on who you were and what your friends were up to. If some random stranger called you, the phone might not ring. But it might shriek at top volume when a friend called or texted with important personal news, like an engagement, a new baby, or a photo of truffle pizza. When you wanted to communicate with a friend, you would just express that, and the phone would figure out the best way to contact that person, maybe even by checking your friend's calendar and location. If she was in a meeting, for instance, it would text her. When you shopped, it would suggest options based on your Likes. If you went to a friend's birthday party, the photos you took would instantly be posted on Facebook. (Palihapitiya notes that the design specs did specify easy ways the user could tinker with privacy settings to rein in excessive sharing.)

Working with Foxconn—the humongous Taiwanese manufacturer that also builds the iPhone—Facebook built a prototype. But as the date for a go-ahead approached, Facebook became concerned about the investment it might take to actually build the product. Taking advantage of this hesitancy were internal foes of the product, chief among them Ondrejka. When he first heard of the project, he told Bret Taylor that it should be killed. Taylor told him to talk to Zuckerberg. "This began a four-month argument with Mark," says Ondrejka, who tried to convince his boss that since the mobile ecosystem was boiling down to two competing operating systems, Facebook had no need to create its own. Neither Google nor Apple was going to screw with Facebook, which was becoming the most popular app in the world. But Zuckerberg still liked the idea of having the Facebook phone as a hedge.

What resolved the issue was a compromise: instead of Facebook making its own phone, it would create an alternate version of the Android operating system to create a scaled-down GFK experience. Facebook would license this to other phone manufacturers. While it retained some aspects of the original idea—these "Facebook Home" devices would be running Facebook even in the "lock screen" mode, before the owner even picked up the device—it fell far short of a direct assault on the current mobile powerhouses. Palihapitiya left not just the phone team but Facebook itself, and started his venture capital company.

Facebook Home finally appeared in April 2013. HTC manufactured the first handset, with Samsung lined up to follow. "We want to make it so that we can turn as many phones as possible into Facebook phones," Zuckerberg told me shortly before it launched. But Facebook Home was a dud. While Facebook was by far the most popular mobile app by then, very few people wanted a phone where Facebook was operating even while the phone was sleeping. There was no second version.

FACEBOOK'S MOBILE CRISIS came at the worst possible time. At the same moment that its future was endangered by a paradigm shift in the way people used technology, Facebook was seeking to float its value in the stock market.

"Would I advise a company to go public in the middle of that shift?" says Sheryl Sandberg now. "No! It would've been way better if we had gone public two years before or two years after."

But Facebook did not really have a choice. As early as 2007 journalists and analysts started asking, *When will Facebook have its IPO?* Every year the questions became more insistent. While Zuckerberg believed in moving fast and breaking things as far as individual products were concerned, he maintained a parallel mindset with a long horizon, patiently plotting Facebook's course for five years, ten years. He spoke often about delaying profit in favor of growth, and complained that the very smell of an IPO might draw people "who want to join the company because they think that they're going to make money very quickly." He did not look forward to the exercise demanded of every CEO: justifying results on a quarterly basis.

Zuckerberg had stalled as long as he could. Since the 2007 Microsoft investment, Facebook had raised big private funding rounds, notably a $200 million influx of cash by Russian mogul Yuri Milner in 2009. Almost every one of those late-round investments had been ridiculed as an overreach. Yet each had proved wise.

But it was always a question of when, not if, there would be an IPO, and there was only so much Zuckerberg could do to forestall the inevitable. In 2010, Facebook quietly began making moves to smooth its shift to a public company. It beefed up its board of directors to include—besides existing members Don Graham, Netscape co-founder Marc Andreessen, and Peter Thiel—Netflix CEO Reed Hastings and Erskine Bowles, who had been Bill Clinton's chief of staff. Bowles made a deal with Zuckerberg: he'd chair the board's audit committee if Zuckerberg would read a stack of finance books. *Look, you'll be a CEO of a public company—you have to understand this,* he said when he dumped the books on Zuckerberg.

In the fall of 2011, Facebook's CFO, David Ebersman, who had previously held the post at Genentech, began interviewing some of the banks. The process had quietly begun for what would be the biggest tech IPO ever.

Facebook, unsurprisingly, chose Morgan Stanley to lead the effort. Its top banker, Michael Grimes, had been snaring the juiciest offerings with regularity. His office was not in New York City or even the financial district in San Francisco but on Sand Hill Road in Menlo Park, where the big VC firms made their bets. He had worked on the Google IPO and recently on LinkedIn's. And he was friendly with Sheryl Sandberg. (As usual, other investment banks and advisers joined in, including Goldman Sachs and JPMorgan.)

Zuckerberg had firm ideas about the way Facebook's stock structure would operate. The key factor was keeping himself in control, presumably forever, by creating two levels of shareholders, with the top level—the one where he had the overwhelming majority of shares—given dominance in any vote. It was similar to schemes that let family-owned newspaper companies, like that of his mentor Don Graham, control the company for decades while owning a minority of the company. It had also been adopted by Larry Page and Sergey Brin of Google. But Facebook's plan topped theirs in how much control a single founder had. Holding 56 percent of the voting shares, Zuckerberg himself would have veto power over anything that other shareholders, or the board of directors, might order.

Likewise, he mimicked the Google guys when he personally wrote a letter to shareholders in the S-1 prospectus that laid out the terms of the offering when it was announced on February 1, 2012. (He wrote the first draft on his phone. Mobile first!) "Facebook was not originally created to be a company," he began. "It was built to accomplish a social mission—to make the world more open and connected." And then he expounded at length on the five values that he'd had Molly Graham and Lori Goler help him codify the year before. It was almost as if he were reading off posters on the walls of Facebook's offices for his bullet points. (Yes, he did write "Move Fast and Break Things" in the formal offering for shareholders to invest in his company.)

Zuckerberg also called Facebook's operating methodology "The Hacker Way." He admitted that the word had "an unfairly negative connotation," but asserted, "The vast majority of hackers I've met tend to be idealistic people who want to have a positive impact on the world." The Hacker Way, he explained "is an approach to building that involves continuous improvement and iteration. Hackers believe that something can always be better, and that nothing is ever complete. They just have to go fix it—often in the face of people who say it's impossible or are content with the status quo."

None of this geek talk dampened the anticipation for the IPO. Something else did—the company's sluggish response to the mobile wave.

Facebook itself stated it directly in the S-1: "We do not currently directly generate any meaningful revenue from the use of Facebook mobile products, and our ability to do so successfully is unproven." About half of Facebook users were now accessing the service on their phones. Since Facebook had yet to monetize its mobile applications, that meant that the company was losing opportunities to serve ads. If that trend continued, the company's revenues would tank.

After the S-1 was published, Facebook found even more people were using mobile, dragging down its financial performance. "Everything here is going really badly," Zuckerberg wrote in a text to Chan that was later exposed in a lawsuit.

Convening one night in a New York City hotel room, Zuckerberg, Sandberg, and Ebersman were actually considering whether to scrap the IPO. "We are making a decision tonight," Zuckerberg texted Chan. Later he reported that the offering would go forward. "Yay," she responded. Yet its financial team was in a quandary about how to deal with the revenue drop. If the company didn't share the information, it might be open to suits or regulatory sanctions. So Ebersman and Grimes decided it would have to amend the prospectus to include five more sentences cautioning of this recent trend. They also felt it necessary to contact the key analysts and tell them individually. It risked the appearance that they were giving information to Wall Street insiders while the public was in the dark.

Because of SEC rules segregating investment bankers from analysts in such situations, Grimes would have to maintain arm's length from the process. This was tricky, since Grimes was at the center of the planning, even writing the script that Facebook's treasurer read to the analysts. So when the treasurer made the calls from their hotel room command center, Grimes went down the hall and sat on the floor. (The SEC was not amused, and cited this incident in a complaint that cost Morgan Stanley $5 million.)

Those calls came a little more than a week before the IPO was slated to take place on May 18. By then news outlets had something critical to say almost every day about the upcoming event. Critics were questioning Zuckerberg's maturity, noting that he wore a hoodie to the "road show" to entice investors. Despite the pushback, Facebook raised its opening price range from $28–$35 per share to $35–$38 per share. That would make it a $100 billion company, worth 100 times its previous year's profit, which seemed a stretch to critics. On May 15, another blow: General Motors told *The Wall Street Journal* that it no longer believed that Facebook ads were effective, and would be scaling back its spending. Though GM was not one of Facebook's bigger advertisers, the claim that Facebook might not be an effective use of ad dollars cast further doubt on its long-term prospects.

So the tea leaves were not well arranged as the day of the IPO approached.

Zuckerberg and his top executives did not make the traditional journey to New York City for the bell-ringing ceremony on the NASDAQ floor, commemorating the moment when company shares go on sale. Instead, employees gathered in the mall of Facebook's new Menlo Park campus, where Zuckerberg would ring the bell remotely. It was just as well he stayed in California.

At the moment the stock was about go on sale, NASDAQ, which prides itself on being the tech-savvy alternative to its more prestigious rival, the New York Stock Exchange, had a computer meltdown. Despite several test runs in the previous few days, the volume of requests overwhelmed its system. NASDAQ postponed the opening, but even when the stock went on sale more than an hour late—to hugs and whoops of joy in Menlo Park—transactions were still delayed. That meant that smaller investors, who had reserved shares at the opening price, were unable to confirm the trades, or to bail out as the stock price took a dive.

Stories abounded of humble investors, excited about participating in the next sure thing, putting in orders and not hearing whether they went through. Typical was the tale of the widow who dumped half her life savings into Facebook and tried, unsuccessfully, to cancel as the order stayed in limbo. As the stock dove, so did her retirement hopes.

With the benefit of hindsight, the IT mess wasn't necessarily all that terrible. If those small stock-pickers had simply held the stock and endured their short-term losses, their investments would have multiplied several times. But investing is a risk because there is no benefit of hindsight. So one cannot easily dismiss the complaint of that widow, who finally wound up selling at a deep loss.

The loss came because when Facebook's stock finally got its listing, investors balked. The stock opened at Facebook's optimistic $38 price, and closed the day barely up. This was only because Facebook's underwriters themselves bought back shares before the market's close to avoid ending the day at a loss, a technique known as "the greenshoe option."

Without similar tactics to prop up the price thereafter, Facebook's stock sagged. After a week the price was down to $32. By September, a twenty-dollar bill could buy you a share of Facebook, and you'd get back two bucks and change.

Even early investor Reid Hoffman, who made hundreds of millions of dollars from his $37,500 investment, would call Facebook's IPO "an egregious fuck-up." In the next few months, courthouse steps in multiple jurisdictions were pounded by the feet of aggrieved investors filing suit against Facebook, NASDAQ, and the underwriters. Over the next few years, Facebook, bankers, and NASDAQ would pay millions of dollars to settle them.

Two things happened immediately after the IPO debacle.

The first was celebratory. On the day after the event, Zuckerberg had invited about one hundred people close to him to attend a backyard gathering, ostensibly to celebrate Priscilla Chan's graduation from medical school and his own twenty-eighth birthday. Those who sensed misdirection had their suspicions confirmed when Zuckerberg showed up in a suit. Later that day he changed his relationship status on Facebook to "married." The status change garnered more than a million Likes.

By exchanging vows, Zuckerberg was ignoring a warning from a fellow billionaire on CNBC, made on the eve of the IPO. "They get married, and then for some reason over the next couple of years they get divorced and then she sues him for $10 billion and she hits the jackpot," said Donald Trump, who had at that point met neither party.

The second thing Zuckerberg did was address the reason Facebook's stock had crashed. The belated move to native apps assured Facebook that its service would run well on the technology that was on a path to ubiquity. The native apps were well received. When people used Apple or Android phones, an average 20 percent of their time was spent on Facebook. (The nearest rivals had 3 percent.) But Facebook had yet to create products that would make money on mobile.

The dive in the stock price was affecting morale. Dan Rose decided to give a pep talk at an all-hands. He recounted his experience at Amazon in the dot-com bust. The stock had crashed from $120 a share to something like $6. Rose's personal plans to buy a house for his family had been shelved. Some people were leaving the company. But Amazon and its leader, Jeff Bezos, hung in there and now ruled the commerce world. Same deal with Facebook. *The world doesn't know it,* Rose said, *but we know what we're doing.* Mobile won't kill Facebook—it will bring it to new heights. People use Facebook more on mobile. And despite the tiny screens on phones, Facebook *will* make money from it. *The product is built for mobile,* he told his colleagues, *We just haven't built the products yet.*

Left unspoken was the flip side of the situation. If it didn't build those products, Facebook would die.

AT CRISIS TIME, Zuckerberg tends to lean on those he knows well and trusts. His choice to head the mobile advertising product team, the effort that would reverse the company's fortunes, was someone he'd known for years, someone who'd literally imprinted a word on his body that signified steadfastness.

Boz.

Not long after the IPO, Zuckerberg and Andrew Bosworth took a walk around the old Sun campus, past the barbecue shack, under the giant Hacker sign, in front of the Analog Research Lab. For the past few months, Boz had been in what he called a "lame-duck period," floating from one group to another to work on various projects. He was about to take the sabbatical Facebook offers after six years of service. When Zuckerberg suggested that he head the ads engineering team, Bosworth thought the idea was terrible. Advertising just wasn't who he was.

Zuckerberg persisted, insisting that if Boz took the post, amazing things could happen. *Boz*, he said, *I think you can unlock four billion-dollar businesses in the next six month*s. He ticked them off. One was a mobile-optimized ad product. Another was a different "premium" ad product. The two others he mentioned are lost to history—neither Zuckerberg nor Boz can recall them.

Bosworth again objected that he wasn't the best person, but Zuckerberg said he absolutely needed someone intimately familiar with the News Feed and Facebook's consumer products, who could come up with solutions based on the addictions of Facebook's current 800 million users. It wouldn't have to be forever. Could Boz commit to six months?

Bosworth agreed. In part, he was moved that Zuckerberg was asking him to take such a key role because by that point, Bosworth had alienated a lot of people with his aggressive style. Kate Losse wrote in her book that he would "jokingly threaten other engineers that he would punch them in the face if they displeased him." At one point he had to make a public apology. "I could be a bit fiery at times and not always the best collaborator," he admits.

He asked to site his team inside the mobile group, not the ad organization. Every week he gathered a small team whose responsibility was to develop mobile-oriented ad products. It included Mark Rabkin, an engineer named Will Cathcart, and Margaret Stewart, a designer whom Facebook had lured from Google, where she had revamped the look of

YouTube. They would meet in a conference room in Building 16 of the old Sun campus. Essentially they were designing Facebook's business from that point. They called their group the Cabal.

The first thing the Cabal came up with was a short-term means to make money on phones. A product manager had an idea based on the Growth team's seemingly omniscient People You May Know feature. PYMK already had a prime location in the News Feed, and people couldn't resist peeking at it to see whom Facebook was recommending (and sometimes freaking out that Facebook had made those obscure connections between them and those faces). Why not mix in some ads among those faces? Some people might want to pay to get their pages inserted in that physiognomic carousel. So Facebook did a product called (sponsored) Pages You May Like. "It was one of the few areas where we could put an ad on mobile," says Bosworth.

The downside was that basically this new product was a scheme to sell Likes, which had dangers for Facebook. If too many of people's Likes had been manipulated, instead of reflecting organic behavior, Likes would be of limited value to advertisers. Facebook had to take the risk. In a few months, the lockdown preventing Facebook employees from selling their stock would end. If the company's revenues were still low when this happened, the stock might wind up even worse than its dismal post-IPO level. The paid product Pages You May Like rolled out that August.

The Cabal next took on what would be a longer-term solution: ads in the News Feed. In the months after Beacon, ads had been purged from Facebook's flagship feature, shunted to their traditional place on the side. But on mobile displays, there *was* no side rail—the screens were too small.

"We built a really nice business on desktop where we didn't have to worry about ads in News Feed because we just had this column of ads," says Zuckerberg. "That business scaled to billions of dollars and we had a separate team that was off to the side; they didn't have to talk to the

News Feed team and it was wonderful. But in mobile, that's not the reality. There's no room for a right-hand column of ads on mobile."

To that point, Zuckerberg had been adamant that "sponsored stories"—ads that looked indistinguishable from the organic stories posted by or about one's friends—must be socially generated. If Pepsi, General Motors, or your local nail parlor wanted to get in your News Feed, it could only do so if one of your friends liked their page, and thus was implicitly recommending it to you (a dubious favor). But Bosworth and others felt that this approach was too limited. The idea of sponsored stories was that people would get ads in low volumes but high quality. But because there were too few ads that could be delivered to a given person—your "friends" could only like so many pages!—the inventory was low. Because there were so few ads to choose from, chances were the actual quality of the available ads wasn't going to be relevant to the recipient.

If advertisers could place stories directly in the feed, targeting them to their specified audience, the quality would rise.

"I love it when people see ads and don't realize they're ads," says Margaret Stewart, who joined Facebook the day before the IPO. "When an ad is great, it's just a great piece of content and you don't necessarily categorize it as an ad, which is usually shorthand for something that's not relevant or valuable."

Zuckerberg resonated with that sentiment. From the early days of the product, he didn't want unwelcome advertising to break the spell of his product. Now, with the data that Facebook had, and the expertise of a product team Bosworth recruited, he felt that Facebook could start to develop ads that were as welcome as organic posts. He gave his thumbs-up to ads in the News Feed, and the mobile team set about implementing them.

"I think it was kind of a cop-out for the first five years at the company to just have the ads on the side, because we weren't tackling the hard problems of figuring out how to actually make the ads good enough to integrate with the user experience," he told me after the switch. Not that he had a choice—there was nowhere else to put ads on Facebook's phone app.

The ads would be intermingled in the feed to sell clothes for Macy's, sundries for P&G, albums for Warner Music, and the wares of millions of small businesses using Facebook's self-service system, where only algorithms oversaw the process. They worked very much like any other post—they could be "liked" and shared in the same way as non-paid posts. Advertisers would pay only for the initial post, and further distribution would be free. Thus a modest ad placement would have the possibility, however remote, to get wide distribution.

Mobile ads in the News Feed were wildly successful, and would push Facebook's annual revenue into the realms of tens of billions.

Of course, no one envisioned that the creative dynamics of News Feed ads might one day be exploited by state propagandists to affect the election of an American president.

Like 2006, a year of controversy, crisis, and stress, 2012 laid foundations for new dimensions of success for Facebook. The smartphone wave that threatened to engulf the company turned out to be its biggest boost since the News Feed. While Facebook Home had been a flop, its native apps for iOS and Android were by far the most popular on each platform. As for Facebook's own Platform, thousands of developers were happy to use Facebook Connect as a way for people to sign in to the apps they created. This allowed Facebook to gather even more data for the digital warehouses it was building around the world—billion-dollar data centers in Oregon, Texas, and North Carolina.

In the next few years, rolling on this momentum, it looked like nothing could stop Facebook. MySpace was a memory. The money was pouring in. The company logged 1 billion people, and then 2 billion. Investors noticed: in the summer of 2013, FB's price finally matched the $38 of the original offering. From there, it ascended steadily. Eventually Facebook's valuation would top $500 billion, ranking Mark Zuckerberg's dorm-room creation among the top ten companies in the world by that measure.

"The people who invested in our IPO did very well if they held on," says Sandberg. "And we were very honest about it. We said, *We have no mobile ad revenue, so we are going to have to build it.* It took us time. We said we'd make a shift to mobile. We did make a shift to mobile."

What Sandberg and Zuckerberg did not see in 2012 was that the compromises made in the last half dozen years—among them the disregard for privacy, the data bartering with developers, the reckless international expansion, and the countless concessions it made to its hunger for growth—had planted the seeds for a series of explosions that would shake not only Facebook but the entire tech industry.

And the first big bomb would go off on Election Day 2016.

13

Buying the Future

KEVIN SYSTROM RECALLS the moment he began dreaming of becoming a founder. It was 2005, and he was talking to Mark Zuckerberg at the original Facebook office in Palo Alto. Systrom, who grew up in a well-off suburb of Boston and entered Stanford after attending boarding school, was interviewing for a job but not really sure if he wanted to leave school. He and Zuckerberg went onto the roof deck, bending down and climbing out the window to get there. This required a deep bend for Systrom, who was six feet five inches tall. (Zuckerberg, at five-seven, got through the window easier.) Over beers, Zuckerberg talked about how starting a company was the hardest thing ever. But all Systrom could think of was how cool it was to build something like Facebook. *I'd really like to do this someday,* he thought.

Systrom didn't take the job. But he kept in touch with Zuckerberg as he methodically climbed the rungs of the Silicon Valley ladder. First came an internship at a podcasting company called Odeo, headed by well-known entrepreneur Evan Williams. His starting week was the same as that of an Odeo employee named Jack Dorsey, who would, when the podcasting idea fell apart, concoct an alternative that became Twitter. Then Systrom took a job at Google. He was hoping to get hired as an associate product manager, which meant inclusion in a fast-track program of people favored for big things at the company. But APMs needed a computer-science degree, so Systrom got into a consolation-prize program called Associate Product *Marketing* Managers. "It was the lowest rung of marketing," he says. "But I wanted to be there."

He learned a lot, but felt drawn to start-ups; he joined one co-founded by ex-Googlers, a site where you got travel recommendations from your social network. Meanwhile he kept thinking of ideas for his own company. It was early 2010, and he knew it had to be centered on those smartphones people carried with them and constantly consulted.

In his spare time, he began writing code for a social app he would call Burbn, after his favorite adult beverage. It was a way to signal to friends what you were up to and where you were. It wasn't an original idea. Twitter already let you report your status by text, and Burbn's flagship function, "checking in" to the bar or restaurant or zoo you were visiting at that moment, was already a popular feature of an app called Foursquare. Still, Systrom wound up getting $500,000 in seed money, half of it from the Valley's hottest VC firm, Andreessen Horowitz. The money men were banking as much on the founder as the idea; Systrom ticked the boxes that VCs look for—Stanford degree, ex-Googler, a quiet intensity that impressed the partners at the pitch meeting.

Systrom quit the social-travel site. Not long afterward, Facebook bought it. Once again, Systrom had just missed working for Mark Zuckerberg.

Marc Andreessen suggested that Systrom find a software-guru partner. That turned out to be Mike Krieger, a Brazil-born engineer who'd majored in Stanford's Symbolic Systems program. At first, Krieger wasn't blown away by Systrom's description of Burbn, but after he agreed to join the beta test, he became intrigued, if not hooked. It wasn't just checking in to a place that drew him in, but the ability to add rich media—photos, video—to what his friends saw. Usually he'd watch movies on his Caltrain rides—he was making his way through the AFI Top 100 films—but he found himself taking pictures out the window and sharing them on Burbn. He joined up with Systrom.

Over the next few weeks, the Burbn beta testers became a small but loyal community. Underline *small*. "It wasn't exactly setting the world on fire," Krieger would later write in an account of Instagram's beginnings. "Our attempts at explaining what we were building were often met with blank stares, and we peaked at around 1,000 users." The founders noted that photo sharing, which was envisioned as a slideshow in the app, seemed to be the most popular feature. The most popular photo-sharing sites, like Flickr or even Facebook, displayed photos as if they were objects in a gallery or scrapbook. On Burbn, people were using them as a form of communication. Systrom and Krieger decided to rewrite Burbn to concentrate on that aspect. The app, written for the iPhone, would open to a camera, ready to capture and transmit a visual signal to the world that showed not just where you were and who you were with, but who you were. It would be primal, pre-linguistic, and lend itself to endless creativity. The photos would appear in a feed, a constant stream shared by people you chose to "follow." It also nudged users into a performance mode, as by default any user could see your photos. It was much more Twitter-like than Facebook-ish.

Shifting Burbn into a camera-first app delighted Systrom. He'd always loved photography. He also had an affinity for old, funky things. He was the kind of guy who'd buy an old Victrola and display it as a piece of art. He was also a craftsman at heart; his standards for detail were Jobsian, without the snide insults to those who dared give him work that fell short. He and Krieger would spend hours on the tiniest detail, like getting the rounded corners right on the camera icon.

It was the antithesis of "Move Fast and Break Things."

One of the key breakthroughs on the revamped app came when Systrom was on a Mexican vacation with his girlfriend, Nicole. To his dismay, she told him she would be reluctant to use the product he was building 24/7 because she'd find it hard to match the quality of photos a certain friend of hers took. Systrom told her they looked good because the friend used filters to make the images more intriguing. So Nicole suggested maybe he should use filters in *his* product. Systrom would later marry Nicole.

He quickly added a filter to the app and used it the next day when the couple were at a taco stand to take a picture of a puppy with Nicole's flip-flopped foot in the corner. The colors had an eye-catching burnt aspect. That was the first picture he posted to the beta version of Burbn's successor, which came to be called Instagram, a portmanteau of "instant" and "telegram." Eventually he and Krieger would create a set of different filters, letting people transform their pictures into virtual tintypes. Later, Instagram snobs would let people know they were past that by eschewing such cosmetics, and branding their photos with the hashtag #nofilter. Instagram wasn't the first app to use digital filters. But it was unique in making filtered photos a form of expression in a social network. In keeping with the company's mission, the filters were less about what viewers were looking at than what they signified about the person uploading the image.

After an all-night programming binge on October 6, 2010, Instagram finally went live. The response was immediate and startling. There was already a built-in demand, generated in part by some notable Valley people who were raving about the beta version. In fact, the small beta group testing Instagram was already producing more content than the entire population of Burbn had. None was more enthusiastic than Jack Dorsey. Twitter's creator began using it compulsively, tweeting its praises to his huge following.

The clamor threatened to overwhelm the one server cage they were renting, requiring Systrom to put in an emergency call to one of the smartest engineers he knew, Adam D'Angelo, who spent a half hour on the phone with him talking through solutions for the deluge. Twenty-five thousand people signed up the first day. They wound up making an emergency transfer to Amazon's cloud service to handle the load.

"I don't know how big this is going to be," Systrom said to Krieger, "but I think there's something here."

Within weeks, Instagram had hundreds of thousands of users, who began using the app's simple rules to create innovative ways of sharing. Using the hashtag function, they would label a photo based on a concept (like #circlesinsquares), and an instant themed gallery of shots based on that topic would emerge. The idea of taking pictures of the food in front of you took hold, especially in high-end restaurants with risk-taking chefs or with interesting street food. In the aggregate, Instagram began a prosaic visual diary of people's peregrinations and observations.

When pop singer Justin Bieber posted a candid shot, Instagram quickly became an important promotional tool for celebrities—drawing attention away from the previous go-to vanity amplifier, Twitter. This didn't bother Jack Dorsey, who kept Instagramming away. Eventually, Instagram became the go-to branding engine of a generation of pop stars, fashion models, and reality-TV insta-figures.

Barely six months old in February 2011, Instagram did its "A" funding round, led by Benchmark Capital. The partner doing the deal was Matt Cohler. Smaller participants included Dorsey and Adam D'Angelo. They invested at a valuation of $20 million.

EARLY ON, MARK Zuckerberg noticed that Instagram was sharing photos in a new way. Since photo sharing was the most popular feature on Facebook, he realized that this tiny start-up was doing something that Facebook wasn't. He saw Systrom a number of times over the next couple of years, making it clear that Facebook had interest in the app. But so did Twitter. Instagram fanatic Jack Dorsey had returned to the company after its board installed Dick Costolo as CEO.

By 2012, Instagram was growing exponentially and needed more funding. The company had no revenue, having adopted the standard practice of focusing on product and growth without bothering to come up with a business plan. Instagram's proposed valuation for the new funding was set at $500 million, and they were having no problem finding investors. The round was shaping up for Sequoia to lead, with other investors chipping in, including Thrive Capital, a VC firm run by Josh Kushner in New York City.

Neither Twitter nor Facebook wanted that to happen.

This was the first test of Zuckerberg and Facebook in trying to acquire a must-have app. Zuckerberg prided himself on an ability to see beyond the horizon. He looked for anything—whether it was a company or a shift in technology—that would threaten his plans. When Google launched its own social network product in 2011, Zuckerberg put the entire company on lockdown for weeks, keeping the cafeterias open on weekends. In an all-hands address, he invoked one of his ancient heroes, Cato the Elder, who ended his own speeches by crying, *Cathago delenda est.* Carthage must be destroyed! The Analog Research Lab fired up the presses to churn out posters with the Latin phrase. (Zuckerberg didn't need to worry: Google Plus was a flop.)

Now, a year later, he understood that if someone else snatched the future of social photo sharing, Facebook would suffer. The best course would be for Facebook to buy Instagram.

But Twitter had the inside track.

Systrom's close connection with Dorsey led Instagram to the brink of a deal that would value Instagram at well beyond the half a billion that the investors in the funding round would have agreed to. Costolo got the board to sign off and thought it would happen. Systrom and Krieger were about to head off to South by Southwest and wanted to think about it first. "We couldn't get it over the finish line with them," says Costolo.

Dorsey especially had reason to worry that he and Costolo had missed their chance. When his own company, Twitter, had gone to SXSW in 2007, the hipsters at the conference swooned over it, ending fears that the app could not make it on its own. Indeed, the Instagram guys were treated like rock stars in Austin, and after the conference, Systrom told Dorsey and Costolo that he felt Instagram had a real shot as an independent, and was going to do the Sequoia round. Dorsey was disappointed but wished his protégé luck, and told him that if circumstances changed, he was up for renewing the talks.

A few days later, Dorsey got on a bus to go to his job at Square, a second company he co-founded. He was the only passenger, which seemed a perfect opportunity to fire up Instagram, which he used constantly, and snap a photo. "A simple morning pleasure: an empty bus," he wrote on what would be his last Instagram photo ever. Because as soon as he got to work, he saw Mark Zuckerberg's post that Facebook was buying Instagram for $1 billion. Costolo was in Tokyo at the time and was frustrated that Twitter had no chance to respond. Instagram, he felt, was worth draining all of Twitter's cash and then some. "We didn't get a heads-up, or I would have obviously borrowed the money from a bank," he says.

What happened? Mark Zuckerberg. When Systrom and Krieger told him Instagram was not for sale, he did not wish them well and tell them to send a postcard from Austin. He summoned Systrom to his house in Palo Alto to give him an offer he could not refuse.

Facebook had bought about twenty companies by then, but the price tags had generally been in the low millions. The biggest, a mobile-app development company in 2011, had been $70 million. Instagram would far eclipse that level of deal-making. But a couple of years of experience in buying companies had given Zuckerberg a playbook for big acquisitions, which he would put to the test with Instagram.

The first principle was direct involvement from Zuckerberg himself, flattering and smothering his quarry with attention. The second was a promise of independence. Co-founders were promised that they would continue to make the creative decisions for their companies—it was their genius that made them so attractive to Facebook!—with Facebook providing all the dull stuff like infrastructure, security, office space, and marketing.

Facebook also had a secret weapon in its arsenal. A year earlier Facebook had hired a head of corporate development who had become its ace closer. His name was Amin Zoufonoun, and he had come to Facebook from Google's business development team. Previously, he had been an intellectual-property lawyer. That compressed résumé does not do justice to Zoufonoun's most-interesting-man-in-the-world bona fides. He was born in Iran to a fabled musical family; his father, Ostad, was a well-known violinist, and himself the heir of a legendary sitar player. The family fled Iran just before the revolution that overthrew the shah, and moved to the Bay Area, where the Zoufonoun home became a center of

Persian music. The family played together as the Zoufonoun Ensemble, with Amin playing the sitar. Amin went to law school while other family members pursued music, but he continued to perform with his family.

After a stint as an IP lawyer for a mobile firm, Zoufonoun joined a pre-IPO Google in 2003 to master a different kind of music: the seductive melodies of a large company well-schooled in wooing founders to sell their start-ups. Broodingly handsome and a deft negotiator, he lubricated the purchase of a number of key acquisitions. He was the second most experienced mergers and acquisitions (M&A) executive at Google when Facebook poached him as part of the continuing talent war between the two companies.

Zoufonoun was unflappable, a master of details, and cool as linen. One founder who was the target of an acquisition vowed that if his company remained independent and got as big as Facebook, the first thing he would do is hire someone like Zoufonoun to buy other companies.

But Dan Rose, who as head of partnerships was Zoufonoun's boss, clarifies that while the M&A head was indeed a killer, the credit for Facebook's game-changing acquisitions should rest almost exclusively with Mark Zuckerberg. "He identified these companies as having synergy and he identified the potential for them before anyone else and he personally led the charge to convince them that [Facebook] was the right home and that they would be able to achieve their vision and their mission for their companies better here than anywhere else, including on their own."

And that's what happened with Instagram. So Instagram was completing a round that would value the company at $500 million? Fine! Now here's a *billion* dollars!

Maybe the most unusual part of the acquisition was that Facebook was weeks away from its IPO. Now it was making its biggest acquisition by far, paying a billion dollars for a tiny company.

The sum was large enough to require a review from the FTC, and this was no exception. The procedure is to do a preliminary review and, if it appeared that the merger presented antitrust concerns or consumer harm, it would move to a second-stage review. One commissioner was sufficiently concerned about Instagram bolstering Facebook's current social-media dominance that he wanted to move to the second stage. But he could not get the majority of the five commissioners to join him, and that was that. (To be sure, it would have been a hard case to win, because of Instagram's zero revenues.)

After the deal was done, Instagram withdrew its support of Twitter, ending the popular practice of seamlessly cross-posting photos to both systems.

ZUCKERBERG HAD BARELY taken possession of Instagram when he perceived another threat. Yet another application, again largely based on photos, was taking off among teenagers and young adults. It had a few twists that hit like punches to Facebook's underbelly. For one thing, the photos were ephemeral—they disappeared after a few seconds, unable to haunt the users in decades to come. This and a few other features—like a nonintuitive interface that anyone over twenty-one had difficulty understanding—made the app, Snapchat, a love object for young people. Its co-founder CEO looked to be a rising star.

Evan Spiegel grew up privileged, his father a successful lawyer in LA. He attended the exclusive Crossroads School, and then Stanford, where he tooled around in a Cadillac Escalade. In his sophomore year, he finagled an invitation to an entrepreneurship course for second-year MBA students, fueling an existing desire to start his own company. The idea that made it happen was a one-sentence pitch shared in a dorm room in April 2010 by his buddy Reggie Brown: *What if you could send disappearing photos?*

Spiegel had the vision to turn that concept into a company. He had a penetrating view of the territory where technology and human behavior merged. He understood not only why Facebook became popular, but reasons why it was increasingly unloved. As a college student, he had witnessed firsthand how the News Feed had shifted from the latest dope on your friends to an avalanche of outside content. Though Facebook had once been more integral to Stanford campus life than beer, now people hardly used it.

With a third co-founder, Spiegel's best friend, Bobby Murphy, they began creating an app around their idea, called Picaboo. Its first year was eventful as the team iterated the product—and Spiegel and Murphy tossed out Brown. (There would later be litigation reminiscent of Eduardo Saverin's suit against Facebook, with a similar outcome—a lucrative settlement for the permanently exiled co-founder.) But by early 2012, the app, now called Snapchat, was beginning to take off. Its refreshing ephemerality made it addictive and intimate: without the weight of knowing you were establishing a permanent record, you could be silly, or unload a secret. (You could also send a nude selfie, but that aspect was always overestimated.)

Snapchat's success drew attention. "When Snapchat started out, I thought it seemed trivial—I was wrong," Chamath Palihapitiya, now a venture capitalist, gushed to *BusinessWeek*. "At worst they are the next-generation MTV. At best they are the next-generation Viacom."

So it was probably inevitable that Mark Zuckerberg would want to own Snapchat. On November 28, 2012, he emailed Spiegel with a baited hook. "Hey, Evan," he wrote, "I'm a big fan of what you're doing with Snapchat. I'd love to meet you and hear your vision about what you're thinking about it sometime. If you're up for it, let me know and we can go for a walk around Facebook HQ one afternoon."

The casual tone masked the seriousness and careful planning that had gone into the overture. Just as Zuckerberg had concluded with Instagram, Snapchat was a threat that would best be neutralized if he owned it. Then he could leverage Facebook's assets to make it grow more quickly.

Spiegel's reply topped Zuckerberg's insouciance by including a text emoji in the message. "Thanks :)," he wrote, "would be happy to meet. I'll let you know when I make it up to the Bay Area." In other words, *I'm not going to drop everything and meet with the great Facebook CEO.*

Zuckerberg had signed his email, "Mark." Spiegel didn't sign, a habit of people younger than Zuckerberg.

Point: Spiegel.

Zuckerberg chose to ignore the diss. In his next message he said he just happened to have an LA trip coming. He arranged to meet Spiegel at a location outside the office.

Spiegel had reason to be cautious. When Facebook bought Instagram earlier that year, most people were stunned at that billion-dollar price tag. Not Spiegel. He thought that the Instagram guys had made a catastrophic mistake. Sure, Facebook's infrastructure might make scaling the product easier. But he didn't respect Facebook's product sense.

At the meeting, Zuckerberg touted the benefits that would come from a merger. It would allow Spiegel and Murphy to grow Snapchat much more quickly with the rocket fuel of Facebook's infrastructure and expertise at global scaling. Facebook would take care of the annoying stuff, allowing them to concentrate on making a great product. And, of course, they would become rich. In addition to the selling price, they would be in line for huge stock bonuses that vested over time.

That was the carrot. Zuckerberg also had a stick. He shared that Facebook was working on a project the two Snapchat founders might be interested in seeing—a chat feature with disappearing messages! He said he was thinking of calling it Poke.

The Snapchat founders turned him down.

On December 21, Zuckerberg sent a message to Spiegel. "I hope you enjoy Poke," he wrote. That was the entire email.

Launching a copycat product after a rejected acquisition offer was a familiar Zuckerberg gambit by now. True, he had not offered to buy Adam D'Angelo's company, Quora, when he created Questions to compete with it. But two years earlier a spurned offer was indeed the impetus for a product meant to kill off Foursquare, a mobile app that used geolocation and game techniques to help people find things to do, and find each other.

Foursquare's GPS technology was superb, and Facebook had wanted to buy the company, especially since other tech companies were pursuing it as well. Its co-founder and CEO, Dennis Crowley, met with Zuckerberg multiple times, taking walks in Palo Alto and near Foursquare's headquarters in New York City's Union Square. Eventually, an offer came in the $120 million range. Crowley had prior experience in having his company sold—years earlier, Google had bought his earlier start-up and let it fester—and he was wary. "I wasn't sure if he wanted us because we were a real shiny thing—everyone was talking about us, we were going to be the next Twitter," he says. So Crowley decided to gauge Zuckerberg's interest on price—would he go up to $150 million? The companies were still dickering when Crowley conferred with his team and decided that instead of selling, they would make a go of it as an independent company. Crowley called Zuckerberg to tell him. "It stuck with me how gracious he was," says Crowley.

The two CEOs kept in touch, with Crowley visiting the Facebook campus occasionally. At one point, Zuckerberg put him together with some Facebook engineers. By that time Foursquare had advanced its technology to work with an array of different sources—Wi-Fi, cellular, GPS—to pinpoint location, and Facebook was struggling with location technology. Crowley generously explained to the Facebook crew how Foursquare operated. Not long afterward, he got a heads-up that Facebook was doing its own location app. "It was like, *Okay, if Foursquare doesn't want to sell to Facebook, we're going to do this too, because you're building something that people want,*" says Crowley.

Sure enough in late summer 2010, Facebook introduced its own location application, called Places. It wasn't *exactly* like Foursquare, but did allow users to "check in" to a business or location, which was the signature activity on the app, as associated with Foursquare as the Like button was with Facebook. But the really galling thing for Crowley was the logo that Facebook used for its Check-In feature. The standard red teardrop-shaped pin signifying location was sitting on top of a square. The square itself had a pattern on its surface. The pattern outlined the number four.

"We laughed," says Crowley. "We were like, *They're trying to kill us. And they're trying to make fun of us.*" Crowley says that the snub motivated his team, and though business was down for a while because customers felt that Foursquare could not compete with Facebook, his company did survive.

Places was a failure. So in 2011, Facebook bought Foursquare's main independent competitor, Gowalla.

Spiegel and Murphy felt that Poke was a pale imitation of Snapchat, and laughed it off. Perhaps they felt a little queasy when immediately after launch, Poke reached number one in Apple's App Store. But they felt a lot better when Poke did a nosedive in the ratings over the next few days.

Not only was Poke a failure for Facebook, but it was a boon for Snapchat. It had legitimized Snapchat's product vision.

Snapchat kept growing, making it even more attractive to Zuckerberg. In 2013, he resumed his hunt, visiting Snapchat's Venice Beach headquarters with his chief dealmaker, Amid Zoufonoun, in tow. He was clearly a pro, smoothly breaking down the numbers and explaining the advantages for Spiegel and Murphy.

Not that he changed Spiegel's mind. But Zuckerberg kept pushing. In May 2013, Zuckerberg wrote an email outlining all the great things that would happen if Snapchat joined the Facebook family. If Snapchat sold to Facebook, he said, Facebook had a playbook to raise the user base to a billion people. There were private APIs that Facebook didn't share with developers. What's more, Zuckerberg wooed Spiegel personally with promises that the younger entrepreneur not only would run Snapchat with some degree of autonomy but would have an opportunity to make an impact on Facebook itself.

> *So even though you'll spend your time on Snapchat it would be fun to work together closely to figure out how Facebook should evolve as well. I have no doubt you could play a broader leadership role over time in addition to your leadership on Snapchat. On top of all this, I*

think it would just be personally fun to work together and build a deeper relationship. I've enjoyed the time we've spent and I think we could both learn from each other and build some great things together.

He also invited Spiegel to do some karaoke at Herb Allen and Company's Sun Valley Conference for media moguls.

Facebook's offer this time around was raised to what was widely reported as $3 billion—though the actual amount was fuzzier, based on contingent sums paid to the founders depending on how long they remained. This seemed to be another part of Facebook's strategy to land unicorns—structuring the deal so that a bigger chunk of the price would go to founders, and less to their investors.

The offer was so large that Spiegel and Murphy had to give it serious consideration. Ultimately Spiegel didn't feel that Snapchat would thrive in Facebook's culture. Though Facebook had impressively emerged a survivor in its move to mobile, in Spiegel's view the company still had very much a desktop mentality. Spiegel and Murphy turned down the offer. They didn't go to the karaoke either.

In a sense, Spiegel's rejection was akin to Zuckerberg's 2006 refusal to sell to Yahoo! In both cases, the founders felt that the bigger company would screw things up. A decade earlier Zuckerberg had been a teenager who'd grown up on the Internet, an advantage that enabled him to slay the aging dragons dominating tech at the time. But now a new generation had mobile mentality baked into their tech DNA. Younger people understood that the website world that spawned Facebook was obsolete. Spiegel had no desire to spend his time teaching Mark Zuckerberg what was cool. He and Murphy would build their own cool products and Zuckerberg could eat their dust. And Spiegel wasn't losing sleep over Facebook copying him anymore: Zuckerberg's Poke debacle had convinced him that Facebook was lame when it tried to emulate Snapchat.

Spiegel was underestimating Mark Zuckerberg. Maybe he never heard Zuckerberg explain he never makes the same mistake twice. In the case of Poke, the mistake wasn't copying—it was making a *bad* copy, one that didn't dovetail into Facebook, didn't take advantage of Facebook's unprecedented user base. Facebook would eventually figure out how to copy products more skillfully.

BUYING COMPANIES LIKE Instagram was not the only way Facebook could create new franchises to shore up its domain. It could create them as well. Over the years, the company has tried multiple times to gin up new properties, generally to duplicate the work of others. Usually they have failed, like the spectacular crash of the Facebook phone. But one success came from cannibalizing Facebook itself: Messenger.

Texting was becoming a core use of phones, and potentially a platform of its own, competing with social-media products like Facebook for time and attention. Facebook, as with all things mobile before its big shift, was late in adapting.

Early in 2011, Facebook had bought a small start-up called Beluga, created by three former Google engineers who were working on a group-chat application. Facebook killed their product, charging them to develop a prototype for what would be a separate app for Facebook messaging. They set about doing that, but Zuckerberg worried about the difficulty of drawing Facebook's users to a new application just for messaging. How could it compete with Apple or Google? So instead of making the new Facebook Messenger totally different, he hedged, and directed the team to graft the new product onto Facebook's current infrastructure for messages. Users could send messages either from this new app or the mobile Facebook app. "The key sticking point was that nobody knew how to grow this new thing from zero," says Ben Davenport, who had been CEO of Beluga. "Mark's thought was you tie it into this existing thing which has five billion messages a day already and that's how you grow the new thing."

That's what Facebook did, but by being just a front end to Facebook, Messenger had built-in problems, especially when someone sent a message to a Facebook user who hadn't installed the separate app. "The notifications were getting lost in the noise," says Javier Olivan. "So when someone sent you a message, yes, there was a notification in the Facebook app, but that might be the seventeenth notification you see."

Also, since it had to accommodate both people who were using it from the main Blue app and those using the stand-alone app, Messenger was limited in how it could innovate. Facebook couldn't add cool new features because it might break the communication between users on different apps.

The app's growth was sluggish. "It was linear, but not hypergrowth, not even exponential," says Davenport. At Facebook, *linear* is the same as *flat*. After a year in the marketplace, the line became more of a plateau, at about 100 million users. That was barely one-tenth of Facebook's user base. "We were struggling for a long time," says Davenport. "We became really concerned that we were not hitting the mark."

All of this set off alarm klaxons within the Growth team. If people became used to the standard SMS text system that came with carriers, or alternatives made by Google or Apple, messaging would be a staging ground for new Facebook competitors. This was especially a worry in regions of the globe where people primarily used the Internet via phones. Since those countries were the area of Facebook's biggest expansion hopes, the messaging issue became a problem for Growth. And when Growth is invoked in any Facebook matter, it takes control of the issue.

Its solution: *force* people to download Messenger by pulling messaging out of Facebook's mobile app. "Javi and I were looking at it and going, *We need to break these apart because we'll be able to control growth better, this will be a better user experience, notifications will work better,*" says Cory Ondrejka, head of Facebook mobile at the time.

The decision defied all rules of serving the user first: if people using mobile Facebook didn't download the Messenger app . . . *no messages for them!* Every time someone using the mobile app wanted to send a message to someone via Facebook a directive popped up warning them that the ability to do that would soon go away, and the remedy was to download the Messenger app. And Facebook made good on its threats.

"People ended up hating us at the time," says Ondrejka. But users had no alternative. What were they going to do, stop using *Facebook*? So out of nothing, Facebook was able to create a valuable property that it didn't have to spend billions of dollars to buy.

Once Messenger was established, Zuckerberg lured a big-name executive to run the franchise. David Marcus was the president of PayPal, then a subsidiary of eBay, but actually the most exciting part. Many thought that Marcus was overqualified to step down from the top post of PayPal to run Facebook's bastard offspring of its chat function. He understood that Facebook had already undertaken the dirty work of prying the feature from the Blue app. It had temporarily infuriated its users, but—like News Feed and other hard-to-swallow pills that the company had rammed down its users' throats—people were warming to the product. "And so I'm glad the team took that step because now we have a product that we control entirely," Marcus told me in 2015. "We control every single pixel and every line of code of that experience." And now he was free to grow Messenger into a business. And even make money from it.

Messenger expanded from a service between friends to something where people could also message with businesses, primarily through the use of automated "bots." Marcus would boast that the future of business communications would be through Messenger—why spend time on the phone or a website making a restaurant reservation when you could quickly do it via a Messenger bot?

Messenger was on its way to joining Facebook itself as an app with more than a billion users. Instagram would join that club as well. Zuckerberg had by now made it a habit to identify prospective new members of that society. He would soon find another prospect.

In 2013, Facebook's Growth team spurred the acquisition of a small Israeli "mobile analytics" company called Onavo. Co-founded in 2010 by Guy Rosen, the company had two interlocking products. The first was a consumer "mobile utility" app that promised to enhance smartphone performance by compressing data, conserving battery, and other tricks.

The second was "mobile analytics," which took information drawn from the behavior of those users, like what sites they visited and what apps they downloaded—and sold it. "We built an app that was very valuable to people," says Rosen. "And then we had a service called Onavo Insights, that provided high-level aggregated analytics into the different kinds of applications or usages that people are doing." Basically, users paid for the first app's benefits by letting Onavo snoop on them and sell the information.

Facebook packaged the merger as part of its Internet.org effort to help connect the world. "We hope to play a critical role in reaching one of Internet.org's most significant goals—using data more efficiently, so that more people around the world can connect and share," wrote Rosen.

But Facebook's motivation wasn't really providing an app to improve phone performance in developing countries. It maintained Onavo's business model, which was gathering data from deceptively "free" apps to inform its money-making business intelligence operations. When the mobile performance tool no longer served its purpose, Facebook created a different honey trap for user data, Onavo Protect, which delivered what seemed like a bargain: a free "Virtual Private Network" (VPN) that provided more security than public Wi-Fi networks. It takes a certain amount of chutzpah to present people with a privacy tool whose purpose was to gain their data.

Facebook now had a powerful way to monitor the mobile activity of thousands of users. The Growth team would study the data carefully, and post results in their regular meetings. Onavo paid special attention to Snapchat. Evan Spiegel's company had security features to block intruders, but according to one Facebook executive, Onavo used a "man-in-the-middle" attack to get past the wall and gather data. Snapchat discovered this and put in protections to thwart the intrusions. With Onavo, a Facebook executive confirmed to me, the company was "able to inject code into Snap and could see how people were actually using the product internally." (According to *The Wall Street Journal*, Snapchat would add this episode to a file it kept of Facebook's actions, calling it "Project Voldemort," after the Harry Potter villain whose name cannot be spoken.)

A visitor to Facebook's campus in this period, Procter & Gamble's Marc Pritchard, recalls seeing a chart of rising start-ups, and getting an explanation about the difference between magnitude and momentum. "They showed all these different companies that were coming up and what they tried to look for," he says. "The ones that were really rocketing versus the ones that were the steady build."

One thing became very clear: a messaging company called WhatsApp was growing at a tremendous rate. So much so that Facebook felt it had to do something.

WHATSAPP PROBABLY NEVER would have existed if Facebook had been less finicky in its hiring practices. Neither Jan Koum or Brian Acton had the kind of résumé that resonated with Facebook's recruiters in 2008. Both were engineers at Yahoo!—not a disqualifier on its own—but neither of them had the kind of check-the-boxes CV that so many Facebookers had and that Chamath Palihapitiya so vividly despised. Rejected from Facebook, they would create a product that was so compelling that Mark Zuckerberg had to back up the Brink's trucks to buy it.

Koum had been sixteen years old when he and his mother had fled the anti-Semitism of their native Kiev for Mountain View, California. It was 1992. Koum had been poor in Ukraine—his school didn't have indoor plumbing—and his small family struggled in the New World, living in subsidized housing and subsisting with the help of food stamps. America had its own challenges, especially when Koum's mother got cancer. Koum, never a big fan of authority figures, got interested in computers and joined an online hacker group. He studied programming at San Jose State University, helping make ends meet by working as a security auditor at Ernst & Young.

Acton was born in Florida, teaching himself computing by typing program listings from magazines into his RadioShack machine. When he went to Philadelphia for college, he'd never even heard of Stanford University. But after his smartest classmates complained about how they had failed to get in, he applied to transfer there, correctly predicting he would be in "nerd bliss." It was also a great pipeline to hot start-ups to get hired—in 1996 Acton joined Yahoo! as the sixth engineer at the company.

Acton did the data processing for Yahoo!'s first generation of advertising, and one of his tasks was to work with auditors to validate the company's claims that it was correctly counting the impressions garnered by its ads. One of the auditors was Koum. The two got along, and reunited some months later, when Koum himself got hired at Yahoo!

But after almost a decade at an Internet giant that was helplessly shrinking, both of the engineers became increasingly bored. They left the same day: Halloween 2007. Their vested stock options were enough to live on for a while, but not forever. After touring South America, Koum returned to the States. That was when Facebook rejected his job application.

Koum hung out often with friends from the expat Russian/Ukrainian community and was a regular attendee at an informal gathering at the home of his friend Ivan Fishman. It was called TDMS, for Thursday Dinner Movie Sessions, though often the talk was so engaging that the movie was never screened. After the iPhone opened up to app development in 2008, the TDMS chatter about it made things particularly unfriendly for cineastes in the group. One night standing at the kitchen counter at the end of the evening, Koum told Fishman about an idea he had for an app that would attach a temporary status update to the names in your address book. It would tell others whether you were unavailable for a call, or maybe that your battery was low and you couldn't take calls. Fishman helped connect Koum to a programmer in Russia to help him create the app.

In February 2009, Koum met up with Acton, who was visiting the Bay Area while temporarily living in New York, for an Ultimate Frisbee session. He excitedly told Acton he had filed the papers for a new start-up that he called WhatsApp. Acton thought it sounded cool but wasn't particularly interested.

Indeed, Koum's original idea was clunky—you would have to go into the app, look at your contact list, and then, if your friend was open to a call, leave the app and make the call. But that June Apple introduced push notifications, which allowed actions inside apps to be sent to the user at any time, even if the app wasn't open. Koum's beta testers began using their status posts to reply to someone else's status. Almost as if they were texting each other by way of notifications. "Nobody was using WhatsApp as intended," says Fishman. "People were trying to use this as a messenger."

This was a revelation. Instead of leaving WhatsApp to start a direct conversation with someone, Koum realized that you could do that *inside* the app.

At the time the only way people could send texts to each other was to use the feature owned by the mobile carriers. Even though the cost to Verizon or AT&T was infinitesimal, those businesses charged their captive customers $5 a month for a limited number of SMS texts, with each additional text costing 10 or even 20 cents. Some teenagers were racking up hundreds of dollars in monthly texting costs; most people simply avoided using it. But Koum understood that if people used an app that sent texts from one user to another, you could bypass those costs entirely.

Koum decided that the phone number itself would be the user ID. Since phone numbers were unique, it was the most direct path to the person who owned that number. Increasingly, a phone number was becoming who you were, a private version of a social security number. On WhatsApp it would *literally* be who you were.

Koum spent the next few weeks revamping WhatsApp to focus on messaging. Meanwhile, Acton was having difficulty figuring out what to do after Yahoo! That summer, he, too, interviewed for a job at Facebook. His tweet of August 3 reported the verdict: "Facebook turned me down," he wrote. "It was a great opportunity to connect with some fantastic people. Looking forward to life's next adventure."

That next adventure turned out to be Koum's project. Acton thought the pivot to messaging was a brilliant idea. So when Koum asked Acton that September if he wanted to partner, he signed on. He kicked in some money, in the low six figures, and they agreed that Acton would thereafter be a co-founder. And for the next five years they worked like dogs.

They realized early that the biggest opportunity would be overseas. The first big burst of growth came in Western Europe. While the costs of SMS were already high, sending pictures on a cellular network could cost from 50 to 90 euro cents. WhatsApp cut that to free. Also, while the EU had generally eliminated tariffs, the mobile world was still balkanized, so when people sent messages from, say, Germany to Austria, the carriers tacked on a charge. WhatsApp eliminated that.

Furthermore, Koum made sure WhatsApp ran not only on Apple and Android, but on other handsets popular around the world even if they belonged to a generation of devices that were seen as being eclipsed by smartphones. Most American companies didn't bother with them. *Ignore the United States,* Koum would tell Acton, *everyone in the world is using Nokia phones!* "That gave us huge growth potential in Latin America, Central America, India," says Acton.

The WhatsApp founders had firm views about their company's business model. They wanted to generate revenues early so they wouldn't be beholden to funders. They lit on charging a monthly fee. "We were building a communication service," says Acton. "You pay forty bucks a month to Verizon for their service. I figured a dollar a year was enough for a messaging service."

Advertising, Acton would later say, "left a bad taste in my mouth." As he came to see it, supporting a business with ads warped incentives, and led a company to create a suboptimal product for actual users. *We're whoring ourselves out!,* he would rail at his boss at Yahoo! They had vowed that WhatsApp would never go down that wicked path. In 2011, Koum tweeted, "Advertising has us chasing cars and clothes, working jobs we hate so we can buy shit we don't need." In June 2012, they published a blog item laying out this philosophy.

> *When we sat down to start our own thing together three years ago we wanted to make something that wasn't just another ad clearinghouse. We wanted to spend our time building a service people wanted to use because it worked and saved them money and made their lives better in a small way. We knew that we could charge people directly if we could do all those things. We knew we could do what most people aim to do every day: avoid ads.*

"Remember," they wrote. "When advertising is involved *you the user* are the product."

BY 2013, WHATSAPP was thriving. Koum and Acton did make one concession to their purist approach: they took venture money. The clever VC who managed to pull off the investment was Jim Goetz of Sequoia. Sequoia had a tool it called Early Bird to help its VCs identify sleeper prospects. It came up with a prominent outlier: WhatsApp may have been a mediocre performer in the United States, but in 35 of the 69 countries tracked by Early Bird, WhatsApp was number one or two. You couldn't say it was a stealth operation—it clearly had millions of users. But its founders were virtually unknown. It was supposedly based in Mountain View, but no one knew where. Goetz actually wandered the streets, looking in vain for signage that might point him to WhatsApp's offices. The search was pointless, because WhatsApp had no sign.

Finally Goetz got to them through connections at Yahoo! He met Koum in a coffee shop. This began a courtship that led Koum and Acton to set aside their doubts and accept an $8 million funding round.

With that bankroll, WhatsApp had the wherewithal to operate and grow independently, and resist the inevitable offers to sell. At first, in part because the company consciously tried to avoid the Silicon Valley buzz machine—it was notoriously press-shy—there weren't many.

Google had made a couple of desultory acquisition efforts. In 2012, the executive making the pitch was Marissa Mayer. But Koum and Acton didn't find it encouraging that when they came to Google's Mountain View offices for the meeting, Mayer's participation was videoconferenced, even though she was somewhere else on the actual campus. "Why sell?" says Acton. "We were having too much fun."

In contrast, the first contact from Facebook came in 2013 from Mark Zuckerberg himself. Like so many things at Facebook, it sprang from the Growth team. Though WhatsApp had been decidedly under the radar, especially inside the United States, Facebook deeply understood how popular it was, because of the private data its subsidiary Onavo had been stealthily gathering for years. In a sense, flagging WhatsApp for attention justified the entire price of the Onavo acquisition.

Zuckerberg proposed that Koum meet him somewhere where they wouldn't be seen together. They decided on Esther's Bakery in Los Altos, a decidedly non-chic coffee shop. The conversation was friendly, not at all a hard sell at that point, with Zuckerberg sharing stories from Facebook's start-up days. It ended cordially, and over the next few months they kept in touch.

Throughout, Koum and Acton felt confident that they could withstand any pressure from a predatory would-be buyer. When *Wired UK* visited their Mountain View offices in December 2014, Acton laid out the case against selling. "I worry about what [an acquiring] company would do with our population," Acton told the reporter. "You're never hands-off in the long term. To have someone come along and buy us seems awfully unethical. It goes against my personal integrity."

In retrospect, that sounds like a young prizefighter's bluster before entering the ring against a cold-blooded belt-holder—things change dramatically at the first monster blow to the head. In February 2014, only weeks after that declaration, a roundhouse punch landed. It was in the form of a report by Morgan Stanley investment banker Mike Grimes. It was a convincing analysis of the same data that led Sequoia and Facebook to identify WhatsApp as one of the most valuable acquisition targets in technology. Somehow it leaked, and was circulated around Silicon Valley. (Speculation still abounds as to who shared the confidential report and why—but it should be noted that Grimes had been a key banker in Facebook's IPO. He denies leaking it.)

If the leak of the Morgan Stanley deck was intended to create a mini-frenzy, it worked. Facebook's Growth team, which had continued to track the remarkable Onavo data, recognized immediately the danger of WhatsApp in enemy hands. Zuckerberg's new priority was now buying Koum and Acton's messaging company. The acquisition machinery cranked up for what would be its biggest and most expensive quest.

Meanwhile, Google reached out again. This time it was CEO Larry Page offering the meeting. It went no better than Google's previous effort. The enigmatic Page was a half hour late. He did ask that if they ever did go on sale, to allow Google to make an offer.

Mark Zuckerberg wasn't going to let that happen. The Onavo numbers told him that WhatsApp was becoming a global powerhouse, possibly blocking Facebook's own messaging efforts around the world. It had 450 million users, including 40 million users in India, 30 million in Mexico. In some countries, it had two-thirds of the market. Only two years earlier, Zuckerberg had shocked the world with a billion-dollar bid for Instagram when the company was closing a round for half that valuation. Now he was prepared to spend much more for WhatsApp. When Koum and Acton told him that they demanded that he value their company on a level with Twitter—which at the time was around $20 billion—he didn't flinch. Which was shocking. WhatsApp back then had only around fifty-five employees. Most Americans had never heard of it.

"Their scale was just so dramatically underestimated by people in the US and a lot of the mainstream press because the US is actually one of their smaller markets," Zuckerberg told me a few weeks after the chase. "But when you look at the growth rate, it's crazy. This is an experience and a network that seems extremely likely to get to a billion people using it, and if you look at the number of things that have reached a billion people, they all end up being incredibly valuable and important things."

Zuckerberg, who had complete control of Facebook and could do whatever he wanted, was ready to pay. Meeting at Zuckerberg's house on Valentine's Day—nibbling on chocolate-covered strawberries possibly intended for Priscilla Chan—they agreed on an arrangement worth $19 billion. (Later changes in Facebook's valuation would nudge the price to around $22 billion.)

The WhatsApp guys had given Facebook a number so high that it seemed inconceivable. And Facebook had called their bluff.

"Mark Zuckerberg checkmated us," says Acton. "When a guy shows up with a big suitcase of money, you have to say yes; you have to make the rational choice." It's one thing to turn down a billion or two, but twenty billion is not just higher—it set the process on an entirely different planet. How can you tell your investors, your employees . . . your mother, that you spurned twenty billion?

Acton also admits that he was tired. When he and Koum started WhatsApp, they hadn't paid themselves, and were living off their stakes in Yahoo! When Sequoia made its investment, the co-founders began taking small salaries. But by 2014, Acton was married and just had his first child. He had been working eighty- or ninety-hour weeks for five years.

And in the future, if WhatsApp didn't take the offer, it would have to fend off Facebook as a competitor. That threat hung over the negotiations like a giant spiked pendulum.

WhatsApp's banking representative on the deal was Mike Grimes of Morgan Stanley, who had created the memo that acted like a starter's gun in the pursuit.

The WhatsApp founders did try hard to get one thing in writing—a promise that Facebook would never force WhatsApp to adopt an advertising-based business model. That would be the true end of their dream. Facebook resisted, arguing that while it had no intentions of doing that, the language was too broad. All finally agreed to a clause that said that if Facebook forced "additional monetization initiatives" on WhatsApp—including advertising—the founders could quit and still collect their full option package, which otherwise would vest over four years. It really didn't address Koum and Acton's fears that advertising would creep into WhatsApp. But by that point, they were worn down. "You say whatever you have to say to get the deal done and then [figure it out] when the dust has settled," says Acton.

Acton, Koum, and Goetz signed the deal outside the Social Services office that once supplied Koum with food stamps. At the announcement everyone said the right things. WhatsApp would be run independently by the founders. Koum would even be a director on Facebook's board, a perk that not even Kevin Systrom of Instagram had been granted.

When sharing the news, Koum and Acton tried to reassure their users that things would be no different:

> *If partnering with Facebook meant that we had to change our values, we wouldn't have done it. Instead, we are forming a partnership that would allow us to continue operating independently and autonomously.*

But as a wholly owned subsidiary of Facebook, all of that was at the pleasure of its CEO. At any moment, WhatsApp could be transformed to a Mark Zuckerberg production, as the founders would learn.

FACEBOOK SEEMED SET to dominate mobile messaging. Now Zuckerberg began to look at what might threaten Facebook after phones were no longer the thing.

Virtual reality had been a hot technology in the early 1990s, the subject of dozens of rhapsodic feature stories. But the hype around it proved unfounded, and nothing much had happened. Until Palmer Luckey.

In 2012, Luckey was a nineteen-year-old Southern Californian whose passion was miniaturizing old game consoles. He had long wished he could be *inside* the games he loved. But virtual-reality rigs were expensive, and even then they did not have the processing power or the software wizardry that made for a fully satisfying experience. (A VR rig included a headset that usually resembled a blacked-out diver's mask, with built-in screens to supply a vision of a computer-generated world. It would be attached to a powerful computer.) So he began to piece together his own. Somehow this savant of Southern California managed to produce something that worked better than the hardware produced by NASA-funded minds with PhDs and research groups. He shared his progress on an Internet news group populated by 3-D video fans.

One of those readers was John Carmack at his home in Dallas, Texas. Carmack was the closest thing to Elvis in the gaming world. A galactic-class coder, he was the brains behind legendary titles like Doom. He had been exploring the VR world and was shocked at how little had been done since the 1990s. One of many frustrations was that the experimental systems at the time had a limited field of view. This was lousy, because the appeal of being in a virtual world was the immersive feeling. You couldn't get this if the illusion ended when you turned your head. Luckey's rig seemed miraculous. "I had been thinking about buying a fifteen-thousand-dollar VR headset that would've had a sixty-degree field of view," Carmack says. "Palmer was putting together something with a ninety-plus-degree field of view, with off-the-shelf parts cobbled together in basically a cardboard box, in the three-to-five-hundred-dollar range."

Carmack took Luckey's headset—bound in duct tape—to a conference, and the demo was a sensation. The news reached a game entrepreneur named Brendan Iribe, who was based in Maryland, near the state college campus. He took some of his longtime colleagues, Michael Antonov and Nate Mitchell, to meet Luckey in a high-end steakhouse in Los Angeles. Luckey showed up in shorts, flip-flops, and an old Atari T-shirt. But the moment he began talking, it was clear that Palmer Luckey was a technology wunderkind. "You could ask almost any question about anything tech- or electronics-related, and he knew the back history stories around why it worked, why it didn't work often, how the whole product was put together," says Iribe. Before the dinner was over, Iribe and his friends pitched him on starting a company.

Luckey wasn't sure. Over the next few weeks, Iribe pursued him, a difficult task, since Luckey had no smartphone. At one point, Luckey said he was thinking of just doing his headset with some friends as a Kickstarter project. (Kickstarter was a site where people could conditionally pay for new products that would only exist if funding goals were met.) But he didn't even have any working prototypes of his own technology anymore. Iribe wrote him a check for $3,700 for the parts. That impressed Luckey, who shook hands. They were in business, along with Iribe's friends. By July 4, Luckey had built the headsets to give Iribe his first demo. The demo made Iribe nauseated, but he was famously prone to motion sickness.

The Kickstarter run would be more professional, manufactured in bulk in China. The Kickstarter went live on August 1, 2012. If the sum from prospective buyers reached $250,000, the project would proceed. Oculus made its goal in two hours. A few days later, when the amount hit $2,427,429, they finally stopped taking money. By then professional investors were making bids. Oculus wound up raising $16 million in its "A" round.

By late 2013, Oculus was coping with classic start-up problems of developing a product while trying to manage a ballooning workforce. It had only thirty employees but was planning to triple its size in order to fulfill the now-delayed Kickstarter orders. The costs would require another funding round, this time for $75 million. But the technology had been progressing well. A former Microsoft engineer named Michael Abrash, who had been working for the game technology company Valve, had combined Oculus technology with a screen-display technique called "low persistence" that minimized motion sickness. For the first time, Iribe could use VR without getting nauseated. Abrash would eventually join Oculus and head its research lab.

The company figured that a great board member might provide guidance, and the logical choice was Marc Andreessen, whose VC firm, Andreessen Horowitz, had been a lead investor in the "B" round. Since Andreessen was also on Facebook's board, he suggested that Iribe get a reference from Mark Zuckerberg. On November 13, Andreessen wrote an email to Facebook's CEO with the subject line "Have you seen Oculus?" Andreessen told Zuckerberg that Oculus "blew my brains out."

On the phone call, Zuckerberg affirmed that Andreessen would be a great board member (duh). Then they started talking VR. "Do you see this being used for more than games?" asked Zuckerberg. "Absolutely!" Iribe said. "You have to see it to believe it!"

On January 23, 2014, Iribe and a small team flew up to Facebook. Since Zuckerberg's glass-walled conference room was exposed (Zuckerberg found it a pain to draw down the shades that could provide privacy), they set up in Sandberg's office. Zuckerberg put on the headset and began exploring a strange landscape with critters running around. One part of the demo particularly impressed Zuckerberg. It depicted a villa in Tuscany, Italy, and allowed the user to roam around, exposing beautiful vistas of the countryside. *This is really cool,* Zuckerberg thought. *I'm clearly not in Italy—I'm in Sheryl's conference room. But I really feel like I'm in Italy because everything I'm seeing makes me feel I'm there!*

The next day, Zuckerberg emailed Iribe. "I was a little dizzy after taking off the headset," he wrote, "but it's clear where it's all heading, and it's amazing." He wasn't offering to buy Oculus yet. But five days later, he flew to Irvine himself for a more elaborate demo.

The second demo clinched it. In the space of a few days, Zuckerberg had concluded that virtual reality was not merely a cool potential feature, but something way larger. *It was the next platform.* Missing this would be like missing out on mobile. Zuckerberg was only two years removed from what he had considered a near-death experience when Facebook almost screwed up that pivot. Virtual reality, he figured, might be ten years out, but here was a company that was building the foundation. If Facebook owned it, and poured in money to make it happen, Zuckerberg not only would be ready for the next big paradigm shift. He would *own* it.

A day later, he had dinner with Iribe and told him he wanted to buy the company. A couple of days after that, Zuckerberg sent his pitch email, with unmistakable similarities to the ones he had sent to Instagram and Snapchat, and had just sent to WhatsApp. *Sure, you will do well on your own*, it essentially said, *but we will fuel your rapid growth, recruit the best people, provide our infrastructure, and help you scale.*

The offer was under $1 billion. Iribe wanted several times that. He politely turned down Facebook.

But Zuckerberg was still obsessed with VR. He conferred with his deal master, Zoufonoun, and agreed that a higher price would be worth it. He invited Iribe to his house on Sunday, March 16. "I'm not going to waste your time," Zuckerberg promised.

Iribe showed up at Zuckerberg's house with John Carmack. Zuckerberg had ordered a pizza. They sat on his porch and got to business. Though Zuckerberg didn't need convincing, Carmack's authoritative account of the technology only made Oculus seem more attractive. When Carmack departed, Zuckerberg made his offer: $2 billion for Oculus itself and an additional $700 million for "earn-outs." That would be just the beginning of billions more that Facebook would be spending to develop the technology. Iribe agreed.

Eight days later, the deal closed. Zuckerberg had done Facebook's second biggest acquisition in a week, just after completing its biggest acquisition, WhatsApp, with similar velocity.

Within weeks, Facebook learned of complications because Carmack's former employer claimed ownership of some Oculus technology. This would lead to a lawsuit where Facebook paid $500 million. Zuckerberg, wearing a suit, testified. There was one moment in the trial where the event transcended the messy arguments about intellectual property and elevated itself to a nerd's dream. After the plaintiff attorney's grilling, one of Zuckerberg's own lawyers asked what his vision for buying Oculus was. Zuckerberg, who had been parsimonious and snarky with the rival attorney, suddenly burst into articulation. Virtual reality, he said, was going to bring people together in the same way Facebook did. His example was that first moment when a child begins to walk. When young Mark took his first steps, he said, Ed and Karen Zuckerberg recorded it in a baby book. Years later, when his older sister's first child took her initial steps, Randi Zuckerberg captured a picture with her smartphone.

When her second child walked, she taped a video. "And when Max, who is my daughter, took her first step just a few months ago," he said, "I recorded the whole scene in virtual reality so I could send that to my parents and share that with the world, and people could just experience that as if they were in our living room with us."

That was what Mark Zuckerberg wanted from Oculus. A social experience big enough to routinely make magic. He couldn't wait to begin showing it off. Even if it would take ten years and a number of to-be-determined breakthroughs for it to happen. When it did, no other company in the world would be close to Facebook in its ability to deliver it.

Zuckerberg's infrastructure was complete. In another five years all those independent companies would all be one big happy Facebook family. And the founders would be gone.

PART THREE

14

Election

NED MORAN WAS part of Facebook's Threat Intelligence team, most of whom worked out of Facebook's DC office, which was mostly staffed by people in policy and communications. While the wonks and lawyers lobbied Congress, dealt with regulators, and spent endless time in videoconferences with their counterparts in California, Singapore, and Dublin, Moran and his colleagues pored over lines of code and Internet links to detect digital intruders and malfeasants. It was from his workstation in the Washington office that Ned Moran first noticed that Russians were using Facebook to tamper with the US presidential election.

The Threat Intelligence team consisted of computer security experts with experience in tracking espionage threats like malware or spear-phishing, which meant tricking targeted people into clicking a link that would give malefactors access to private information, or even Facebook's code. Facebook was concerned that skilled operatives, perhaps even working for a foreign power, might try to use Facebook to find their targets.

Such forces certainly were known to be afoot. In recent years, the security firm CrowdStrike, among others, had been tracking the activities of teams the firm nicknamed Fancy Bear and Cozy Bear. The stuffed-toy nicknames are deceiving. These were two separate groups of digital marauders based in Russia. Intelligence officials knew them as Units

26165 and 74455 of the Main Intelligence Directorate of the General Staff (GRU), roughly the Russian equivalent of the CIA. "Their tradecraft is superb, operational security second to none, and the extensive usage of 'living-off-the-land' techniques enables them to easily bypass many security solutions they encounter," wrote CrowdStrike's cyber-espionage specialists.

Facebook knew that some of its active accounts were associated with the GRU. Instead of shutting them down—they weren't doing anything illegal anyway—the Threat Intelligence team monitored them, to keep track of potential security concerns. In early 2016, the team noticed that whoever was behind those accounts started searching on Facebook for people in government posts, journalists, and Democrats involved with the Clinton campaign. It looked like a prelude to a spear-phishing attack on those people. Facebook alerted the FBI, which accepted the reports and did not follow up with Facebook.

The Bears became more assertive. In June of 2016, CrowdStrike reported that the GRU squads had conducted a series of spear-phishing hacks on the Democratic campaign that included candidate Clinton's emails and the emails of her campaign manager, John Podesta. (The attacks were launched from the victims' Gmail accounts.) In June, a Fancy Bear member calling himself "Guccifer" claimed credit for breaking into the Democratic National Committee and stealing its emails.

That's around the time that Moran found more activity from the GRU-related accounts. This time they weren't phishing or searching for targets. The Russians instead were engaging Facebook *users*. In essence they were using Facebook the way its engineers designed it to work—as an engine for sharing. It was an activity that the Threat Intelligence team—and Facebook itself—had not anticipated and were unequipped to handle.

In the 2008 and 2012 elections, Facebook had gotten a boost as a civic plus. It had proved a valuable tool for campaigns, and its co-founder Chris Hughes had a role in the first Obama win. Beginning in 2008,

Facebook had begun co-sponsoring debates, with its users posing questions to candidates. But Ned Moran's discovery was the first signal that 2016 would be different for Facebook, in a terrible, terrible way.

ALEX STAMOS BECAME Facebook's chief security officer in June 2015. A barrel-chested engineer in his mid-thirties, he had come from a troubled, yearlong tenure at Yahoo!, where he had held the same title. Stamos was known for his deep ties to the white-hat hacker community, allying with them on key issues, like the support of strong encryption. He was also something of a wild card. The outspoken Stamos was no fan of the tradition that dictated that CSOs should stay out of public view. The popularly held theory was that the more a company spoke of security vulnerabilities, the less trust people would have in that company, or in technology at large. He thought the opposite—that silence on those issues just led to more vulnerabilities.

Stamos's first run as CSO, at Yahoo!, was his first job at a major technology company after years of working for a number of security firms. Before he arrived, Yahoo! had suffered several serious breaches, exposing the information of more than a billion users. Stamos, leading a group nicknamed "the Paranoids," launched multiple initiatives to bolster security, but clashed repeatedly with his bosses about implementing even stronger measures. When he left for Facebook in May 2015, his new job would be even more challenging, protecting the information of 2 billion people and an infrastructure that sprawled across the globe. Plenty could go wrong.

What Stamos didn't realize was that from the minute he had been hired, something already had gone wrong: the company had reorganized, and the CSO and his team were now in Sandberg's organization. His was the only team in her domain where virtually everyone (well over one hundred people) was technical, and had actually done the coding part of Facebook boot camp. Stamos didn't even report directly to Sandberg—his boss was general counsel Colin Stretch. So the man in charge of Facebook's security was a hop from the chief operating officer, with no routine access whatsoever to her and certainly not to Zuckerberg and the top lieutenants running the company.

Now Stamos was seeing the first signs that Facebook was unprepared for some issues it would be encountering in the 2016 election. It wasn't so much that the Russians were interested in Facebook—that was expected, and Stamos's team had been on high alert for attacks foreign and domestic. Cyber-espionage villains like these were what the Threat Intelligence team was always looking for: dark-side hackers potentially trying to take over accounts and steal information.

Facebook itself was not utilized in the actual spear-phishing attack the Russians successfully launched on the Democratic National Committee and other players involved in Hillary Clinton's presidential campaign. But in the summer of 2016, the service was being used to spread selected stolen emails that would embarrass the Democrats and potentially discourage voting for Clinton. That spring the Russians, masking their involvement, had created a website called DCLeaks as a staging ground for distribution. Around June 8, when the website went live, the Russians started a Facebook page with the same name. (They also opened a DCLeaks Twitter account.) The Russians had identified Facebook as an excellent venue to conduct activities.

The Threat Intelligence team took a close look at the troublesome DCLeaks page. On its face, it seemed legit. It had been started by someone calling herself Alice Donovan, who had a preexisting Facebook account. The page was promoted by two other accounts with Anglo-sounding names, Jason Scott and Richard Gingrey. From there, hundreds of thousands of unsuspecting Facebook users would eventually be exposed to the content. Threat Intelligence analyzed the data of the page's origin and interconnections. By matching it to the known GRU-related accounts he was tracking on the system, Ned Moran discovered that the DCLeaks page was in fact connected to the Russian hacking organization. (Later the US special investigator's team led by Robert Mueller would charge Aleksey Aleksandrovich Potemkin, a supervisor in Unit 74455, with actually starting the page. The irony of the surname—which was real—is unbearable.)

"We saw them start to push stories with journalists to try to get stories printed about the information from the Hillary emails, the DNC emails, and the Podesta emails," says Stamos. His group passed this information to Facebook's lawyers, who presented the problem to the policy team.

It should have been an easy decision for Facebook to take down the page. But such decisions were not simple at Facebook in the 2016 election year. Internally, the company was at odds with whether, or how far, it should bend its own rules to maintain neutrality, or even to acknowledge the difference between truth and destructive lies.

FACEBOOK HAD BEEN burned by charges of political favoritism in the past. Remember Ben Barry's last-minute implementation of the "I Voted" button in the 2008 election? For the 2010 midterm election, Facebook expanded the program, with a prominent button proclaiming "I Voted" visible to users. But not all visitors. Facebook used the midterms to conduct an elaborate experiment. Two of Facebook's top data scientists, working with researchers at the University of California at San Diego, decided to test whether the voter button actually affected voter turnout. If you saw that your friends voted, would it influence you to do the same?

Cameron Marlow, who headed Data Science for Facebook at the time, says the experiment was an innocent exercise: "We had a product that had run in every single election and we were starting to run in other countries' elections—the goal was to get people out to vote." He says that when the UCSD scientists proposed the experiment, it seemed like a natural one to run. The Data Science team was part of the Growth operation; it was always looking for ways to learn about user behavior that might increase engagement.

The study, "A 61-Million-Person Experiment in Social Influence and Political Mobilization," created controversy when it was published in

Nature in 2012. It indicated that Facebook could be a factor in political behavior. Indeed, the study contended that Facebook's power could affect elections. "The results show that the messages directly influenced political self-expression, information seeking and real-world voting behaviour of millions of people," wrote the authors, later adding, "The Facebook social message increased turnout directly by about 60,000 voters and indirectly through social contagion by another 280,000 voters, for total of 340,000 additional votes. That represents about 0.14% of the voting age population of about 236 million in 2010." In a nation of close elections, such numbers could be decisive.

Even more disturbing was the idea that Facebook could use that power to manipulate people to get the results it wanted. The study itself was an example. Facebook cooperated with the study by splitting part of its population into two groups: those that could see the "I Voted" button and a control group that could not. Then, using voter records, the authors compared. Since it turned out that voting increased with the experimental group, then it was possible Facebook affected the election by masking the buttons in the control group. And what if Facebook decided to withhold the button in Republican districts and feature it prominently in Democratic strongholds? Essentially, Facebook was in some small way determining turnout by its geographic selections. (Marlow says the selection of who saw the button was random.)

The study horrified observers. A typical headline was "Facebook's 'I Voted' Sticker Was a Secret Experiment on Its Users." Like many actions coming out of the Growth organization, it was one more instance where the company had moved fast, got slapped for recklessness, and pulled back. Thereafter the company was cautious to avoid reminders that a bias on its part could affect an election. *Everyone* would see the voting reminders, with no control groups.

Facing the 2016 election, though, the concept of Facebook's partisanship had once again become a center of controversy both within and outside of the company.

The head of Facebook's Washington, DC, office, as well as the VP in charge of global policy, was a former George W. Bush aide named Joel Kaplan. As a young lawyer, he clerked for Supreme Court Justice Antonin Scalia, and then worked on the recount that put Bush in the White House. His last job in the Bush administration was succeeding political Svengali Karl Rove as deputy chief of staff for policy. In 2011, he was an energy lobbyist when Sheryl Sandberg tapped him to be a VP of policy for Facebook. He was the ultimate Friend of Sheryl Sandberg—he had dated her at Harvard and had maintained ties to her despite their differing political affiliations.

At the time Sandberg's friend Marne Levine—a Democrat—had been running the office. She and Kaplan created a natural balance of party power. Things changed in 2014 when Levine moved to California to become the chief operating officer of Instagram. Kaplan became head of global policy and did not seek out a counterbalance on the other side of the aisle. In 2015 he hired a friend from the Bush-Cheney campaign, Kevin Martin, who had headed Bush's FCC. Martin's tenure at the agency had ended with an investigation that resulted in no criminality but noted that his "heavy-handed, opaque, and non-collegial management style . . . created distrust, suspicion and turmoil among the five current commissioners." According to the report, at one point Martin had ordered his staff to rewrite a report that he disagreed with. (Martin would later ascend to head of Facebook's US public policy.)

Sandberg herself chafes at the entire FOSS narrative. "There's a long history of great people I worked with and thought were awesome," she says. "They have now worked with me job to job and I have become very close with them."

To some in the office, Kaplan seemed to act as if his role were specifically to make sure that Facebook was not favoring liberals. "That was Joel's role at the company—to find out what conservatives wanted and to make them happy," says a Facebook official who worked with him. At one point, the official says, the staff intended to promote a voting initiative by the Obama administration. This was in line with Facebook's ongoing belief, even after that unfortunate study, that increasing turnout was a good thing. But Kaplan nixed the idea, saying that it would be partisan for Facebook to aid anything the president of the United States was doing. After all, Obama was a Democrat. "I think his reason was, *Republicans don't like voter registration*," says another Facebooker working in DC at the time.

That was only one way that Kaplan seemed hypersensitive to concerns that Facebook was tilting the field against Trump. In December 2015, the Trump Facebook page posted a video with the candidate's demand to ban Muslim immigration. "It was a clear violation of our policy," says a Facebook person working in policy. But Zuckerberg, who was a passionate defender of immigration, was questioning whether the post should be removed.

The issue came up in the weekly "Sheryl meeting," a videoconference involving policy executives in DC and Menlo Park. Kaplan argued to leave the post up; others noted the difficulty Facebook would have in explaining why it didn't act on something that violated its own rules against hate speech. According to *The New York Times*, Kaplan said, "Don't poke the bear," referring to Trump. But the Community Standards team was also fine with leaving it up. "It was one of those ones that was right on the borderline," says Monika Bickert, who heads that team. "The person is a major candidate in a major election in the world; this is of course very newsworthy."

Sandberg usually made the final call, but since it involved a presidential candidate she said she would take it up with Zuckerberg. Despite his own feelings on immigration, the CEO allowed it to stay. Facebook settled for a tortured compromise. It would leave the video on the Trump page. But if anyone else posted the same video, Facebook would take it down. (Twitter had been grappling with the same issues with Trump's hateful tweets, with Jack Dorsey finally deciding that the newsworthiness of a candidate, and later a president, would outweigh a policy violation.)

The partisan tap-dancing continued in the spring as another political controversy erupted, this one centered on a Facebook feature called

Trending Topics. It seemed innocuous on its face—just a list of news subjects, sited on prime real estate to the right of the News Feed, that were getting a lot of attention on a given day. A small group of journalists, technically working for the outsourcing firm Accenture, acted as admins for the lists. (Facebook, as well as other companies, employed that technique to avoid hiring more expensive full-time employees.) Trending Topics used algorithms to initially detect stories with high circulation; the journalists would filter out bogus posts and make sure the trends reflected actual news. They might remove content if it wasn't timely, or was insufficiently sourced, or was satire or an outright hoax.

In May 2016, a story in *Gizmodo* reported that, according to a politically conservative former "news curator" on the team, some of the journalists were suppressing conservative content on Trending. In addition, curators would "boost" content from liberal sites.

For years, right-wing conservatives had been complaining that Facebook—run by those liberals in Silicon Valley—discriminated against them by down-ranking their posts. The claim was unsupported by data, and by many measures conservative content was *overrepresented* on Facebook. Fox News routinely headed the list of most-shared posts on the service, and even smaller right-wing sites like the Daily Wire were punching above their weight.

Nonetheless, Republicans suddenly became huge *Gizmodo* fans. GOP senator John Thune, who chaired the committee overseeing the FTC, demanded an explanation. Facebook checked its data, and in its twelve-page response Facebook confirmed that Trending Topics treated content in a nonpartisan manner, regardless of political tilt. "Boosting," it explained, was done when several of a list of high-quality publications, like *The New York Times* or *Wall Street Journal,* were heavily covering a national topic. Kowtowing to the right, Facebook said it would stop the boosting. But it retained the Trending Topics feature.

"The joke of it was that Trending Topics wasn't a big deal inside the company," says Andrew Bosworth. "It was not a big investment, it wasn't important."

Bosworth and others urged Zuckerberg to shut it down, but the CEO resisted for a while. He liked the idea of showing people current information. But by the summer, the election was obviously polarizing the country, and the last thing Facebook wanted to do was to be seen as biased. The fact was, Facebook *did* have a bias—it wanted a standard of quality in the news sources it cited in Trending Topics. The human beings used to cull the choices from the most popular ones on Facebook omitted trolling, inaccurate, or downright fabricated sources. It just happened that some of the flagship publications of the conservative movement weren't too picky about accuracy.

But Facebook felt it had to do something. In August it de-contracted—fired—the human operators and turned the job over to an algorithm. Previously, Facebook had used artificial intelligence to sort out stories for the human curators, who were able to further cull ridiculous or outrageous posts or links. Without that human oversight, the algorithms rewarded the types of posts that thrived on the News Feed—attention-getters, without regard to truth, good intentions, or newsworthiness. On the day of the announcement a CNN reporter noted that the topics on her feed included Go Topless Day, Al Roker, and rapper Yung Joc's haircut. A few days later, topping the Trending list was a fabricated story from a website called endthefed.com claiming that Fox News anchor Megyn Kelly had been fired for supporting Hillary Clinton. Endthefed's source was another obscure website, which got *its* information from a right-wing blog that, according to a *Washington Post* story, "read like anti-Kelly fan fiction."

Trending Topics was no longer biased against conservatives. It was biased against journalism. Amazingly, Facebook retained the feature until quietly pulling the plug in 2018.

AFTER THE TRENDING Topics fiasco, Kaplan suggested that Facebook invite a slate of right-wingers to Menlo Park so the company could convince them that it was giving them a fair shake. Some Facebookers found this an insulting contrast to the virtual snub Facebook had given Black

Lives Matter some weeks earlier. Members of the civil rights group had asked for a meeting to air out issues, including Facebook Live's streaming of violent crime and police killings. Another sticking point: that February at Facebook HQ, employees had crossed out the words "Black Lives Matter" from a graffiti wall and replaced it with "All Lives Matter," a response associated with racism. Zuckerberg had condemned the action. But neither Zuckerberg nor Sandberg attended the session in the DC office. Nor did Joel Kaplan, the top policy person in the office. Facebook sent Monika Bickert, who was in charge of Content Standards; a Facebook policy director who works with Democrats; and an African American staffer whose work wasn't relevant to the subject matter. (Higher Facebook executives, including Sandberg, have met with Black Lives Matter at other times.) In contrast, Facebook was treating a motley lineup of right-wing pundits like rock stars, flying them out to Menlo Park to listen to Zuckerberg and Sandberg's explanation of how respectfully their posts were treated, even the shrill conspiracy charges from Rush Limbaugh and Glenn Beck. Facebook officials registered as Republicans staffed the event, with policy people on the Democratic side asked to stay out.

In a sense, the meeting was a cynical exercise. According to *Wired* magazine, the thought was that the conservatives—who were sprinkled along the right-wing spectrum from principled to wingnut—would turn on one another. Or be bored to death from PowerPoint explanations of how the News Feed worked. Indeed, some conservatives bickered amongst themselves, some demanding special favors like a quota for conservatives on Facebook's workforce. But Glenn Beck thought Zuckerberg was sincerely listening. "I sat across the table from him to try to gauge him," he says. "He was a little enigmatic, but I thought he was trying to do the right thing."

Despite whatever good feelings the meeting engendered, after leaving Menlo Park the conservatives returned to complaining about Facebook's treatment of them—while piling up millions of views because of their skill in exploiting Facebook's algorithms.

. . .

THAT MAY MEETING was the backdrop for the June discussion of whether to take down the DCLeaks page, which smelled very much like an attempt to disseminate stolen emails from the DNC. To some at Facebook, Kaplan seemed more concerned about offending the Republicans than whether the page violated terms of use.

Kaplan's boss, Elliot Schrage, strongly takes issue with the perception that Kaplan was carrying water for the party he identified with. Schrage says that the decisions on DCLeaks—and, for that matter, all the decisions that later were criticized because of Kaplan's alleged bias—were reached by vigorous debate, with his participation. Schrage has a background as a human-rights activist and describes himself as "a First Amendment advocate in the Brandeis tradition," which leads him to give the benefit of the doubt to free speech. In a famous dissent in a First Amendment case, Louis Brandeis wrote that "the fitting remedy for evil counsels is good ones," though it's not clear what he would have thought of the News Feed. "I am hard-pressed to recall a single one of these debates where Facebook's conservative public policy head and his liberal boss disagreed on how to proceed," says Schrage. Of course, Zuckerberg, while not necessarily a student of Justice Brandeis, also tilted toward a free-speech approach.

As for DCLeaks, whether due to lofty principles or political calculation, Facebook's original decision was that the page itself *didn't* violate any policies. Or maybe it *did* violate policies, but its newsworthiness (like Trump's hate posts) overruled it. The page remained.

It was an explanation that only an engineer would love—*Yes, it's terrible but these are our rules!*—and far from the last time that Facebook tried to retain a post that seemed indefensible. And the decision did not go well with the press and public. Facebook came under immediate pressure to explain why it was okay to support the brokering of information stolen by Russian hackers.

Finally, Facebook figured out a reason to take down the page. In violation of the company's rules, DCLeaks was exposing the private information of some individuals—namely wealthy financier George Soros, who was supporting the Democrats. It was like taking down Al Capone for underpaying his income tax. The removal was part of a pattern that would repeat endlessly over the next few years: after a lame attempt to defend something indefensible on its service, Facebook would cave to pressure and suddenly discover a reason to take it down.

"We found the broken taillight and wiped it out," says Stamos. "But effectively there was no policy and the Policy team did not want to be seen as getting involved in the election. They absolutely did not want it."

By the time Facebook made the decision, the DCLeaks page was irrelevant anyway. The GRU did not need a Facebook page to distribute the leaks because WikiLeaks, the distributor of secret information that had inspired its name, had posted selected stolen emails, and the American press feasted on them, just as the Russians hoped they would.

THE REAL TAKEAWAY of Trending Topics was the degree to which misinformation and cynically wrangled outrage dominated the News Feed. This came to be called fake news. In part, fake news was a consequence of the Growth team's victory in producing high engagement. Even if you had very few friends, Facebook would rope in the friends of your weakest ties to deliver the stories you were most likely to react to—to comment on, click on a Like button, or even to stop scrolling for a few seconds to take in the post. When your eyes lingered on a story, Facebook took this as interest too.

Compounding the problem was that when users shared links from web publications, Facebook visually presented them in the same style regardless of whether the source was a newspaper with a hundred-year reputation for truth, or a phony site that sprang up two weeks earlier. Users rarely checked the source. Over the years, fakesters discovered that between Facebook's engagement practices and lack of filtering, the platform was a gold mine to reap ad revenue (delivered when someone clicked on a story) or to push their sometimes radical ideas. The problem grew with little intervention from Facebook, which took its cue from its leader, an advocate of free speech who often said that Facebook wanted no part of determining the truth behind what was posted.

"Fake news, maybe it's always existed, but no one I know was thinking about it in 2015, right?" Sheryl Sandberg says to me in a 2019 interview. "I mean, it really became a thing in the last few years."

But people *were* thinking about it in 2015. Facebook just wasn't listening.

Renee DiResta was a researcher and writer who was also involved in start-ups. She had her first child in 2013 and became active in the pro-vaccination movement. Around that time there had been a measles outbreak at Disneyland, and some state legislators were trying to get a bill passed mandating inoculations. DiResta started a Facebook page called Vaccinate California. She discovered that it was possible to see the performance of potential competitors to the page and was shocked to find that anti-vaxxers had been building audiences for years. What's more, when you searched on Facebook for vaccination information, it was the anti-vaxxers—with bogus science and conspiracy theories—who dominated the results. Though they were only a tiny minority in a huge state, the fringe owned the discussion.

When the Trending Topics curation ended, DiResta realized that the anti-vax problem was indicative of a much bigger Facebook problem. "It just exploded into the most batshit-crazy, insane conspiracy theories," she says. *Oh my God,* she told herself, *this stuff is viral everywhere on the platform.* She saw the phenomenon as inevitable given Facebook's growth and advertising practices. The company had marketed itself as an influence machine—*Reach your audience, change their hearts and minds, and sell them your T-shirts.* But there was no fundamental difference between commercial persuasion and political persuasion. Facebook, she felt, had built an engine to push propaganda. She managed to get a meeting with a News Feed director, who conceded that some groups were problematic but that the company did not want to hamper free expression. "I wasn't asking for suppression," DiResta says. "I was saying your recommendation engine was growing this community!"

. . .

IN FACT, HALFWAY around the world, there was terrifying proof of those fears. In the Philippines.

By 2015, nearly all inhabitants of that Pacific island country of 10 million had been on Facebook for several years. A major factor in making this happen was the Internet.org Facebook program—hatched from the Growth team—known as Free Basics. It was designed to increase Internet activity in poor countries where many people could not afford to pay for data charges. Free Basics allowed people to use Facebook—for no fee. While the program had run into trouble in India, in its 2013 test bed in the Philippines, it was a "home run," said Zuckerberg in a 2014 conference appearance. (A couple of years later, Zuckerberg heard that 97 percent of Philippine Internet users were on Facebook. His joking reaction was, *What about the other 3 percent?*)

The country also got the bulk of its news from Facebook. This is why when one of the country's leading journalists, Maria Ressa, started a publication called *Rappler* in 2010, she designed it specifically to run on Facebook. "I had always thought that this technology would help to solve problems bottom-up," she says. "And it did for a while, up until 2015."

That was when a candidate for the Philippines' May 2016 presidential elections, a populist authoritarian named Rodrigo Duterte, spread misinformation about his opponents and misrepresentations about conditions in the country at large. A body of pro-Duterte bloggers flooded Facebook with horrific posts that took full advantage of the viral power of the News Feed. Its design visually treated marginal or unscrupulous "news" sites the same as the most highly vetted publications. And because those dicey operations commonly dealt in sensational content that was hard to ignore, Facebook rewarded them.

"Newspeople don't tell lies, but lies spread faster," Ressa says. She had bet her entire publication on Facebook but now was being eclipsed by the false information from Duterte bloggers. The country was inundated with posts like a fake sex tape where the head of Duterte's female opponent was digitally grafted onto the body of a porn actress. Facebook also was empowering the Duterte mob to use the platform to attack his critics, putting them in danger from his angry supporters. Ressa was personally targeted.

And despite her multiple complaints, Facebook was doing nothing to stop this.

Ressa thought that after Duterte won the election in May 2016, things might calm down. But then he began using the same tactics on Facebook to push his governance platform of strong-arm tactics.

Ressa understood that the Duterte forces were drawing a road map for future political abusers around the globe to use Facebook. She pushed for a meeting to warn the company. In August 2016, she met with three senior Facebook officials in Singapore. She had identified twenty-six fake accounts that were able to amplify their hateful and false information to 3 million people. "I began showing them lies, the attacks against anyone who attacked [violent acts by Duterte supporters]," she says. One example was a post from the Duterte campaign spokesperson, showing a photo of a girl he claimed was raped in the Philippines. "We did a check and it showed that the photo was a girl from Brazil," says Ressa, speaking to me in 2019. "And yet that post was allowed to stay up. It's still up there today."

It seemed to Ressa that the Facebook officials were in total denial of what she was pointing out with clear evidence. "I felt like I wasn't talking to people who use Facebook as well as I do," she says. Despite her handing over the names, Facebook did not act on it for months, even after Ressa published a three-part series about the misinformation, and she was personally targeted with thousands of hate messages. (Facebook says that when it got the necessary information, it acted on the accounts.) Later she would recall a moment in the meeting when, frustrated, she reached for the biggest hyperbole she could think of to portray what might happen if such practices continued. "If you don't do something about this," she said in August 2016, "Trump could win!"

The Facebook people laughed, and Ressa joined them. It was a just a joke. Nobody thought *that* could happen.

In the fall of 2016, Facebook still wasn't thinking of the News Feed as a propaganda machine. But the Trending Topics experience had made it impossible to ignore how many low-quality stories and outright hoaxes were being spread on Facebook. Every Monday, Facebook's top managers—the small group—gather in Zuckerberg's conference room for a long meeting. The first hour is devoted to the topic du jour, and the rest of the time the focus turns to specific projects. That first hour is freewheeling—a time where anything goes. The fake-news topic came up one Monday as the election approached. While the company surely had to address it, the small group decided it was too risky to do so in the heat of the contest. "We didn't want to overreact to it and create a political snafu for ourselves," says Bosworth. "We're worried about taking action and spawning a big flame-up. We were aware of our natural perception as being aligned to Democrats. So we assume that's the bias. We didn't want to interfere with an election. We figure anything that looks like we're playing one side of this against the other is off-limits."

So in order to avoid interfering with the election, Facebook effectively gave a green light to misleading, sensationalistic posts that themselves arguably interfered with the election.

The ultimate justification for this could be attributed to the engineering mentality that Mark Zuckerberg celebrated in his company. It was a matter of metrics. Compared to the number of total posts hosted by Facebook, the disputed content was minuscule. Those on the product side viewed it from a data perspective and noted that fake news comprised a tiny percentage of the billions of stories posted to Facebook every day. The numbers did not indicate the urgency of the problem.

"These people had all the power," says one Facebook executive. "All their metrics were better ad metrics, more growth, more engagement. That's all they cared about. And [on the Sheryl side] they're dealing with the downsides of all that. And that was effectively how the company ran."

In short, Zuckerberg's inner circle had no clue that misinformation was thriving in their system because, well, where was the data? "We do a lot of work to understand what the top twenty-five things are that people are concerned about or the things where people are having bad experiences," says Chris Cox. "We asked them what are the bad experiences you're having and then we rate the bad experiences and then we get things like sensationalism, click bait, hoaxes, redundant stories, and stuff like that. But as a practical matter, [misinformation] wasn't on our radar. We missed it."

"The dirty secret no one talks about is that the stuff was really small," says Bosworth. "So we're just like, *How can we deal with this? Can we build good policies that we think are uniform?* And so we're talking about it but it's not urgent. I would say honestly it was just business as usual until the election, which all of us thought Hillary would win. Like a lot of other people, I assume."

IN LARGE PART, Facebook's rationalization for not addressing the outrageous misbehavior on its platform was just that: Clinton was going to win anyway, so why needlessly alienate the losing team?

In 2008 and 2012, the Obama campaign had been masters at Facebook. Rather than building on that experience, the Clinton campaign acted as if social media was some fringe unproven medium. The Clinton-Kaine team stuck to traditionalist media buys and seemed to take almost a perverse pride in being clueless novices at Facebook. When Facebook offered to provide on-site guidance on how to run campaigns on Facebook, the Clinton team spurned the opportunity. "The Hillary camp just did not understand the value," says a Facebook official. "They didn't see it." They spent only a fraction of the Trump team's Facebook budget. And the few Facebook ads they did use were woefully misspent. One example was a painstakingly produced two-and-a-half-minute ad about her campaign—a mini-documentary of sorts—that for some reason the Clinton media people felt was appropriate for Facebook. The ad resonated more with women, and as a result, the Facebook algorithm delivered it to female users. Because the Facebook ad auction rewards advertisers who target people who most would want to see the ad, it also cost less to show it to an audience of exclusively women. But the Clinton team wanted it shown to both men and women, even though it would bust the budget to show the ad to men as well.

"Hillary's team looked at that and said, I see the problem, so we'll increase the budget to get more men," says a tech executive familiar with the ad. "They are effectively paying more money to get the ad to people who don't want to see it!"

The Trump team also began as novices on Facebook. But they learned quickly, hiring a hitherto obscure forty-year-old website designer named Brad Parscale to run their digital campaign. Parscale got the job by playing a long game. Years before the election he'd hooked up with the Trump family by underbidding rivals to get the job of designing a Trump corporation site. His work impressed Trump's son-in-law Jared Kushner, who tapped Parscale in 2016 to help with the election.

Parscale understood that a traditional campaign would not work for this nontraditional candidate. He also understood that the microtargeting tools of Facebook—and the expertise of Facebook's free consultants—could compensate for a gap in spending between Trump and his opponent. Parscale *did* accept the professional guidance Facebook offered all big advertisers, and several Facebookers worked virtually full-time advising the Trump team how to maximize their ad spending.

"I asked Facebook, I want to spend $100 million on your platform, send me a manual," Parscale said to *Frontline*. "They said we don't have a manual. I said send me a human manual then, and that's pretty much it." The advantage of having people on site was, when a glitch arose, the Facebook person could contact the engineers immediately to address the problem. "If I would have chosen the way Hillary's campaign did it," said Parscale, "I'd have to send an email and make a phone call, wait a couple of days, and then have it fixed. I wanted it fixed in thirty seconds."

Parscale started with a budget of $2 million to build up a database, and sunk it all into Facebook, eventually spending, as he noted, much, much more. Like Facebook itself, the Trump campaign's Facebook team was a giant testing machine, treating every ad like an experiment, and sifting through the results to figure out which groups responded to which ads. They took Trump's stump speeches, cut them into fifteen-second slices, and ran them to a variety of demographics. The ones that Facebook delivered were repeated and further refined. The ones that didn't work were junked. By October, Trump was running hundreds of thousands of different video "creatives"—meaning styles of ads—with algorithms testing almost infinite variations. One Trump campaign official told *Wired* that the campaign once ran 175,000 variations of an ad on the same day.

The targeting was made possible by specific Facebook tools designed to ostensibly make ads relevant and welcome to its users. Parscale began directing ads to groups defined by Facebook as Custom Audiences, where advertisers could mix and match traits like gender, race, residence, religion, and other interests (BMW owners! Gun lovers!) to isolate groups. When a group was found to be a particularly fertile ground for sowing Trump affinity, the campaign used a tool called Lookalike Audiences to expand its targeting to non-obvious but algorithmically like-minded populations. This strategy—known as "uplift "or "brand lift"—had been pioneered by the Obama campaign.

Furthermore, Parscale used multiple creative agencies that competed with one another to deliver the best Facebook ads. Each of these teams would wake up at 6 a.m., start a campaign for a new region, and at noon would rejigger the budget to spend on what had worked best. The agency with the best ads got the money, and the losers would then look for a different demographic to win the next day.

By the end of the campaign, Trump's team had a database of age, gender, region, and other demographics, and which messages resonated for each one. Facebook's worry had been that its targeting infrastructure would encourage politicians to deliver different messages to different groups—pro-immigration to one region, anti-immigration to another. That was tempting because Facebook ads, unlike, say, radio or television ads, aren't generally exposed—they go straight to the News Feed streams of targeted users. But Trump's campaign didn't *have* to do that because it used Facebook to figure out which of its many messages would drive a dagger into the brain stem of each individual. "They were just showing only the right message to the right people," says the tech executive familiar with the techniques. "To one person it's immigration, to one person it's jobs, to one person it's military strength. And they are building this beautiful audience. It got so crazy by the end that they would run the campaigns in areas where he was about to give a stump speech and find out what was resonating in that area. They would modify the stump speech in real time, based on the marketing."

Because of the News Feed's tendency to promote sensational content, Trump's wild experimentation found that the most salacious ads would be generously shared by the targets to their friends—and the resulting "organic" distribution was utterly free.

And what did the Trump people do when they found an audience for whom *nothing* resonated, implying that they weren't likely to vote for Trump? To those people, they ran anti-Hillary ads, hoping to discourage anti-Trumpers from voting at all. According to a *Bloomberg* article written by Joshua Green and Sasha Issenberg, who were granted access to the Trump digital campaign in the late stage of the election, Parscale and his team identified three groups among those who were never going to vote for Trump: "idealistic white liberals, young women, and African Americans." The liberals got ads tweaking Clinton for the misdeeds uncovered by the hacked emails of her campaign aides (conveniently stolen from Democrats' inboxes by Russian military operatives). Young women would be reminded in lurid terms about Bill Clinton's sexual misdeeds and the candidate's unsavory treatment of the White House intern at the center of the scandal. African Americans would be reminded that Clinton once referred to some criminal black men as "super predators." (The Trump people, of course, did not remind African Americans that Donald Trump had taken out a full-page ad calling for the execution of the Central Park Five alleged rapists, who had falsely confessed.) The explicit goal was voter suppression.

If you had an audience for whom *everything* resonated, you'd send them more donation ads, which were crucial because Trump, taken aback by his shocking triumph in the primaries, went into the general election with an empty wallet.

Parscale worked his database—which he called Project Alamo in tribute to his San Antonio headquarters—overtime in key states like Florida, Michigan, and Wisconsin. States that would tip the Electoral College to Donald J. Trump.

"Fucking beautiful!" says a tech executive who followed the campaign. "They ran the greatest digital marketing campaign I've ever seen, completely by accident. They just did very commonsense things in a new era."

Lots of people at Facebook knew that Trump's people were playing the platform like a Stradivarius while the Clinton team was banging it like a broken tambourine. As a matter of course, the ad team had weekly meetings where they discussed big spenders and whether their budgets were increasing or being cut back, and what could be done to better serve them. As Election Day approached, the disparities were more and more stark. Not only was Trump outspending Clinton, but his campaign was simply better.

"In every way, the way they used the product was different," says Facebook advertising VP Rob Goldman. "The degree to which they measured their outcomes, the kinds of creative they used, the timing of their spend, the way they did their targeting. They took our best practices and operationalized them."

Yet even as people in Facebook's ad organization saw the tremendous imbalance in the way each campaign used the platform in both quality and quantity, they regarded the disparity as a curiosity, not something that could become a factor in electing a candidate most of them fervently opposed. "Even having seen all of that [Trump ad activity], nope, I didn't think Trump was going to win," says Bosworth, echoing a sentiment widely held inside Facebook. "It was so unthinkable to me that I had ruled it out."

. . .

IN THE BLUEPRINT of the company organization that Mark Zuckerberg and Sheryl Sandberg drew up when she joined the company, policy issues, security, and communication were all in the domain of Sheryl World. While he would weigh in on major decisions, Zuckerberg was happy masterminding product at Facebook. Sheryl would oversee the rest.

But by many accounts of Facebookers, as 2016 loomed, Sandberg herself was not operating at peak performance.

On May 1, 2015, she and her spouse, Dave Goldberg, were visiting a high-end resort in Punta Mita, Mexico, celebrating the birthday of Marne Levine's husband with several other couples. That afternoon, Goldberg went to the gym to exercise and didn't return on time. Sandberg and Dave's brother found him on the floor by the treadmill, blood around his head. He wasn't breathing.

Dave Goldberg was dead at forty-seven.

Sandberg had lost her rock. Goldberg was a perfect partner and foil for Sandberg. Though he was a formidable CEO on his own, building the SurveyMonkey site into a Valley success story, he had also partnered equally in domestic duties with his high-flying bride. He also had taken pride in his wife's success, both as an executive and as the bestselling author of a book, *Lean In*, about the empowerment of workingwomen, as well as the foundation intended to organize women around the movement.

Which made it even harder when he died. As she later recounted in a second bestselling book, Sandberg had spent a lifetime controlling her surroundings and overcoming problems with preparedness and hard work. Grief could not be erased with an A+ performance.

Some at Facebook say that even a year later, during the critical 2016 election, Sandberg was not the same. In part, the impact on the company was mitigated because some of the toughest work she had done at Facebook had paid off—the teams she had built were able to operate on their own. Other areas, though, could have used a firmer hand.

Sandberg could be a challenge to work with: despite her public persona as a corporate goddess of sympathy, she was prone to yelling at subordinates when they did not live up to her considerable demands. She could be obsessed with her public image. The meticulously planned 2013 rollout of *Lean In* had been soiled by a hostile prepublication article that set the agenda for criticism: Sandberg, read the indictment, was a super-rich executive not in touch with the problems of everyday women. It was especially devastating because the article came from a *New York Times* writer known as a pro-feminist. The article corralled feminist commentators who were happy to label Sandberg as an elitist—"double Harvard degrees, dual stock riches . . . a 9,000-square-foot house and a small army of household help"—falsely providing hope that women could Have It All. "It set up a counternarrative," says Brandee Barker, Facebook's early PR person, whom Sandberg had hired to help promote the book. "She was shaken." What made it worse is that the *Times* doubled down on the criticism with a snarky Maureen Dowd column. "Sandberg has co-opted the vocabulary and romance of a social movement not to sell a cause, but herself," wrote Dowd. Despite the book's huge success, Sandberg still bristles at the jibes. "I did this because it had to be done," she told me. "When I meet with people I will talk about Facebook and I will almost always talk about women. And sometimes they like it and sometimes they don't, but I will keep doing it."

When Facebook came under scrutiny after the election, Sandberg became even more conscious of her image. Sheryl World, says an insider, "is comms [communications] driven. That is the center of gravity. She thinks in the lens of *How is this story going to be written, what are the headlines?* She knows everything going on from a comms perspective, down to the lowest person. She regards herself as the best comms person in the company. And she's quite good at it." But maybe she was losing her touch. Sandberg had commandeered part of the policy budget to hire an outside firm, TSD Communications, led by a former Treasury colleague, at what one official with knowledge of the terms pegs at $30,000 a month. (Sandberg's office disputes the figure.) According to two people close to the situation, TSD reviewed anything that went out under her name. Now that Facebook was under scrutiny, she micromanaged even harder for every press appearance; once she confided to a colleague her pre-interview strategy of telling the reporter she was nervous, in hopes of an easier interrogation.

Throughout her tenure she had been close to Facebook's head of policy and communications, Elliot Schrage. But in the months after her husband's death, Sandberg seemed to sour on him. People around her report there were screaming matches in her conference room. (Sandberg professed puzzlement when asked about this, saying that her behavior during mourning might have been misinterpreted.)

In any case, going into the election year her energies seemed focused more on the business side than policy. "If you think about who was actually running a bunch of this [policy] stuff, who was making the real decisions," says one executive in Sandberg's organization, "it was Joel Kaplan and Elliot Schrage because Sheryl was not managing those guys." (Sandberg says this is not the case.)

Sandberg, along with Zuckerberg, would later admit that Facebook was too slow to attack the fake news problem. But by the time the election season began, the platform itself was an ideal misinformation distribution machine. Due to the design and algorithms of the News Feed, fake news was essentially a product problem, under Zuckerberg's domain. But there was no appetite to apply engineering to fix the problem. In part, misinformation persisted because Facebook, from Zuckerberg on down, believed in free speech, even when people didn't tell the truth. He held a Panglossian view of the goodness of humanity, and felt that people would sort out for themselves what was true. In addition, he viewed with pure terror the idea that Facebook would be drawn into becoming the arbiter of truth.

"Facebook at the time had no interest in actually making decisions about quality and not quality, truth or not truth," says Andrew Anker, who joined Facebook in 2016 to help with its news strategy. "That was a very dangerous area."

But in the last weeks of the campaign, fake news spiked dramatically, and some on the policy team came to understand that the company's inaction was leading to disaster. When news outlets and researchers began to question them about why some of the most popular posts on the entire network were bogus lies, Facebook had no compelling answers.

The press had little problem identifying the offending posts. One can't-miss technique among the fakesters was to create a phony news source with a legitimate-sounding name, make up stories that hurt Hillary Clinton, and publish links to the bogus story on Facebook. Even people who didn't follow up with a click would see the headline and a brief description. The classic example of a fake-news outlet was *The Denver Guardian*. A website of that name was registered on July 16, 2016, and it was largely dormant until November 5, when a faux news article appeared with the headline FBI AGENT SUSPECTED IN HILLARY EMAIL LEAKS FOUND DEAD IN APPARENT MURDER-SUICIDE. So many people assumed that it was reporting by the major news source in Colorado that the real newspaper, *The Denver Post*, ran a story proclaiming, "There is no such thing as the *Denver Guardian*, despite the Facebook post you saw." It pointed out that the newsroom address of the *Guardian* was a tree in a bank parking lot. Nonetheless, the story got more than 500,000 shares and its headline was viewed 15 million times.

Later an NPR reporter discovered that *The Denver Guardian* was a creation of a forty-year-old man living in the suburbs of LA. A Democrat, he'd been running a stable of twenty to twenty-five writers cranking out totally fictional stories appealing to conservatives. "We've tried to do similar things to liberals," he told NPR. "It just has never worked, it never takes off. You'll get debunked within the first two comments and then the whole thing just kind of fizzles out."

In the weeks before the election, a shockingly high percentage of Facebook's top stories appeared to be made-up news emanating from a small town in Macedonia. In early November BuzzFeed tracked more than one hundred top US political sites, many with huge Facebook pages, to Veles, population 45,000. As with *The Denver Guardian*, the motivation was purely financial. "These Macedonians on Facebook didn't care if Trump won or lost the White House," wrote a *Wired* reporter who visited Veles after the election. "They only wanted pocket money to pay for things—a car, watches, better cell phones, more drinks at the bar." The cottage industry of Veles was plucking out an anti-Clinton story somewhere in the conservative blogosphere, circulating it on Facebook, and cashing in on ad views when people clicked to see the article, which was usually a complete fiction. But on Facebook it looked like real journalism. That was the case with Veles's biggest hit, "Hillary Clinton in 2013: 'I Would Like to See People Like Donald Trump Run for Office; They're Honest and Can't Be Bought.'" In one week, Facebook users engaged with it 480,000 times, and viewed it millions more times. In contrast, BuzzFeed noted, the *New York Times* mega-scoop on Trump's finances generated 175,000 Facebook clicks in a month.

Indeed, in the last three months of the election, the engagement from fake news stories on Facebook exceeded those from mainstream media outlets. And people were noticing.

Roger McNamee, one of the company's early investors, was upset enough to take the company to task in an editorial that *The Verge* had asked him to write. Before he sent it to the editors, he shared it with Zuckerberg and Sandberg. "The cover note was, *Guys, I'm really concerned that there's a systemic problem. This is an op-ed that I've been invited to publish but I really want to talk to you about it, okay?*" They both wrote back assuring him everything was in hand, and sent him to Dan Rose, who spoke to him a few times before the election. McNamee didn't publish the op-ed. But he'd be back.

AS FACEBOOK EMPLOYEES arose on November 8, 2016, they had every reason to believe that the troubles of the election season would end with the victory of Hillary Clinton.

But during the day some Facebook policy people who were monitoring real-time conversations as well as who was sharing the "I Voted" button began to suspect otherwise. "Trump always had the lead in conversation on the platform, and people were like, *That makes sense, a lot is negative*," says one Facebooker watching the results. "But around the time Florida turned, I knew there was going to be a new level of scrutiny."

Donald Trump was the president of the United States. Facebook employees were not alone in their shock and grief. But not many companies had questions of culpability like this one. Almost immediately, the question came up: Did *Facebook* have a role in this?

In the office the next day, employees straggled in like they had been beaten up in a bar fight and woke up without their wallets. People were crying openly. Zuckerberg led a town hall meeting of stunned engineers, designers, PR people, and policy wonks. Groups sprang up on Facebook's internal, company-only pages with names like "Facebook (the company) Is Broken" or "Refocusing Our Mission." The introduction to the latter read, "The results of the 2016 Election show that Facebook has failed in its mission."

In Facebook's policy sphere, some people were angry at Joel Kaplan, who had protected the conservative cause throughout the process. Kaplan had to address the issue. He told them the results were a shock to him as much as anyone. Even though he was a Republican, he hadn't voted for Trump, he said. But now Facebook had to get used to the fact that Trump was elected, even if much of its core leadership didn't like the guy.

Two days after the election, Zuckerberg was scheduled to be interviewed onstage at the Techonomy conference. The feeling among his staff was that addressing the issue early might be a good way for Facebook to get past it. It did not work out that way. When the interviewer asked Zuckerberg about the election, he answered, as he often does, in a long-winded explanation of Facebook's mission and how its system worked. Finally, he addressed the issue of fake news.

> *I've seen some of the stories they're talking about around this election. Personally I think the idea that fake news on Facebook, of which it's a very small amount of the content, influenced the election in any*

way, I think is a pretty crazy idea. Voters make decisions based on their lived experience. . . . We really believe in people and you don't generally go wrong when you trust that people understand what they care about and what's important to them, and you build systems that reflect that. . . . There is a certain profound lack of empathy in asserting that the only reason why someone could have voted the way they did is because they saw some fake news.

I was in the room during that interview, and can report that the remark in question was greeted with equanimity. Zuckerberg seemed to be reasonable in discussing it. But once out of the room a single remark survived: *Facebook's CEO said it was crazy to think that fake news had an impact.*

Months later, he would apologize. Given what was to be unearthed, he had no choice.

WHILE FACEBOOK WAS making a belated attempt to see where the fake news came from, outsiders were also focusing on it. In the weeks after the election, misinformation on Facebook became a favorite target of finger-pointers devastated by the Trump victory.

Even allies of Facebook, like early designer Bobby Goodlatte, acknowledged that Facebook's algorithms supercharged fake news. "Sadly, News Feed optimizes for engagement," he wrote in a post on Facebook's internal discussion forum the day after the election. "As we've learned in this election, bullshit is highly engaging." Loyal Facebookers, including Boz himself, disputed this. One of the policy PR people responded with a comment that the presence of fake news was *good*: allowing users to share misinformation is part of Facebook being "humble as a company . . . the last thing we need to do is define 'truth.'"

Among the critics was the outgoing president of the United States. Before the election, speaking at a Clinton rally in Michigan, Obama inveighed against the "outright lies" plaguing the candidate. "As long as it's on Facebook and people can see it . . . people start believing it," he said.

"And it creates this dust cloud of nonsense." In an interview with *The New Yorker*'s David Remnick, he diagnosed the problem that Facebook was unable to justify as Election Day neared: "An explanation of climate change from a Nobel Prize–winning physicist looks exactly the same on your Facebook page as the denial of climate change by somebody on the Koch brothers' payroll. And the capacity to disseminate misinformation, wild conspiracy theories, to paint the opposition in wildly negative light without any rebuttal—that has accelerated in ways that much more sharply polarize the electorate and make it very difficult to have a common conversation."

Now, post-election, he was continuing to express his concerns. At a November 17 appearance in Berlin with Angela Merkel—a victory lap in Europe soured by the election result—Obama bemoaned how "very well-packaged" misinformation appears on Facebook to be genuine news. "If everything seems to be the same and no distinctions are made, then we won't know what to protect," he said, twice saying that fake news threatened democracy itself.

In mid-November, Zuckerberg was scheduled to appear at a summit in Peru where the president would also attend. Obama requested a small, off-the-record meeting with him. Obama staffers told *The Washington Post* that the president intended to give Facebook "a wake-up call," urging Zuckerberg to be more aggressive in dealing with fake news. The Facebookers say they were there in part to brief *Obama* about fake news and the steps they were (belatedly) taking to address it. "I actually asked for the meeting because he had made some comment publicly and I wanted to make sure that he knew everything that we were doing to work on it," says Zuckerberg.

When things go wrong, Zuckerberg always prefers to couch a mea culpa inside a prescription for fixing things. In November, the News Feed team had set about what would be a long process in trying to deal with the problem. Adam Mosseri, who was in charge of the News Feed, convened a meeting in his conference room (named Dunder Mifflin, after the hapless firm in *The Office*) to come up with ideas. True to its engineering mindset, Facebook was trying to fix its problem by product tweaks. The team came up with a number of approaches to minimizing fake news, like helping people identify the sources of a story, fact-checking questionable stories, and more aggressively weeding out bogus accounts that spread toxic posts. All of these were on the table now that the election was over. But what was not on the table was the idea of outright banning misinformation from the platform. That would be a violation of Zuckerberg's core belief about granting his users free expression. A platform of censorship would mean the end of his dream. The goal would be minimizing those lies, or burying them in the low-ranking sub-basement of the News Feed scroll.

On the plane to Peru, the Facebook team worked on an announcement Zuckerberg would make on his page when they landed that evening. He admitted it was an unusual post in that he was expressing what Facebook was planning—the News Feed tweaks discussed at Dunder Mifflin—instead of what Facebook was shipping. "Some of these will work well and some will not," he wrote, "but I want you to know that we have always taken this seriously, we understand how important the issue is for our community, and we are committed to getting this right." In the meeting with Obama the next day the two sides appeared to talk past each other. Obama seemed unaware of the announcement, and just repeated the talking points he expressed in Germany.

The Facebook people would come to wonder: *If Obama's people knew so much, why didn't they tell us?*

ON ELECTION DAY, Alex Stamos had been in Lisbon, Portugal, scheduled to speak at a big web conference the next day. Though he felt it would have been fun to follow the results, he wanted a good night's sleep, so he popped an Ambien and turned off his phone. The next day, stunned by the outcome, he hurriedly added a line to his talk: "We are coastal elites here. We're the class of people that are famously surprised by the election." When he checked his inbox afterward, though, the expressions of disbelief overwhelmingly were deep questions about the causes of the misinformation people had been seeing. Had this been an organized attempt to use Facebook to tilt the election?

Stamos vowed to look into it.

Over the next few weeks, he and his team pursued an investigation into where fake news was coming from and how it might be identified in the future. He came to feel that Facebook, and Zuckerberg in particular, still hadn't grasped the enormity of the problem.

He had a completed report in December. It found that most fake news stories were in the Macedonian vein of sensational stories pitched at willing viewers, in hopes of getting lucrative clicks. These had been easy to find, by following the money. The "landing pages"—websites that you visited when you followed the links—did not resemble real publications but were jammed with low-quality ads.

But Stamos also wanted to make it clear that the foreign involvement itself was a serious problem. And Facebook had yet to figure out the extent of it. The report went into detail about the GRU's involvement, admitting that Facebook had yet to learn how deeply the Russians had cracked the code of using Facebook to spread propaganda. It included a number of screen shots of pages that his Threat Intelligence team believed were of Russian origin, not just involving the election but earlier forays into Ukrainian-related disinformation, and even propaganda regarding the Olympics. To emphasize this point, Stamos put the logos of the Russian intelligence agencies on his report.

Attacks from a hostile super-state were not something that could be addressed by tweaking News Feed signals. It required deeper understanding—and direct involvement from Zuckerberg. That was tricky. Because of the organizational structure split between Zuckerberg and Sandberg, Stamos, the chief security officer, had never even had a one-on-one with his CEO.

So Stamos did something unusual with his report: he essentially crossed the border of Sheryl World and engaged the product side. He sent it as an email attachment to the people whom Zuckerberg listened to most closely—Chris Cox, Adam Mosseri, Naomi Gleit, Javier Olivan. He understood that those people were the ones who really ran the company, working on the things that Zuckerberg cared about. Those who were on the phone at midnight during every crisis. That was the only way he felt he could bound past the Policy team's negativity and truly get Zuckerberg's attention.

After those leaders got the report, Stamos met with them in Chris Cox's conference room. As head of product, Cox was probably the second-most important person at Facebook—more than Sandberg, some insiders believed—and he was unhappy that this was the first he was hearing of it. All agreed that Zuckerberg should hear it, too.

The next day, about twenty people met in Zuckerberg's Aquarium to discuss the report. It seemed that Zuckerberg, like Cox, had not been previously briefed on the Russian issue. He pelted the team with questions, many of which no one had answers to. Zuckerberg ordered Facebook's leaders to form a committee to examine what could be done. They called it Project P.

The P stood for propaganda.

"I think at that time we still didn't have a great way to systematically understand this," says Naomi Gleit, who was in charge of the project. Largely guided by Stamos's team, they dove into analysis of the problem. Gleit, one of the earliest employees besides Zuckerberg himself, felt like a film had been lifted from her eyes. In close conjunction with Facebook's growth czar, Javier Olivan, she worked with data scientists to produce their own report.

But once again, the tyranny of metrics dictated the outcome. The Project P team found that of the top one hundred fake news stories, none of them originated from the suspected Russian groups. Fake news, they concluded, was really a problem of cutting off the money supply of malefactors, like the Macedonians in Veles, who were gaming the system. In a sense, they saw it as a similar situation to Facebook's problem with spammy developers. Dealing with fake news, under that view, was not really much different from taming the excesses of Zynga's Mark Pincus.

"We understood some of the disinformation efforts had been traced to Russia [but] the fake news problem seemed larger," says Colin Stretch, Facebook's general counsel. So, belying its name, Project P went hard on financially motivated fake news and gave propaganda a pass.

While not disputing Project P's conclusion, Stamos still felt that Facebook should sound an alarm about foreign interference on Facebook. He felt the public should know about the GRU's meddling, as such activity was an ongoing concern. With two members of his Threat Intelligence team he co-authored a white paper that would be openly distributed. Once again he found himself at odds with Facebook's policy heads. The December report had included screen shots of some of the pages that Stamos and his team had traced to Russian origins, along with other explicit mentions of GRU activity on Facebook. The policy heads of Facebook—particularly Joel Kaplan, according to sources—did not want any such information included. Whether intentional or not, the argument had a political implication. By then, Donald Trump was vociferously denying that he was helped by the Russians during the election. Why taunt the new president?

So while the thirteen-page white paper talked at length about how foreign interference might operate, it did not single out Russian involvement. In fact, the word "Russia" did not appear. "Facebook is not in a position to make definitive attribution to the actors sponsoring this activity," the authors wrote. They also cautioned that state disinformation is but a small part of "false news" on Facebook. The one concession to acknowledging Russian involvement was a note about how nothing in the white paper contradicted a recent report from the US director of national intelligence—which *was* explicit that the Russians tried to screw up our election. But you would need a sharp eye and a lot of knowledge to pick up that clue.

"The compromise was that we point to the DNI report and don't say *Russia, Russia, Russia,*" says Stamos now.

It was later reported by *The New York Times* that Sandberg herself approved the omission of Russian activity in a paper whose point was exposing, well, Russian activity. She denies it vociferously. "I knew there was a white paper being written, like, vaguely, but no one asked me about whether Russia should be in the footnote or not," says Sandberg. "I wasn't involved in that."

Facebook released the white paper on April 17, 2017. Despite its enforced timidity in identifying who was attacking the company (and the United States), Stamos did not consider the white paper to be a whitewash. "In the end, I was okay with the compromise because we got it out," he says. "We had to get something out." They felt that, unlike other parties (Twitter, YouTube) who were victims of interference, Facebook took a step to warn the public and the authorities of the danger of foreign meddling on social media. Editing the Russians out of the paper, goes this theory, was simply caution.

Later it would become clear that the report was incomplete. Despite Stamos's investigation, despite Project P, Facebook still had no idea of the degree to which Vladimir Putin had played Facebook.

But they would soon find out.

15

P for Propaganda

On February 9, 2017, Zuckerberg summoned me to the Aquarium in Building 20, the hangarlike headquarters across the street from the former Sun complex, now called the Classic campus. Designed by Frank Gehry, Building 20 was an extreme example of the just-about-to-fall-apart ethos of the blue-chip architect's work, with exposed pipes, wires hanging from the distant ceiling, and walls that looked like temporary plywood barriers. ("He did not want it overly designed," said Gehry of his client.) Stuck on those walls was the latest crop of posters from the Analog Research Lab, including newly silk-screened pleas to "Be the Nerd." A quarter mile long, Building 20 contained under its twenty-two-foot ceilings a seemingly random jumble of long tables with young people staring at their personal cluster of screens. Distributed haphazardly throughout the 430,000 square feet were conference rooms, no-fee cafés, and even a premium coffee shop that took credit-card swipes for its chai tea and Americanos. The roof had been sodded to support regional flora, with dirt paths winding through the foliage to support work-break mini-hikes. (Later, an attached twin structure—Building 21—would allow for longer perambulations.) People who'd worked in the building for months still had to consult the ubiquitous display monitors on the walls to find their next meeting. They had nicknamed the two main thoroughfares through the building, parallel paths through the silicon jungle, after the two highways binding Silicon Valley and San Francisco: Interstate 280 and Route 101.

Roughly in the center of Building 20 was Zuckerberg's glass-walled conference room, with couches and chairs arranged around a center table, giving it a casual feel. Whiteboards and display screens guarded the edges of the room where one might sketch the next big product—or, increasingly, a fix for a product causing havoc in the social graph.

Just as foot traffic in Building 20 was split between paths 101 and 280, Facebook itself was moving along two thoroughfares in 2017. One was a high road of good intentions and blockbuster earnings reports. The other was a nightmarish downhill boulevard of damaging revelations. In 2017, its onetime wunderkind founder was now husband, father, billionaire, and protector and harvester of 2 billion signed-in Facebook users (though almost none of them had plowed through its Terms of Service file, which had grown to *Finnegans Wake*–ish proportions, both in bulk and in comprehensibility). Befitting his status, Zuckerberg was trying to sail above crises with big thoughts. Unlike the earlier days when he would limit these to a private notebook, he now dispatched his views like lightning bolts from his fishbowl Olympus, to be consumed (and "liked") on his Facebook page, which had millions of followers.

In the midst of the discussions at Facebook about its role in the election, fake news, and whether the secret Project P would uncover Russian interference, Zuckerberg contextualized his company's problems as signals of a wider unrest, a divisiveness that split America in the recent election and that was spreading globally, like a viral post on the News Feed. He sensed that in the coming year he would have to admit to some of Facebook's shortcomings. But with his typical instinct to seize opportunity, the onetime addict of the game Civilization had decided to broaden the discussion to a larger conversation about what ailed not just his company but the world at large.

On January 3, Zuckerberg had posted his annual challenge. He planned to embark on a political-style listening tour around his home country, visiting every state on the continent, save the ones he'd already spent a lot of time in. "My work is about connecting the world and giving everyone a voice," he wrote. "I want to personally hear more of those voices this year."

Zuckerberg had also been cooking up a manifesto to frame his thoughts and share his vision. That's what he wanted to talk about on that February day.

While previous blowups at Facebook had been centered on discrete missteps—the News Feed, Beacon, the Terms of Service debacle—the post-election crisis struck at the essence of Facebook itself. All the decisions made in the name of growth, commerce, and a single-minded push toward sharing had created an unhealthily addictive system vulnerable to bad actors. Zuckerberg's now-constant refrain, in a phrase that its officials would intone incessantly over the next few years, was that Facebook *had a lot of work to do.* Zuckerberg was always happy to share the steps it was taking to get the work done—he'd ordered bulked-up teams in safety and security, operating on the premise that the company should *proactively* attack problems rather than apologize and fix them after they surfaced. He now conceded that he'd misspoken when he said that it was a "crazy idea" that fake news affected the election. "I might have messed that one up," he told me once I had been offered a beverage from one of the well-stocked mini-kitchens in the building. (That visitor offer was as formalized at Facebook as tea service in corporate Japan.) But while he could tick off by rote the measures Facebook planned to mitigate fake news, he framed the issue as just a symptom of the global shift toward divisiveness and ill will. He felt that Facebook—*he*—could do something to turn it around.

"This is kinda my thesis here," he said. "That there's this infrastructure that needs to get built for our society and civilization to reach the next level and transcend the current tribalism that we have of 'we're a bunch of countries,' to really feeling like we are a world that can get things done together."

While he conceded that the Trump victory had unnerved people in and out of Facebook, his new crusade wasn't about a single person but a global movement. Facebook was in a position to address the problem as a builder of *communities*. That was to become his buzzword for 2017. (Facebook had long transcended his previous view circa 2007, when he was telling journalists like me that "We're not a community/social site at all.")

Indeed, when his 5,700-word essay ("Reading Time: 27 minutes") appeared on Facebook a week after our conversation, it was called "Building Global Community." It tacitly admitted that simply "connecting the world" was no longer the unalloyed blessing Facebook had envisioned. The data were in: things were more complicated. *Should we keep connecting or reverse course?* he asked.

Unsurprisingly, he chose the former. As he had shared with me earlier, the answer was *communities*, and Facebook's role would be making them supportive, safe, informed, civically engaged, and inclusive (in that order). In each of those "pillars," Facebook had . . . work to do. But Zuckerberg focused on the positive. In discussing the challenges, he hit on instances where Facebook had helped things, as opposed to breaking them. Yes, Facebook made mistakes, he admitted, but these were not because of ill intent or a poisoned business model but things like differing values among communities, or "operational scaling issues." The manifesto looked beyond those mistakes, inviting the world to join Facebook in creating a new world order of understanding and comity.

Toward the end of the essay, Zuckerberg quoted the words of Abraham Lincoln, in an 1862 speech: *The dogmas of the quiet past, are inadequate to the stormy present. The occasion is piled high with difficulty, and we must rise with the occasion. As our case is new, so we must think anew, act anew.*

Zuckerberg did not cite the words that came after that passage in the Lincoln speech, but he may as well have done so. *We . . . cannot escape history,* the sixteenth president wrote. *The fiery trial through which we pass will light us down, in honor or dishonor, to the latest generation.*

. . .

IN JULY 2017, Ned Moran and his colleagues on the Threat Intelligence team began a process that would lead to another alarming discovery. Someone from the Legal team had passed on a tip it had gotten from the government:

Look at the ads.

The question was already in the air. Only a month after the Stamos white paper that April, a *Time* magazine cover story reported that intelligence officials found that part of the Russian 2016 propaganda campaign was targeting Facebook ads to susceptible users. "They buy the ads, where it says 'sponsored by'—they do that just as much as anybody else does," a "senior intelligence official" told *Time*. An angry Senator Mark Warner of Virginia had visited Menlo Park that summer, demanding that Facebook look deeper into the source of fake news. Warner, a member of the Senate Intelligence Committee, had been increasingly critical of social media, and Facebook in particular. After the election, he had pushed Facebook to more closely examine Russian interference. He later told *Frontline*, "I was pretty disappointed with the initial pushback from Facebook, saying, in effect, 'This is crazy; Warner didn't understand what he's talking about.'"

But to that point, Facebook hadn't thoroughly checked into the role of ads when trying to identify fake news intended to disrupt the election. This omission didn't stop it from deflecting charges that ads were involved, even as the Threat Intelligence team was investigating. "We have seen no evidence that Russian actors bought ads on Facebook in connection with the election," a Facebook spokesperson told CNN on July 20.

Looking at the ads was not an easy task. Facebook had 5 million advertisers at the time, creating hundreds of millions of ads every day. Moran began winnowing these down. He was joined not only by his own team but by part of the ad organization called Business Integrity. They chose a three-month window before the 2016 election, and started looking for advertisers originating from Russia, or using Russian Internet providers, or posts written in Russian, or ads paid for with rubles. Using signals like that, the team winnowed the field to a few hundred thousand ads. Then they began looking at the ads themselves, determining which had political content. They looked for keywords like "Trump" or "Hillary." It was difficult, because some of the language in the ads wasn't in text format but part of a graphic and not searchable. Still, they began to narrow the field even more.

Then Moran began looking for interconnections between the advertisers, whether through similarities in the ads themselves or shared links. And, like when an image developed in a film photograph becomes clear—a darkroom phenomenon never witnessed by most of Facebook's young population—Moran suddenly surfaced a discrete network of twenty or thirty users. They had one thing in common—they were originating from the Russian city of Saint Petersburg.

That meant something to Moran. He flashed back to a 2015 *New York Times* article by Adrian Chen. It described the activities of a toxic "troll farm" operating out of Saint Petersburg, known as the Internet Research Agency. Its goal was disrupting free states to the advantage of the Homeland. "Russia's information war might be thought of as the biggest trolling operation in history, and its target is nothing less than the utility of the Internet as a democratic space," wrote Chen.

Moran and his colleagues got to work. As best as they could ascertain, the IRA had spent about $100,000 on around 3,000 ads, most bought with rubles. They were used to promote 120 pages associated with the IRA. Those pages posted more than 80,000 pieces of content that reached 129 million Facebook users.

Once Moran understood that Russia's IRA team was placing ads on Facebook, he took a close look at the content. It made him sick to his stomach. There were thousands of ads purporting to come from news outlets or engaging American citizens by making outrageous claims (Hillary Clinton's intimacy with Satan, for example), stirring up racial resentments, and playing on the darkest of fears.

That gastrointestinal response passed through Facebook like an *E. coli* outbreak, as various officials at Facebook viewed the tranche of ads that the Russians had bought and, using networks they had built up over years, further circulated. "We were in a conference room looking through them and it was just revolting," says Stretch. "It just felt so exploitative and infuriating." One in particular stuck with him—someone with a flamethrower shooting at an unidentifiable mass labeled with an offensive term for Muslims and the caption, *Let's burn them all!* "That sort of violence and the idea it was being used to whip up people who may carry certain prejudices was just awful," he says. "It was terrible content, and disturbing to all of us, I think, certainly to me, that we didn't see it."

Left unspoken was how easy Facebook made it to use demographics and interests to target voters with whom these ads might resonate. Often both sides of an issue would be targeted, with one set of ads motivating those on the Trump side to vote and another alienating Democrats in hopes they would stay home. Some were just outright civic stink bombs. Those prone to fear immigrants would be pelted with stories of crime by non-citizens, further stoking divisiveness in a country already at odds with itself.

Instagram, owned by Facebook, was also affected. As cited in the later Mueller indictment, the IRA created an account called "Woke Blacks," urging African Americans to stay home on Election Day. "We cannot resort to the lesser of two devils," said one post. "Then we'd surely be better off without voting AT ALL." Another account, called "Blacktivist," pushed the idea of voting for the ultraliberal third-party candidate Jill Stein. "Choose peace and vote for Jill Stein," said one post. "Trust me, it's not a wasted vote."

Special Counsel Robert Mueller's team looking into Russian involvement was already looking at the IRA. It would later reveal that the IRA operation was known internally as Project Lakhta. (The Lakhta Center was a recently constructed skyscraper that dominated the Saint Petersburg skyline.) Basically, the IRA was acting like thousands of companies that used Facebook as a marketing engine. It monitored metrics with internal dashboards and scolded managers who did not meet quotas. The special counsel's indictment tells how an account specialist for an IRA-created Facebook group called "Secure Borders" was called out for "a low number of posts dedicated to criticizing Hillary Clinton." The specialist was told that for the last few weeks of the campaign, *it is imperative to intensify criticizing Hillary Clinton.*

That indictment was months away. For the time being, Facebook alone had the knowledge that thousands of ads and tens of thousands of posts proved how the Russians had made Facebook part of their attack on the American election. Facebook not only had allowed the Russian poison on its platform but had given the implicit endorsement that ads enjoy on Facebook. (Facebook's standards for ads are more stringent than for the free-speech-driven user posts.)

So how did Facebook miss this as it was happening? One reason was technical: in looking for fake news, the Project P researchers had used English-language words as "classifiers"—terms for its machine-learning algorithms to identify. The ads that the Russians posted typically did not store words as text but superimposed them on images. Intentionally or not, this helped them evade Facebook's fake-news dragnet.

Another reason was the relatively low number of Russian ads. Rob Goldman, the Facebook ad executive who was in charge of Business Integrity, would later try to explain: Every day, thousands of Russian advertisers were spending tens of thousands of dollars for ads shown outside of Russia. The IRA ads were only around $100,000 total, spent over the course of eight months.

Goldman, though, recognizes that those figures, and the technological blind spot, in no way excused the oversight. After the revelation about the IRA, Goldman became obsessed with the subject of Russian disinformation campaigns, which were referred to by its intelligence agencies as "active measures." "I became a bit of a Russian scholar," he says. He read history and shared findings, such as the memoir of KGB defector Oleg Kalugin, with what had become kind of a masochistic book club of Facebook executives belatedly learning what they should have been paying attention to earlier. "The Russians have been doing this for a hundred-plus years," he says. "There are people who knew that they would try to do something like this. They had agents painting swastikas on synagogues in New York in the '70s. It's essentially the same thing they were doing with Facebook ads sixty years later."

Because Facebook had operated on the convenient premise that advertisers were well intentioned, it didn't look for what, in retrospect, were obvious signals. "Is the fact that ruble-denominated accounts were buying ads having to do with US elections unusual? Yeah, that's unusual," Goldman says, noting that Facebook later changed its practices and now watches for such things. But in 2016, the company had been too busy stoking its overseas advertising business to thoroughly consider and monitor what might go wrong. "It was one hundred percent knowable that [the Russians] would use social media in this way," he says. "The fact that we didn't think about it and identify it is a shame."

What made things even more complicated was that even when Facebook identified propaganda from Russia, it had no way to distinguish it from what its rules considered perfectly acceptable content. The team tagged the 3,000 ads and 80,000 posts according to subject: *racist . . . anti-Hillary . . . LGBTQ . . . guns . . . immigration . . .* All were valid discussion topics on Facebook. The vast majority of the propaganda fit well within what Zuckerberg considered the "free expression" of his users. Facebook took down the IRA pages because of *who* posted them, not what was in them.

"Once we finally understood what these ads were about, there's a big question of what do we do about them," says Goldman. "How do we change our policies to [work] against them? The reality is all the ads were against our policies for a weird reason—because they were *fake accounts*. If they had been created by real accounts, we would have had a hard time stopping these ads. There is no standard that you can use to say it's not okay to run that ad about immigration. If we applied something [that said] basically you can't talk about immigration, that would really curtail Americans' ability to talk about immigration, politicians to talk about immigration."

So the pages were removed, because the IRA had used dummy accounts to post them. And considering the volcanic outrage over all the fake news that had appeared on Facebook, one might have assumed that the company would immediately announce what it found. But it didn't.

For one thing, the timing was awful for Facebook. It had been touting its reaction to fake news based on its previous conclusions that most of it was financially motivated. It had released a white paper that didn't mention the Russians. And the public might scoff at Facebook's contention that it didn't know the extent of the Russian meddling until that July. So the company proceeded, depending on one's point of view and whom you talked to, either deliberately or deceptively keeping the revelation out of the public sphere.

Which isn't to say it did nothing. As Moran and Stamos had done with the original GRU news, they told Facebook's general counsel, Colin Stretch. The news made its way up to Sandberg and Zuckerberg. Facebook's legal team told the FBI. They also informed the Special Counsel's Office, which got a subpoena to allow Facebook to turn over the ads to them. And when Facebook started to brief Congress, they were surprised that the questions they got from some of the legislators with access to classified information indicated familiarity with the incursions. "In the hearings everyone was saying that they knew this," says Stamos. "If they did, why didn't they help us?"

Through that summer, Facebook gathered more information. But it still did not mention to the public that its previous declarations of no Russian ads had been superseded by the new discoveries. Alex Stamos has since said that he doesn't believe that his bosses were intentionally deceiving the American people, but were simply being cautious with a national-security issue. "[The policy team] thought we could keep our heads down and make it through," he says. "The playbook was, something bad happened, don't talk about it until we absolutely have to talk about it. But that's not the same as a cover-up." Stretch cites privacy. "We are in this space where we want to be protective of user content, as we always are," he says. "So the idea of just putting it out publicly was something I was very uncomfortable with, just like I'd be very uncomfortable with the idea of putting your private messages out. Right?"

By the end of August, Facebook still hadn't handed the actual ads to the congressional committees that were clamoring for them. Stretch says that this was due to Facebook's customary wariness of voluntarily handing over user information to government bodies. "Once you sort of declare [it's] open season for government authorities, who knows what sorts of requests you're going to get," he says.

Not everyone at Facebook felt that those justifications were well intentioned. "They were covering their ass every step of the way," says one Facebook official with knowledge of the process.

Sandberg now insists that Facebook was just working overtime to uncover the extent of the problem. By then, she says, she was fully engaged. "It's been asserted that I didn't see Alex that much before—I didn't—but then I started seeing everyone [on his team] all the time, talking to the people doing the research myself, like, *Where are the ads? Where is the organic content?* Really trying to find all of it. I was very worried we were going to miss a bunch." She says that she usually takes an August vacation, but in 2017 she canceled it to deal with the issue. "I had plans I blew up," she says.

It was September when the company prepared to inform the public that Facebook had been a staging ground for a Russian disinformation campaign designed to help elect Donald Trump. (A charge made more volatile because of the revelations coming out of the Special Counsel's Office and press accounts that the Trump campaign had frequent contact with the Russians and, as the final Mueller Report concluded, "welcomed" the interference.) Facebook would agree to hand over to Congress—but not the public—the stack of Russian-generated ads. "Ultimately we decided that we're going to give everything to Congress and then they can decide how to release it publicly," says Sandberg.

There was a lot of discussion about how much Facebook should own up to the *degree* of Russian involvement. Stamos wanted to go all out, and his proposed blog post hit on the theme that Facebook itself was a victim, and the government had done little or nothing to respond to what was, after all, an attack on America. But his bosses at Facebook threw out his draft and presented something much tamer: an anodyne report that gave the figures of ads and pages that violated policies, and noted that the ads focused on "amplifying divisive social and political messages across the ideological spectrum." That was a true but misleading statement, considering that, overwhelmingly, the ads were designed to help support Donald Trump. The same goes for the sentence about their origin: "Our analysis suggests those accounts and pages were affiliated with one another and likely operated out of Russia." No mention that this was from a Russian *state agency.* Then the post pivoted to all the things that Facebook was doing to improve.

Despite his reservations, Stamos put his signature on the sanitized blog post. He was essentially on his way out: soon Facebook would reorganize its security operation, eliminating the CSO post and blending the researchers and computer security scientists into other groups. Stamos agreed to stay until the middle of the following year, to help Facebook prepare for the next election. From running a group of 127, he would be in charge of a team of around five people.

Before going public on the Russians, Zuckerberg and Sandberg had to brief the board of directors, due for its quarterly meeting on September 6, 2017. The day before the meeting, Stretch, Stamos, and Schrage met with the Audit Committee—consisting of outside directors Erskine Bowles, Marc Andreessen, and Gates Foundation head Susan Desmond-Hellmann—to brief them on what Stretch and Schrage would present to the full board the next day. The directors were shocked, both at what they were hearing and at the fact that they hadn't heard it earlier. Bowles, who was well schooled in politics and national security, understood right away that Facebook had a huge problem. He wanted to know if there would be more revelations. Stamos said that was possible. *Who knows what else the Russians might be doing?* Certainly the US government wasn't helping Facebook keep the Russians off the platform.

The meeting lasted about an hour. Soon after, directors and executives gathered for their usual dinner the evening before a board meeting. The revelations cast a pall on the meal. The directors were angry that they had not been informed earlier that the company they were overseeing had been a Russian tool to influence the election. They let Sandberg and Zuckerberg know it.

"I don't recall any screaming at any board dinner ever, including that one," says Sandberg. "But people were pretty upset; this was a big deal. And I think we thought it was a big deal too. I think we were upset and they were upset. We were all upset together. I mean, you're really upset to find out that foreign powers or anyone would have tried to interfere in the election, like, really upset."

The next day, at the actual board meeting, the directors continued to take Zuckerberg and Sandberg to task. Sandberg was shaken. She was particularly unhappy to hear of Stamos's remark to the Audit Committee that Facebook might find even more interference. The next day, at a meeting with about twenty people, she launched a high-volume verbal assault on Stamos, telling him that she had never been as angry with anyone at Facebook. Though it was not unusual for Sandberg to scream at someone, this was a humiliating takedown, especially since some of Stamos's subordinates were in the room. After some minutes of her fury, Zuckerberg finally told her to let up.

Reports later came out that Zuckerberg had himself upbraided Sandberg for what had happened in her organization, on her watch. Zuckerberg won't share whether or not that happened. I eventually would ask him directly: *Do you think Sheryl let you down?*

There is a pause, though not as long as the epic silences of the young Zuck. "That's not the way I think about it," he finally says. "I think we all missed that this was going to be a bigger thing that we should have paid more attention to. I mean, certainly there were mistakes that I feel like *I* made directly, in decisions that I made that should have been different."

MAYBE ONE MISTAKE was taking a tone-deaf road trip in a year when Facebook's reputation was going south. While various teams at Facebook had been unearthing Russian ads and meeting with Mueller's minions,

Mark Zuckerberg kept on with his domestic travels and high-minded dispatches. What had started out as a low-key way to meet real Americans was turning out to be a complex mission to pull off, and a hide-and-seek game with a press that was suddenly rabidly interested in the young CEO who alternated between bursts of Ciceronian pronouncements and a travel agenda seemingly drawn from episodes of old TV shows like *Route 66* or *The Fugitive*. On weeks he traveled, he would try to hit a few adjoining states in the same batch. He visited celebrities and local politicians, parachuted into the lives of heartland Americans, and engaged with citizens as they dealt with difficult problems like a broken education or social justice system. One day, Daniel Moore of Newton Falls, Ohio (about fifty-five miles from Cleveland), got a phone call out of the blue, asking if a mystery guest could join his family for dinner. Fifteen minutes before the family meal, Moore learned it would be Zuckerberg, who had been looking for an Obama voter who pulled the lever for Trump in 2016. "We got to know a very cool guy," Moore told the Youngstown *Vindicator*. "Just down-to-earth and real easy to talk to."

In Indiana, he puttered around South Bend with mayor Pete Buttigieg, with whom he'd overlapped at Harvard. He took a much faster drive around a racetrack with NASCAR driver Dale Earnhardt Jr. His scheduled trip to Delaware was shaping up to be much less fun—a court appearance in a civil lawsuit about his attempt to change the corporate structure to keep all his power even if he sold off considerable stock, or ran for office. (Hmm.) But Facebook dropped the attempt at the last minute, so he zipped up Route 95 for a cheesesteak in South Philly. He dipped into the worlds of people in crisis, sitting amidst a group of opioid addicts or convicts. He visited with serving military in Fort Bragg, North Carolina, and the Naval War College in Newport, Rhode Island (the war games they studied, he told the officers in training, seemed a lot like his own favorite, Civilization). He tried his hand at building a Ford truck on an assembly line. He threw a group of workers at a North Dakota oil rig off schedule by showing up unannounced for a talk about their work. He toured a dairy farm, an oyster farm, and a wind farm.

As much as possible, he tried to low-key the visits in real time, traveling with only an aide, a PR person, and what had become a standard complement of bodyguards. (Once the visits were complete, though, he'd post to the world.) As his wealth rose over $40 billion, Zuckerberg had become increasingly concerned about safety. Though he only took $1 as annual compensation for his job as Facebook CEO, the company paid for his security, which came to $7.3 million in 2017, and would rise to double that figure the year after. In the videos regularly posted on his Facebook page, though, it would be just friendly Mark, soaking up whatever wisdom his hosts were dispensing.

One foray was built around the Harvard commencement address he would deliver in May. When Harvard president Drew Faust invited Zuckerberg she told him that previous speeches had included General Marshall announcing his plan to rebuild Europe and David Hackett Souter's first speech since retiring from the Supreme Court. So he should not shy away from addressing big topics. The prompt was unnecessary. Zuckerberg mused about his message in another pre-brief I got a week before the event. "The standard commencement trope is *Go find your purpose and go find what matters,*" he told me, "but my point is that Millennials kinda get that instinctively. There is actually a broader challenge for this generation, which is to make sure we create a world where everyone has a sense of purpose and meaning." That was his theme—purpose. Thefacebook did not have a lofty purpose when Zuckerberg launched it thirteen years earlier, but later he had come to see it as a calling.

Returning to Harvard to give the speech would be like closing a circle, he said. I asked if it would be an emotional return, and he brushed it off, returning to the subject a few minutes later. Maybe people won't notice, he told me, but he was structuring the stories in his speech around his own life journey. "So even though I'm making a point about the need to create a sense of purpose, and then talk about big projects and the problem of wealth inequality, and then global community, the emotional arc is my life," he said. The speech, delivered in a drenching downpour, won him plaudits.

Also on his New England sweep, he went to Maine to visit displaced workers in a town where the mill had closed, and checked out a middle school in Providence, Rhode Island.

No, Zuckerberg was not running for president, but was settling in for a stint as a social theorist with the rare power to affect the communications of 2 billion people. No country on Earth had a population as big as Facebook. The presidency would be a step *down*.

As he wandered the nation like some nerd Tocqueville, he also pondered Facebook's woes. As one would expect from an empiricist, he was grappling with the challenge to his core belief that when people connected and shared, the world was a better place. It really shook Zuckerberg that the Russian incursion in the election risked negating that view. His attackers *knew* that the features meant to engender goodness in the world could reliably be exploited for destruction and divisiveness. "I think what was really challenging and upsetting about some of the Russian propaganda was the way that they were misusing the platform they were trying to engage in," he later told me. "They were basically creating communities on both sides of these issues—pro-immigration/against immigration—but they obviously didn't care about immigration, they were just doing it to cause the conflict."

For months, Zuckerberg had been focusing on communities. The manifesto on "Building Global Community" was only the ante. He raised the stakes in a visit to Chicago for Facebook's first "Community Summit." It covered the expenses for 350 or so unpaid administrators of Facebook groups—things like Black Dads, Nomadic Travel, Disabled Veterans, and Fishing Places in Austin—to attend workshops, share tips, and cheer loudly when Facebook executives like Chris Cox and Naomi Gleit trotted out for keynotes, promising that Facebook was going to use the same techniques to boost invites to Groups as it successfully did for ads.

Zuckerberg appeared as an unbilled speaker, and the audience went bonkers. He pumped hands as he made his way to the semicircle stage, like the chief executive before a State of the Union speech. *The meaningful groups like the ones you people administer,* he said, *were the most valuable thing on Facebook.* But of the 2 billion people on Facebook, only 100 million were currently in these meaningful groups. He wanted everybody on Facebook in one of those groups.

Then he shared a surprise. Right then and there he was changing Facebook's entire *mission.* It was no longer just connecting the world. From this point forward it was, "Give people the power to build community and bring the world closer together." The refinement was almost an admission that Facebook's blind chase for growth had created a formless mass, ripe for manipulation. Now he was giving that mass shape.

"I always believed people are basically good," he said. "But I've also found we all need to feel supported. We don't want to feel afraid. But when we don't feel good about [our] lives here at home, it's hard to care about people in other places. Communities give us that sense that we are part of something bigger than ourselves, that we are not alone, that we have something better ahead to work for. . . . We have to build a world where everyone has a sense of purpose and community—that's how we'll bring the world closer together.

"*I know we can do this!*"

In other circumstances, the mission shift would have been a huge story. Did this mean Facebook was going to be *different*? But the announcement had less impact than any of the Facebook scandals that were now arising. Maybe people had manifesto fatigue. Or maybe, in light of what people were learning about the Russians, the shift seemed like a conscious distraction.

Still, if you were in the room full of shrieking group administrators in Chicago, Zuckerberg's message felt sincere and even inspiring. Off-key or not, in 2017, Mark Zuckerberg was devoting much of his mental bandwidth to such lofty concepts. He was consciously cruising the high road.

After the summit, he dashed off to visit a charter school for African American boys. Illinois was now checked off on his list of states.

Behind the upbeat message he conveyed at Chicago's Communities

Summit was Zuckerberg's understanding that Facebook had to change, or at least course-correct. The complaints about Facebook—from fake news to intrusive ads—focused overwhelmingly on its most popular product: the News Feed. The 2006 concept first outlined in Zuckerberg's Book of Change had become a battleground for the company's survival. The popularity of the service, and Zuckerberg's ambitions to make the feed an indispensable source of personalized information, had put an impossible burden on the stream of stories.

Though in theory scrolling could go on for thousands of posts, people only viewed a handful at a time, so the race for top ranking was a vicious contest, determined by a scoring system beyond the grasp of a single human being. Facebook loves to claim that it makes no editorial choices, and each person's preferences dictate what appears in their individualized feed. But the choice of signals and how they are weighed *were* Facebook's choices. For years Facebook evolved its EdgeRank algorithms to determine which posts had the highest ranking, but eventually the system became so much more complex that it evolved into a complicated digital gumbo of more than 100,000 signals. The weights and balances were the result of a never-ending series of experiments, conducted by data scientists working within the News Feed team. They reported to Growth. So it was not surprising that success was measured in terms of the things that Growth valued: building and retaining the user base. Engagement was still treasured.

The drawbacks of such a system were best expressed by former Google interface engineer Tristan Harris. He had earlier taken his own employer to task for addictive techniques. Essentially, his argument was that the traditional methods of maintaining attention—well known in television and even serial novels—had reached a new dimension of toxic addictiveness with the digital tools and artificial-intelligence breakthroughs of the twenty-first century. He considered News Feed and other "infinite scrolls" the worst offenders, Facebook being the worst of the worst. In the United States, around one-fourth of all mobile Internet time was Facebook time. In some countries it was even more. To Harris, these products were not just addictive distractions but an existential threat to humanity. While the movies portrayed the threat of artificial intelligence as Terminator-style robots pursuing us, what we really should fear, he argued, was Mark Zuckerberg, whose algorithms were overwhelming us with irresistible digital junk food. The fake news controversies raised the stakes even higher, manipulating our impulses to check out sensational and destructive content.

"We've actually built an AI that's more powerful than the human mind and we hid it from all of society by calling it something else," Harris says. "By calling it the Facebook News Feed, no one noticed that we'd actually built an AI that's completely run loose and out of control." Harris says that using the News Feed is like fighting an unbeatable computer chess player—it knows your weaknesses and beats you every time. It goes, *Shall I play the pawn, which is like the Trump article, or should I play, Here's your friends having fun without you! Oh, that one works really well! So I'll put that one in front of you.*

"We know what happens when human beings play chess against computers—we lose," says Harris. "It's, like, totally checkmate for the human evolutionary instrument."

Not surprisingly, the News Feed team rejected the idea that their daily labors were steps toward annihilating the human race. But the election and its aftermath forced them to deal with the fact that the News Feed could be . . . bad for users. "We have always been working to identify and reduce distribution for bad experiences," says John Hegeman, who was a VP on the News Feed team. "[But] I don't think anybody expected the extent to which this would become a big focus and an area that we realized that we needed to invest a lot more in."

In other words, *We screwed up and now have to fix it.*

ONE SPECIFIC AREA of focus was the way the News Feed handled journalism.

Beginning around 2010, in large part because of Zuckerberg's efforts to crush Twitter by making Facebook more timely, more and more actual news began appearing in the News Feed. "We started to see that referral traffic to publishers was organically growing," says Nick Grudin, a former *Newsweek* business executive who headed content partnerships for Facebook. "It wasn't necessarily by design, [but] over the course of several years, we saw Facebook becoming a really important source of distribution—20 percent or 25 percent of clicks were coming from Facebook." Many publishers reported even higher numbers. During those years, Facebook almost willfully ignored the impact it was having on the news industry. "You don't read the news on Facebook," Chris Cox told me in 2014. "You *discover* the news on Facebook." Facebook, he told me, was not so much a news source but "a place where you could send caring eyes to journalism."

Facebook didn't *ask* to be a power in the news industry, and was uncomfortable taking responsibility for its role. Its work with news outlets focused on helping them negotiate the News Feed, without giving a thought to changing the product to distinguish between quality and dreck. Facebook's first big product to serve publishers was "Instant Articles," a way to more speedily load articles by storing them on Facebook's servers. (Google had a similar product.) The feature actually helped Facebook more than its media partners, because an Instant Article bypassed a newspaper's website, where the publisher could show ads and collect data on the reader. "It turned its goal quite well in terms of speeding up delivery of articles," says Grudin. "But it was not as widely embraced as we had anticipated it would be." What? Publishers wanted *money*?

Facebook was taking its cue from Zuckerberg himself. He was not a believer that the News Feed should give preferential treatment to the big legacy publishers. To him, that smacked of making a choice of what news was, and being an arbiter was anathema to him. Let the users decide! Newer, data-driven publishers like Upworthy or BuzzFeed, whose content was optimized to exploit the News Feed algorithms, resonated with Zuckerberg. The names of both companies described their approach to news. Upworthy packaged tear-jerking, inspiring stories that begged to be shared. BuzzFeed specialized in goofy and ephemeral stuff that people couldn't stop talking about. When one of BuzzFeed's social-media managers unearthed someone's Tumblr photo of a dress of ambiguous color—some people saw black and blue, others gold and white—she took all of five minutes to whip up a story she shared on Facebook. With the benefit of BuzzFeed's social-media acumen, the story cut through the social graph like Sherman's March. Twenty-eight million views in a day. Then BuzzFeed ran dozens of follow-up posts, flooding the zone as if it were *The New York Times* reporting on 9/11. Such was news in the age of Facebook.

In 2016, the same techniques BuzzFeed used to virally propel "The Dress" would drive big numbers for stories about the pope endorsing Donald Trump, and Hillary Clinton running a child-sex ring in a pizzeria. Meanwhile, the citadels of actual journalism couldn't get traction for well-reported stories that somehow weren't as "engaging" as fake ones or stupid ones. Facebook shrugged. "Facebook certainly plays a unique and important role in the news ecosystem," says the News Feed's Hegeman. "At the same time, that is not the primary reason that people use Facebook." The number he cited was 5 percent—one in twenty posts on the News Feed was a link to news.

Even before the election, some Facebookers knew that the company had to do better. In 2015, a media-savvy executive named Andrew Anker joined Facebook with a loose mandate to help out the news media. After a few weeks he came up with a plan: add a paywall option to Instant Articles. That meant that in order to keep viewing stories on a publication, readers would have to be subscribers. Publishers had been begging for something like that to monetize their stories on Facebook. Anker went into Zuckerberg's Aquarium and began to pitch it. Zuckerberg stopped him about two minutes in. "Facebook's mission is to make the world more open and connected," he reminded Anker. "I don't understand how subscriptions would make the world either more open or connected."

After 2016, Facebook knew it had to grapple with news, even if it just meant eliminating phony news. Post-election, Mosseri and Anker came up with some quick changes to address misinformation. One of them was to authorize and fund fact-checking operations, which would look into potentially bogus stories and call them out if investigation exposed them as phony. (Facebook wasn't fond of the term "fake news," which had become appropriated by the winner of the 2016 election.)

The first iteration of its fact-checking product had a startling result. When Facebook labeled a post as "disputed content"—meaning that professional fact-checkers had found the article fraudulent—people actually engaged with it more! "There was a set of people who believed us marking it false meant it was more true," says Anker. Facebook would later try other techniques to discourage people from gorging on phony tales designed to manipulate them. The most effective action against fake news was to "downrank" it in the News Feed, so it would take a lot of scrolling to see it. The action Facebook steadfastly refused to take, however, was eliminating such content. By not exercising that nuclear option, Facebook didn't have to take responsibility for saying—*This is false and it's out!* Better to avoid that controversy and make the Pizzagate stories less likely to appear in someone's News Feed.

The internal struggles continued after Anker left and was replaced in 2017 by a former *New York Times* digital manager named Alex Hardiman. "The question then was whether or not Facebook had the appetite, given the risks, to define and promote quality journalism," she says.

Also that year, Facebook hired former CNN reporter Campbell Brown to become its liaison with media outlets. Brown was a well-connected New Yorker married to a former George W. Bush official. During the 2016 campaign, she had been disappointed at the way legacy media, in particular cable news, had covered the contest, skating past issues to boost ratings. "To me, Facebook during the campaign wasn't that relevant," she says. "My background was television news, so I was focused on television news." She quickly moved to dispel the idea that she was hired as a celebrity mouthpiece to tout the company, confronting the company's controversies head-on. She tried to thread the needle of being an internal voice for publishers and an external defender of

Facebook. She found headwinds on both sides of the role. Publishers already blamed Facebook for two things: a credibility crisis in journalism and a naked grab for their revenues. Internally, there was still the feeling, taking the cue from Zuckerberg, that publishers deserved no special consideration in the News Feed.

Along with the News Feed's leader, Adam Mosseri, Brown held a number of off-the-record dinners and meetings with editors and media writers. These could become contentious. At one meeting, Ben Smith, the news editor of BuzzFeed, exploded. *Why are these people here?* he asked, referring to super-partisan right-wing sites like the Daily Caller. He didn't endear himself to them by calling them "trash" publications.

The news team found that it was getting harder to actually run anything because of Facebook's fear of alienating the right wing. The seemingly intractable problem was that the media outlets that spent the most on quality were generally perceived as liberal, while a number of popular right-wing outlets thought nothing of twisting the content of an article into partisan fantasies. Determining truth was scary enough for Facebook, but asking it to do so when truth was politicized made it impossible. "You would run an experiment and say, *Okay, we're ready to go*, but you might have one hyperconservative publisher who's doing fairly nefarious things who is going to be penalized and that person has a very loud lobbying body," says one news team official. "Are you okay waging war with a certain type of constituent in our current government?" That was the debate: *Can we actually try to do what is as close as possible to the right thing versus the more politically sensitive thing?*

Could Facebook ever restore the News Feed to its prelapsarian state as a feel-good way to connect with friends and family? That actually was the hope. In one meeting at the end of 2015, Fidji Simo, who headed video efforts at Facebook, presented a plan to double down on video content in the feed, introducing professional content. Chris Cox and some others objected—if video dominated the News Feed, Facebook would lose its unique advantage of being the place where people connect with each other. That led to a wider discussion about the quality of the News Feed.

"That was the start of the conversation that led to meaningful time spent," says Andrew Bosworth. Yes, Facebook itself was now using the terminology of its toughest critics. It began to run focus groups in Knoxville, Tennessee, asking actual human beings what they wanted to see in their News Feed. It asked its Core Data team to dig down on the "time well spent" concept.

Facebook began thinking about changing its priority metric from engagement to something that encompassed quality. "But it was moving along at a regular pace," says Bosworth. "There wasn't an urgency to it." In 2019, the company would change the algorithm to try to temper the velocity of posts designed to go viral. It would also rebalance the signals to highlight posts from close friends and family. Dismayed news sites saw their numbers drop.

Toward the end of that year, Facebook would change direction entirely, announcing a program to pay selected news organizations for their content, which would live on a separate "News Tab" outside the actual News Feed. Industry enthusiasm for the program was tempered by the inclusion of the toxic alt-right Breitbart News.

THE IDEA THAT he might share his personal fortune to help people first hit Zuckerberg in 2006, when he realized that if he took Yahoo!'s offer, he would personally be worth hundreds of millions of dollars. He took a long walk with his then girlfriend Priscilla Chan, who pointed out that that kind of money could be used to help humanity. "That number is shocking, and it makes you realize this is a very real opportunity—you have to really think twice before you turn that down," she told him. But neither of them could light on a great way to spend that money. "I joke lightly that he didn't have any good ideas, so he needed to go back to his day job," she says. "But I think the way he would say it is, he had a vision of where things needed to go and he needed to see that through." Chan, who taught science to elementary school kids before she entered medical school, did want to ground her future husband, and demanded that he do some teaching himself. At first he resisted. "I'm like, *I'm busy, I'm running this company*," he says. "But she insisted. So I taught that after school program and I learned so much from the kids." He kept up with a group of four students, seeing them once a month or so, and took pride that after five years of mentoring them, every one entered college. Their varied backgrounds spurred Zuckerberg's passion for the value of immigration.

That experience didn't help Zuckerberg's first major attempt at philanthropy succeed. In 2010, he went on *Oprah* to announce his $100 million donation to support New Jersey governor Chris Christie and Senator Cory Booker's program to reform Newark's schools. He hoped it would be a model to transform education nationally. The program was something of a fiasco, with the data showing little improvement in student performance. (Chan suggests that long-term results might show otherwise; nonetheless, the Newark foundation shut down in 2016.)

After the Facebook IPO made Zuckerberg one of the richest people in the world, with those millions replaced by billions, he and Chan, now his wife, began to explore philanthropy with more urgency. He tapped his friendship with Bill and Melinda Gates, whose eponymous foundation spent billions on world hunger, education, and social justice. They had been urging zillionaires to sign the Giving Pledge, a non-binding commitment to donate at least half their riches to the social good: Zuckerberg signed up. And then, with the Gateses as role models, he and Chan began their own philanthropic organization. They would focus on health, education, and social justice.

The Chan Zuckerberg Initiative has one big difference from the nonprofit Gates Foundation. CZI is a for-profit LLC (limited liability company). When CZI launched in 2016, critics questioned why. Could starting a for-profit even be considered giving back? But David Plouffe, the former Obama political operative whom Chan and Zuckerberg hired to head policy at CZI, says that the reason is to provide more flexibility. "Let's take a three-sixty view," he says. "Is there an investing play, is there a grant-making play, is there an engineering play, is there an advocacy play, is there a storytelling play, is there a policy play?" By having the freedom to indulge in those activities without worrying about losing charitable status, CZI can do more. Especially since Plouffe was eager to have the initiative dive into policy recommendations, particularly in the field of social justice.

As CZI took shape in 2016, it became clear that Zuckerberg and Chan would have to be deeply involved themselves. For Zuckerberg that meant spending real time at CZI: usually a full day Friday and a few stolen hours on some other day. (In recent years, he has cut back because of the demands of a beleaguered Facebook.) For Chan, who loved her work as a pediatrician, it meant giving up clinical work to become a full-time co-CEO. She had also been in charge of a school that Zuckerberg funded, and to run CZI, she turned that work over to an administrator, almost as if she's a prisoner of big philanthropy. "I miss the patient care," she admits.

She's doing it, she says, for the *future* Priscilla Chans and their patients. "CZI is just an incredible opportunity to be able to change the systems that have prevented me from being successful in the clinic, in the hospital, and in the classroom and in the community," she tells me when I visit her at the organization's headquarters. (With people neatly arranged at their workstations, a well-stocked mini-kitchen, and a free café, it's like a Facebook office on mute.) In a few minutes, she returns to the subject of the life she gave up. "I wanted to be part of fundamentally setting the DNA of the organization with Mark. It's hard, and not as rewarding," she says, "[but] I want to be able to have the path to changing lives of those that I care about."

In September 2016, CZI had made its biggest announcement at an auditorium in the sprawling Mission Bay medical complex in San Francisco (which includes the Zuckerberg San Francisco General Hospital, whose reconstruction benefited from a $75 million donation). Zuckerberg came onstage promising that CZI would be spending $3 billion to "cure all disease" during the life span of his two daughters, roughly by the end of the century. (The Zuckerbergs' second child, August, was born in 2017.) Though it's true that in the last hundred years medicine had made amazing progress, this seems an astounding goal. Part of the approach was starting a "BioHub" that merged the resources of Stanford, UC Berkeley, and UC San Francisco. After Zuckerberg drove that stake in the ground, Chan made her debut in high-stakes public speaking. She invoked her background in a financially struggling family, her desire to go to medical school to serve families, and the care she gave her patients, many of whom were children with life-threatening illnesses. When she told of having to sometimes break that horrible news to parents, Chan teared up, demonstrating an onstage sincerity her husband still had yet to master.

The last speaker was Bill Gates. His remarks were enthusiastic, but I wondered if I caught a bit of a verbal eye roll when he commented on Zuckerberg's goal of eradicating disease by the end of the century. As if he were thinking, *I've spent billions of dollars to get rid of polio and haven't done it—you're going to cure ALL disease with your measly stake?* (A few days later, a commentator in *The Washington Post* pointed out that $3 billion over ten years is but a drop in the $7 trillion ocean of annual healthcare spending.) Gates later explained: "The world needs aspirational goals, and the world needs concrete goals," he says. "In the world of philanthropy, sometimes people confuse which are your concrete goals and which are your aspirational goals. Maybe I'm just a stickler because I manage an organization where we're serious about concrete goals. But we live in a world where the aspiration of getting rid of all disease is absolutely what motivates us."

Ultimately, like Facebook itself, CZI is a reflection of its co-founder's belief in the engineering mindset. CZI's distinguishing feature is an effort to create the digital tools to attack the problems he and Priscilla were addressing. It competes with tech companies, including Facebook, to hire software engineers and AI scientists. Zuckerberg's ambition is another common factor. "When I joined it was January '17; we were twenty people," Plouffe told me in early 2018. "We've now more than [ten times that]. So I will say Mark's experience with Facebook's growth has been super helpful."

Not so helpful is the steady contrast of CZI's noble intentions with the public profile of Facebook, which was taking a dive during Mark Zuckerberg's year of travel and contemplation. No matter how earnest those intentions, there would always be critics charging that the whole enterprise was but a distraction from the mother ship's woes.

The fact was, being associated with Facebook or its CEO was no longer a ticket to admiration. Even giving away his money could not change that, as demonstrated by a request that would come from a San Francisco supervisor, asking that the Zuckerberg name be stripped from the general hospital.

ON HALLOWEEN DAY, Facebook's general counsel, Colin Stretch, and representatives of Twitter and Google raised their hands to testify at a Senate Judiciary hearing. The subject was "Extremist Content and Russian Disinformation Online: Working with Tech to Find Solutions." It was the first of a doubleheader; the following day, Stretch and his two counterparts endured a similar hearing in the House of Representatives. After vigorous negotiation, the legislative bodies had agreed to listen to corporate general counsels defend the tech companies. The committees had hoped to get top executives: in Facebook's case, Zuckerberg or Sandberg. Their appetite would not be sated with these lawyers.

The senators came prepared with graphic evidence of Facebook's accommodation of Russian agents. Using screen images pasted on poster boards (Senate technology being frozen in 1950s science-fair territory), Senator Mark Warner showed how, from the IRA offices in Saint Petersburg, Russians had planned a potentially violent rally at an Islamic center in Houston in May 2016. One Russian page called "Heart of Texas" urged its anti-immigrant followers to protest, and another, claiming to be "United Muslims of America," called for Muslims to stand up for their people. The questioning was sharp, unearthing Stretch's admission that Russian activity on Facebook had reached more than 120 million Americans, though Stretch had to admit that Facebook still wasn't sure it had identified all of the IRA's work.

Stretch's lawyerly caution frustrated the legislators. His response after any question that was even mildly challenging was to thank the senator for asking it. After the session his wife told him, "Honey, I'm not sure anybody's going to tell you this: you don't have to thank them for every question." He told her he was just figuring out what to say. All too often, it was some variation of "I'll get back to you."

The senators made it clear that the hearings were only the start of a new reality where Facebook and its peers—but especially Facebook—were about to be harshly scrutinized. Senator Dianne Feinstein, representing Facebook's home state, said it outright. "You've created these platforms, and now, they're being misused, and you have to be the ones to do something about it," she said. "Or we will."

While Stretch testified, Mark Zuckerberg was 7,000 miles away, on an annual Beijing trip he took as a member of the Tsinghua University School of Economics and Management advisory board, along with other US business leaders like Apple CEO Tim Cook and Goldman Sachs strongman Lloyd Blankfein. Facebook had been blocked in China since 2009, and for years, Zuckerberg had been courting the country's leaders to figure out a way to get around the ban. How can one connect the world if there are a billion human beings blocked in China? During his 2010 challenge to learn the language, he had spent an hour a day studying, and in his 2014 visit to Tsinghua University, he actually gave a speech in Mandarin. Rough translation: *I'm very glad to be in Beijing. I love this city. My Chinese is really a mess, but I study using Chinese every day.* According to *Page Six,* he even asked China's president Xi to select a Chinese name for his then-unborn daughter. (Allegedly, Xi declined. Facebook denied the story. The Zuckerbergs chose their own Chinese name for Maxima—Chen Mingyu.) But no amount of pandering could overcome Zuckerberg's inability to deliver a version of Facebook that would satisfy China's desire to censor online speech and access personal data without due process.

"Every year this trip is a great way to keep up with the pace of innovation and entrepreneurship in China," he wrote on his Facebook page about his 2017 sojourn. Maybe that was not the best year to continue his study of Chinese innovation. As Facebook's woes accumulated, Zuckerberg's onetime dream of getting Facebook into China was no closer to being a reality.

I MET UP with Zuckerberg on the last leg of his US ramble, in Lawrence, Kansas. It was November, with the end of a bruising year in sight. He was leading a town hall–style meeting at the University of Kansas. Metal detectors had been set up to screen the students invited to the event.

It had only been fourteen months since we were in Africa. It seemed like an eternity. Zuckerberg might have maintained his vibe—the friendly guy in a T-shirt who says, "Hi, I'm Mark"—but the jauntiness that once seemed like charming modesty now made people wonder if he was living in a bubble of denial. Zuckerberg seemed strangely impervious to the constant invectives hurled his way in the press and even in comments on his own Facebook posts. After that perspiration-drenched D: All Things Digital event, he would always make sure that the AC in the green room pre-events was chilled to the temperature of a meat locker. A similar chilliness seemed to freeze out hostility directed toward him.

"I don't look back on stuff we've done," he said in answer to a question from a Kansas student who asked him about his attitude. "It's the way I'm wired. I don't get happy. There's always more to do. When you have a platform like this, you have [a] responsibility to help more people. People will criticize you and beat you up when you make mistakes. But it's optimists who build the future."

Those remarks didn't fully sync with a post he'd made a few weeks earlier, on the high Jewish holy day of Yom Kippur. He'd shared a personal note of atonement to his millions of followers: "For those I hurt this year, I ask forgiveness and I will try to be better. For the ways my work was used to divide people rather than bring us together, I ask forgiveness and I will work to do better. May we all be better in the year ahead, and may you all be inscribed in the book of life." When I interviewed him backstage at the university facility the night before his speech, I asked him, Why did he do that?

"Well. I spent the day fasting and then I went to synagogue and was just reflecting," he said, nibbling on the catered dinner of ribs provided to the Facebook team. "And then as I was riding home I wrote that on my phone and posted it. The point of Yom Kippur is to reflect on your mistakes and the ways that you may have hurt people intentionally or not and to try to make good on those for the next year. Of course, the work that we intend to be used positively to bring people together is going to be used in some ways to divide people, especially some of what we know now the Russians tried to do. I do really feel bad about that. And that was what I was reflecting on."

During his visits to the states, he said, he found that people weren't so concerned with national issues as much as ones closer to their day-to-day existence, that political disagreements among neighbors didn't have to be alienating. People can differ on what dogs they love and what sports teams they follow, he said, but that doesn't mean that they don't share basic things in common. That's the glue of *communities*. That's why it's disturbing when outsiders burst in to stir up divisiveness.

But didn't Facebook provide a platform for troublemakers to do just that? I asked.

"We have a lot of work to do," he said.

In early January he would announce his 2018 resolution: working full-time to restore Facebook's reputation. He'd lead Facebook to do better in protecting users from abuse and hate, fending off foreign interference, and rejiggering Facebook so that users' time on it would be—and here Tristan Harris's term spilled off his tongue—time well spent. "If we're successful this year then we'll end 2018 on a much better trajectory," he promised.

It would be the challenge of repairing Facebook—a lot harder than wearing a tie or learning to speak Chinese.

16

Clown Show

I MET ALEKSANDR KOGAN one day in a Starbucks south of New York's Central Park. I had expected someone with dark Slavic looks and an air of mystery. Instead I was meeting a gangly, goofy, honey-haired American in a sweatshirt and jeans, appearing younger than his thirty-two years.

We went into the park and sat on a bench, and he shared some geeky statistical networking theories with me. It wasn't until our next interview that he described what led to his handing over the personal information of up to 87 million Facebook users to a shady political consultancy that would use it to help elect Donald Trump. When news of this broke in March 2018, it was like all the negativity about Facebook that had been accumulating since the election like some ammo dump of recklessness suddenly blew up in a fireball worthy of the climax of a Marvel flick.

The name of the consultancy would live in infamy: Cambridge Analytica.

The fiasco had roots in decisions that Facebook had made years earlier: sharing information with developers on its platform, changing its News Feed in a way that accelerated sensational content, and allowing advertisers to microtarget its users based on the unfathomably wide dossier it had built on each and every one of them. Not to mention Facebook's main boon and burden, its fetish for growth.

Kogan was born in Moldova, a tiny Soviet satellite wedged between

Romania and Ukraine, in 1986. His father taught at a military academy. Fleeing anti-Semitism, the family, including seven-year-old Aleksandr, immigrated to Brooklyn and later moved to northern New Jersey. Kogan grew up as a typical American kid.

He entered UC Berkeley intending to study physics. But after feeling helpless when some friends suffered depression, he found himself drawn to psychology. That led him to the lab of Dacher Keltner, a renowned scientist who studied the other side of depression's coin: happiness, kindness, and positive emotions. The approach resonated with Kogan, and he joined Keltner's lab. His specialty was quantitative data: "If we needed to learn a new statistical technique, I'd be the one pegged to learn it," he says. He went on to get his doctorate at the University of Hong Kong in 2011 and, after a postdoc at the University of Toronto, landed a job at Cambridge University. It was an open-ended offer that could have evolved into a lifelong post.

"That was the plan," he says. He was twenty-six.

In Cambridge's psychology department, professors commonly had their own laboratories, recruiting postdocs and students to do research, ideally paid for by grants. Kogan's lab was called the "Cambridge Prosociality and Well-Being Lab." He lured some grad students and postdocs. "That was my little university," he says. But within a few months, he was also drawn to participate in the work of another lab in the department called the Psychometrics Centre, founded by a professor named John Rust. Rust had come to Cambridge with his wife, who was an academic superstar. As part of the deal to hire her, Cambridge gave him a lab. His lab did a lot of work for outside industry, including test makers.

One of the associates in the Psychometrics Centre was a Polish researcher named Michal Kosinski. His work centered on finding the means to extract useful information from modest inputs, something that could be helpful when facing a chronic shortage of funding for projects—essentially, making something out of not so much. One day Kosinski stumbled upon an online survey called myPersonality, created by a grad student at Nottingham University named David Stillwell. The quiz lived on Facebook, which at that time was a rather unusual setting for an academic study. The study itself was not so remarkable—it was a standard set of questions that helped determine which of seven personality types the user belonged to—things like introvert, extrovert, neurotic, and others—as defined by a well-known personality-identification system called OCEAN.

The novelty of Stillwell's work was its use of the Facebook News Feed, specifically its power to widely distribute stories that people were engaging with. Stillwell's test was cleverly seductive. Once people took the test—and who could pass up a chance to find out something about themselves?—they would share results with their friends, who would then give feedback to determine whether the test actually nailed the takers' personalities. And then, of course, those friends would be tempted to get in on the fun and to take the test themselves. It was the same viral technique used by Flixster and other opportunistic developers.

Kosinski realized that this scheme was a game changer. Previously, social scientists had to struggle to get responses. They would have to pay subjects to fill out forms, and the process involved all sorts of problems. But with Facebook, all you had to do was get the survey in front of people. They couldn't wait to fill it out and share it with their friends. Goosing the process was the News Feed's EdgeRank algorithm, built to indulge such sharing. And since this was still in the early days of Platform, Facebook wasn't finicky about flooding the feed with viral come-ons like game invitations, tossed sheep—and quizzes. "Facebook was brutally sharing anything," says Kosinski. At one point 100,000 people per month were interacting with myPersonality. Eventually 6 million people would take the test.

Kosinski contacted Stillwell and asked if he could share his data. Soon the two were collaborating. Kosinski begged Rust to bring Stillwell to Cambridge, and they quickly became a success story in the Psychometrics lab. But they had trouble creating a successor as popular as myPersonality, because Facebook had changed the News Feed to suppress spammy apps, whether they were Zynga games or personality quizzes.

That's when the two psychometricians realized that they did not need huge numbers of people to take the quiz. Because Facebook had increasingly been exposing information about its users to the public—a practice that the FTC would later cite as a privacy violation—a lot of data was there for the taking. What's more, the recently introduced Like button, beginning in 2009, opened a world of new data to anyone. You didn't even have to sign in to Facebook to get it—just type in a command to the Facebook API and there it was. Unlike responses on a questionnaire, this information could not be tainted by poor memory or fake responses. "Instead of having them fill in a personality questionnaire, your *behavior* becomes a personality questionnaire for me," says Kosinski.

Kosinski encountered some skepticism about this methodology. "Senior academics at that time didn't use Facebook, so they believed these stories that a forty-year-old man would suddenly become a unicorn or a six-year-old girl or whatever," he says. But Kosinski knew that what people did on Facebook reflected their real selves. And as he used Facebook Likes more and more, he began to realize that, on their own, they were incredibly revealing. He came to believe that you didn't need an OCEAN quiz to know a ton about people. All you needed was to know what they Liked on Facebook.

Working with Stillwell and a graduate student, Kosinski used statistics to make predictions about personal traits from the Likes of about 60,000 volunteers, then compared the predictions to the subjects' actual traits as revealed by the myPersonality test. The results were so astounding that the authors had to check and recheck. "It took me a year from having the results to actually gaining confidence in them to publish them because I just couldn't believe it was possible," says Kosinski.

They published in the *Proceedings of the National Academy of Sciences* (*PNAS*), a prestigious peer-reviewed journal, in April 2013. The paper's title—"Private Traits and Attributes Are Predictable from Digital Records of Human Behavior"—only hinted at the creepiness of the discovery. Kosinski and his co-authors were claiming that by studying someone's Likes, one could figure out people's *secrets*, from sexual orientation to mental health. "Individual traits and attributes can be predicted to a high degree of accuracy based on records of users' Likes," they wrote. Solely by analyzing Likes, they successfully determined whether someone was straight or gay 88 percent of the time. In nineteen out of twenty cases, they could figure out whether one was white or African American. And they were 85 percent correct in guessing one's political party. Even by clicking on innocuous subjects, people were stripping themselves naked:

> *For example, the best predictors of high intelligence include "Thunderstorms," "The Colbert Report," "Science," and "Curly Fries," whereas low intelligence was indicated by "Sephora," "I Love Being A Mom," "Harley Davidson," and "Lady Antebellum." Good predictors of male homosexuality included "No H8 Campaign," "Mac Cosmetics," and "Wicked The Musical," whereas strong predictors of male heterosexuality included "Wu-Tang Clan," "Shaq," and "Being Confused After Waking Up From Naps."*

At the paper's conclusion they noted that the benefits of using Likes to broadcast preferences and improve products and services might be offset by the drawbacks of unintentional exposure of one's secrets. "Commercial companies, governmental institutions, or even one's Facebook friends could use software to infer attributes such as intelligence, sexual orientation, or political views that an individual may not have intended to share," they wrote. "One can imagine situations in which such predictions, even if incorrect, could pose a threat to an individual's well-being, freedom, or even life."

In subsequent months, Kosinski and Stillwell would improve their prediction methods and publish a paper that claimed that using Likes alone, a researcher could know someone better than the people who worked with, grew up with, or even married that person. "Computer models need 10, 70, 150, and 300 Likes, respectively, to outperform an average work colleague, cohabitant or friend, family member, and spouse," they wrote.

Kosinski and Stillwell both had good relationships with Facebook before the Likes paper—Kosinski says the company had offered each of them a job. So as a courtesy, he shared the paper with his contacts there a few weeks before publication. The policy and legal teams at Facebook, still smarting from the 2011 FTC Consent Decree, immediately recognized the paper as a threat. According to Kosinski, Facebook called *PNAS* to try to stop publication. It also contacted Cambridge University and warned that the researchers might be illegally scraping data. But, as Kosinski notes, there was no need to do this because Facebook exposed everyone's Likes to the public. At that point there wasn't even an *option* to make them private.

"I think this was the moment when people working for Facebook realized *Hey, we are doing things that might not be entirely neutral to the safety of people and privacy of people,*" says Kosinski. But Facebook knew that all along. In fact, in 2012 the company had gotten a patent for "Determining user personality characteristics from social networking system communications and characteristics," basically doing the same thing as Kosinski and his collaborators went on to do. And because Facebook's work had started before the Like button had taken off, its researchers had been using the text of people's posts for keywords that would provide clues to a user's private traits. It turns out that the Facebook data team had created a secret database called the "Entity Graph," which charted the relationships of every one of Facebook's users not just with other people, but to places, bands, movies, and products and websites—sort of a stealth Like button. "The inferred personality characteristics are stored in connection with the user's profile, and may be used for targeting, ranking, selecting versions of products, and various other purposes," said Facebook's patent application.

After the Kosinski paper, Facebook changed the default settings on Likes so that only friends could see them, unless people chose to share more widely. The exception was Facebook itself, which saw everyone's Likes and could keep using them for . . . targeting, ranking, selecting versions of products, and various other purposes.

. . .

AS A MEMBER of the Psychometrics Centre, Aleksandr Kogan got to know Stillwell and Kosinski and became one of Kosinski's thesis examiners. He was bowled over by Stillwell's initial discovery that Facebook could be a revolutionary way to collect social-science data. "It was the first-mover advantage," says Kogan with admiration. "There weren't a lot of personality quizzes yet on Facebook—now you see a billion of them."

Kogan wanted to do such work himself and asked Stillwell to give him access to myPersonality data, and he began to analyze it. One of Kogan's grad students began looking at the idea of applying the research to an economics question: How does intrapersonal contact between countries affect things like trade and charitable donations between countries?

Answering that question required more data, so Kogan called his mentor Dacher Keltner. He told Keltner about his project and said that he'd love to look at all the friendships of the world, parsing them country by country. Keltner, who was consulting with Facebook, said he'd help him get in touch with the one place in the world that could help him.

"So he makes an introduction to the Protect and Care team at Facebook," says Kogan. "They said, *Cool, we'll give you the data set.*"

BY THE TIME Kogan began collaborating with Facebook, its Data Science team had grown considerably. It was unabashedly part of the Growth organization. While it employed genuine social scientists and statisticians, its aim was not pure research, but studying the behavior of Facebook users in order to fulfill Growth's goal of expanding the user base and retaining current users. One theme of its research was discovering how sharing worked, in experiments like the one that produced the "Gesundheit" paper that showed how posts went viral. Another showed how the social dynamics of sharing affected people's behavior. Mostly, its work was unpublished. Data scientists worked with product teams as they iterated products.

But the team did sometimes publish its work. To social scientists,

Facebook was like the data set of the gods. You had 2 billion people in a petri dish. You could tweak a feature for hundreds of thousands of people and compare the results to an equally huge control group.

Usually the statistics-rich papers were circulated exclusively among the social-science community. But sometimes, an experiment exposed to the general public would raise ethical questions. Or reveal something uncomfortable about Facebook's powers. One example was the controversial voting study—co-authored by researchers at UCSD and Facebook—that made critics concerned that Facebook could affect elections by selective availability of the "I Voted" button.

But the most controversial study in Data Science's history came in Kogan's specialty: emotional well-being. In 2014, a study called "Experimental Evidence of Massive Scale Emotional Contagion Through Social Networks" appeared in *PNAS*. It presented the results of an experiment involving 689,003 Facebook users. The News Feeds of those users were altered to prioritize a small number of posts. For one group the posts had positive content (*here's my cute dog!*); for another the content was negative (*my dog died yesterday*). A control group was served an unmanipulated feed.

Ironically, the purpose of the study was not to see whether the News Feed could depress people. It was specifically conceived to help Facebook thwart that very criticism—and make sure that people kept using Facebook. One gripe about Facebook was that some people used the News Feed to boast how great their lives were, whether it was true or not. Every vacation was fabulous, every baby 24/7 adorable, every Warriors game viewed from courtside. Seeing friends so happy, the theory went, made everybody else feel lousy.

Facebook disagreed that good things made people feel bad, so a researcher in its Data Science team, Adam Kramer, set out to disprove it. As he later wrote, "It was important to investigate the common worry that seeing friends post positive content leads to people feeling negative or left out." In addition, "We were concerned that exposure to friends' negativity might lead people to avoid visiting Facebook."

He asked Jeff Hancock, then a professor at Cornell, to help design the study. Hancock's previous work was about "emotional contagion." In earlier experiments he had gone to extremes to document the effect of negativity, like showing people the terrible scene in *Sophie's Choice* where Meryl Streep has to choose which of her children the Nazis will kill. The Facebook experiment, which would boost or diminish posts that were already in people's News Feed queue, seemed rather tame in comparison. For a week in 2012, Kramer, Hancock, and one of Hancock's postgrad students tweaked the News Feed of almost 700,000 users. They found a tiny impact due to the manipulation—a slight increase in negative posts from those who had been shown the nonorganic downer stories—but because of the size of the experiment, the effect was measurable and significant in the aggregate. The good news for Facebook was that good news from other people did not make people feel bad. *Bad* stuff made them feel bad, but only a little.

That would have seemed like a win for Facebook. As Hancock says, "It said, *Look, however small it is, it's the opposite of what people are saying about us, that hearing good things about your friends makes you feel bad*. And so, *Yeah, baby, I guess I am wrong about Facebook*."

That's not the lesson that people drew from it when the study appeared in *PNAS* in June 2014. The trouble began with a blogger who wrote, "What many of us feared is already a reality: Facebook is using us as lab rats, and not just to figure out which ads we'll respond to but actually change our emotions." The post got the attention of the media, which piled on to what would be called "the mood study." Summing up the general perception of the experiment, *Slate* wrote, "Facebook intentionally made thousands upon thousands of people sad."

Facebook admitted that it had made a mistake in not declaring its motivation for the study, but insisted that its terms of service gave it leave to conduct the experiment. Hancock agrees that was insufficient: "Nobody looks at terms as a form of consent, because nobody looks at terms of service." Hancock himself had to justify his work to the Cornell administration because academic research standards are more rigorous than in the corporate world. *PNAS* had to run an apology. The entire episode played into fears that the world's biggest social network was manipulating what a billion people were seeing—which of course it was.

From that point on, Facebook was dramatically more cautious about its research. "One of the things that came out of that example was, we have a very robust list of things that are sensitive," says Lauren Scissors, a research director at Facebook. "There are certain topics that we believe are not right for our users." The work still continued—after all, research led to Growth!—but the company did not want to be misunderstood again. "I don't think that they stopped doing experiments—they just stopped publishing them," says Cameron Marlow, who headed Data Science but left shortly before the emotion paper was published. "So is that a good thing for society? Probably not." As I found by informal conversations at a Data Science conference on campus in 2019, though, most of its researchers stuck around. They feel that their work is important.

BEGINNING IN 2013, Kogan was visiting the Facebook campus regularly. He ate a lot of free lunches. He did a presentation. Eventually he would provide consulting services to Facebook and work at Menlo Park for a short stint. "I know Building 20 well," he says. Meanwhile, Kogan's lab had grown to fifteen people. A postdoc from Texas named Joseph Chancellor had joined his lab. He shared Kogan's affinity for statistics and his interest in Facebook, and they collaborated, working closely with Facebook's Care and Protect team.

Kogan needed more data for more studies. So he figured he would write his own version of myPersonality—a new survey to snarf up the information of willing participants. To get the most information, he would draw on the generous access Facebook granted developers not just to people using the app, but to their friends as well.

In the fall of 2013, Kogan wrote the app called thisisyourdigitallife. He had been coding since he was an undergraduate, and in any case, Facebook made it very easy to gin up a simple application that used Facebook Connect to suck up data from the service. It took him one day.

"It's not like an app that *runs* anything—it's literally that stupid Facebook log-in button you see anywhere," he says.

Indeed, mining user data with a Facebook app was laughably trivial. Kogan used the Facebook Login protocol, which allowed developers to access data, as Facebook later put it, "without affirmative review on approval by Facebook." Further, at that time Facebook was still using the version of Platform known as Graph API V1—the Open Graph that had caused controversy inside and outside Facebook. Some referred to it as "the Friend API," because it gave developers access to not only someone's information, but detailed data of their friends as well, including a detailed dossier on their likes and interests. It had been the technology behind Facebook's Instant Personalization, the so-called privacy hairball that had given rise to internal opposition at Facebook, which Zuckerberg overruled. To Kogan it was a godsend. "They send you the data," he says. "Done."

The data he was talking about wasn't what people typed into his survey. These were paid subjects whom he hired via Mechanical Turk, a crowdsourced network of freelancers run by Amazon. In exchange for the pennies per hour they received for taking the survey, the "turkers" also gave Kogan permission to access their Facebook data—and also the data of their friends, who had *not* given permission to share their data.

Kogan justifies the practice by saying the people taking the survey were better informed than users of commercial apps. "You know how in industry you put in terms of service and nobody often even clicks that link to even see it?" he says. "Academia, it's front and center. The first page is a terms of service and we really try to make it really intelligible and where we have to spell everything out."

But that's for the people taking the test. Their friends would have no chance to grant or deny permission. They wouldn't even know that their personal information had been exposed. And since each of the informed users—known as "seeders"—has about 340 friends (the Facebook average at the time), by far the majority of Kogan's data set would be totally unaware of their inclusion in his project.

What Facebook did not allow was for developers to take that information and use it somewhere else. Facebook had always set its terms so that information could not be retained, transferred, or sold. But it had done very little to enforce those rules. And despite the promises it had made again and again about how it was policing developers to make sure they did not retain or distribute the data, it still had no way of actually knowing what happened to the data after it left Facebook. Facebook's employees and developers alike agree that if someone were to accumulate a database of Facebook information and abscond with it, there was little Facebook could do.

Kogan says Facebook was fully informed of what he was doing. "Nobody had an issue. We normally collected the demographic information, page likes, friends information," he says. Then he thinks a bit. "We might have collected wall posts," he adds.

Things were going great. Kogan had ten papers in the hopper. Then one of the students in the psychology department mentioned that he had been doing consulting for a company called SCL. Would Kogan be interested in meeting them? He described it as a political consulting firm.

"The hook for me was that they had a lot of big data and they'd be potentially interested in sharing it with my lab," says Kogan.

So the student set up a meeting for Kogan. With a guy named Christopher Wylie.

WYLIE WAS A Canadian-born data nerd who, as an eighteen-year-old, crossed the American border to help ad targeting for the Obama campaign. He immigrated to London in 2010, pursuing degrees in law and fashion forecasting, but, as he later told *The Guardian*, his passions were politics and data. He had excitedly followed the work of Kosinski and colleagues on personality prediction. And in 2013, he met Alexander Nix, director of a company called SCL. Nix, thirty-eight, was from a prominent family, educated at Eton, and had been a financial analyst before joining SCL in 2003. Listed as a military contractor, SCL was actually a consultancy that offered services to candidates, corporations, and governments. Its exploits sounded like something from a Ross Thomas novel: working behind the scenes in places like Uttar Pradesh, Kenya, Latvia, and Trinidad to influence the citizenry, both in voting and in their general attitudes. "Our services help clients identify and target key groups within the population to effectively influence their behaviour to realize a desired outcome," read some of their promotional copy. Nix convinced Wylie to join. "We'll give you total freedom," he promised. "Experiment. Come and test all your crazy ideas." Wylie, at twenty-four, was suddenly research director for the SCL Group. Later, he learned that his predecessor had died in a hotel room in Kenya under suspicious circumstances. It was a hint that SCL might have a shady side.

Not long after, Wylie met hard-core conservative warrior Steve Bannon, then editing the notoriously partisan right-wing news site Breitbart. Somehow the gay nerd and the proto–white nationalist bonded. "It felt like we were flirting," Wylie would later write about their data-wonky intellectual jam sessions. Soon they were hatching a plan for SCL to enter America. Bannon set up a meeting with a wealthy funder of right-wing causes named Robert Mercer. Before making his fortune in hedge funds, Mercer had been a celebrated IBM researcher, so SCL's promise to change voting behavior resonated with him. He agreed to fund the subsidiary. In December 2013, "Cambridge Analytica" was registered in Delaware. The name came from Bannon, who liked the implication that it was involved with the university.

Cambridge Analytica began devising a plan to sell services to Republican candidates, with the flagship project a Wylie plan named Project Ripon. It would require a huge database of voter personality profiles, which it would then match with voter rolls in key states, directing ads to them that would hit hot buttons they didn't even know they had. Or so was the theory.

The pursuit of data for this project is what led Wylie to travel from London to meet with Kogan. They got together in a restaurant in Cambridge. Kogan was impressed with Wylie. While Wylie would later adopt a distinctive look—short-cropped pink hair, earring—he dressed traditionally for the meeting. Wylie first told him about his work for the Obama campaign, and how they had collected all kinds of cool data. Now, he said, he was working for a company that wanted to do similar things. It was, he admitted, associated with the right. Though Kogan's own politics leaned left, this was not a deal-breaker. "Even though I'm an Obama fan, I'm not like, *the Republicans are evil people*," he says.

Kogan says that, at that first meeting, Wylie offered to work with him and share data. "What they initially wanted from me was just some consulting," says Kogan. "Not even Facebook-related consulting, just like how-do-we-write-a-better-survey-question consulting." Kogan was excited. He began dreaming of setting up a data science institute of his own. Instead of studying undergraduates or people on Mechanical Turk, he could gather a broader set. Wylie loved the idea and the two of them began talking about building a society in silico, with voluminous data on every living person.

In the short term, though, Wylie was seeking personality-based information for SCL. Kogan began to sketch out a grand scheme where he would generate data through his Facebook app and then he'd get Kosinski and Stillwell to use their personality-prediction techniques. Then they would send personality scores to SCL. Wylie loved the idea. "So this is kind of aligning for me," says Kogan.

Since mixing paid work with university activities was forbidden, Kogan started his own company, Global Science Research (GSR), to do the consultancy. His colleague Joe Chancellor was a partner. Walking him through the process was Chris Wylie.

In the UK, all applications using private data must register with the Information Commissions Office. Kogan did so in April 2014. That same month, at Facebook's F8 conference, Mark Zuckerberg announced that it was closing the gap that allowed developers to access without permission the information of the friends of people who used their app. Graph API V1 would be sunset, with developers moving to version two. Though motivated by Facebook's desire for "reciprocity" from developers, Facebook spun it as a move toward more user privacy. The first part of his keynote concentrated on new rules for developers that limited the user information they could suck out of Facebook. But not limited enough. Instead of immediately locking down friend-of-friend information, Facebook was grandfathering in existing apps, allowing them to violate user privacy for a one-year grace period. A move it would regret.

Facebook's new rules included an "App Review," where developers had to request permission to access certain user data. Kogan went through the review and was turned down—but because he had a preexisting app, Facebook allowed him continued access to user data during the one-year transition. If Facebook had enforced its new rules immediately, the GSR–Cambridge Analytica partnership would have ended. Without the friend information he accessed during the grace period, Kogan would have been able to provide only a tiny fraction of the population he promised, insufficient to target a significant number of voters.

Kogan's efforts to draw Kosinski and Stillwell into the project did not bear fruit. Part of the problem, he says, was that Wylie kept changing the terms. The first proposal was that Cambridge Analytica would give the money as a grant to the Psychometrics Centre. Then Wylie flipped—the money would go to Kogan's company and *he* would pay the center. But the amount was only $100,000. Previously he had been talking about $1 million.

Kosinski and Stillwell felt Kogan had dealt shadily. "He used our credibility to obtain a grant that was supposed to be funding the work at the university," says Kosinski. "He's suddenly redirecting the grant from the university to his private company and then paying us for some work. And we said, *First of all, one hundred thousand dollars, you cannot even hire one postdoc for a year so that's not enough money.* And second, *This is absolutely unethical, it's outrageous.*"

Kogan himself says he began to harbor doubts about SCL. He met with Nix a few times and thought he was dicey, like a used-car salesman. "He doesn't understand much of the product, but he's just trying to sell a dream," he says.

It turned out that Wylie didn't like Nix either. "He talked about Nix like Nix is the idiot of the world," says Kogan. "And he started revealing this plan that he wants to start his own company."

In fact, later in the summer of 2014, Wylie left and did start his own firm. But not before helping Kogan deal with Facebook's terms of service, which forbade companies like Kogan's to do exactly what he was intending—selling or licensing the personal user data Facebook provided to developers.

According to Kogan, Wylie claimed that as an expert in law and data privacy, he would take charge of getting over that hurdle. His suggestion was that Kogan provide a new terms of service agreement for survey-takers that allowed him to give the data to SCL with no restrictions. Wylie would draw up this new agreement. "He wrote my terms of service," says Kogan. "He's like, *just fill in your name.*" Wylie confirms this, saying that he simply used Google to seek out sample terms of service agreements. When Kogan looked at the document, he says he saw a lot of legalese. Wylie, he says, directed him to one section in particular. "He's like, *This section says you could transfer and sell it.* I think he made that a point to point that out, to make sure, to assure me that we're giving proper authorization."

"It didn't feel very nefarious at the time," says Wylie.

Yet what they would be doing blatantly violated Facebook's own terms of service, which did not permit the transfer of data that Kogan was engaging in. Kogan later insisted that he submitted the new terms to Facebook, but without Facebook's affirmation, his and Wylie's new terms were meaningless. It was as if they were tenants who had rewritten their signed apartment lease to cut the rent in half, and then dropped it on the landlord's doorstep, concluding that they were now that much richer. It's unclear whether anyone at Facebook put eyes on it.

WITH STILLWELL AND Kosinski out, Kogan could not use their prediction system for the data he was gathering for Cambridge Analytica. So he revised his app to gather information for SCL. Instead of harvesting data from Mechanical Turk, he acquired his "seeders" from a commercial company called Qualtrex that both provided survey software and found participants. Qualtrex agreed to recruit about 200,000 people to take the survey, paying them about $4 each. SCL paid for it. The survey-takers agreed to share their Facebook information, which included the personal data of their friends. Kogan then purportedly had his team work up a system to emulate what Kosinski and Stillwell had done in analyzing the data to predict traits. In a May email to Wylie he suggested a couple dozen things that Cambridge Analytica might want to flag in the profiles, from political proclivities to "sensational" interests in subjects ranging from guns to "black magic."

The data-gathering process took about four weeks. The 200,000 survey-takers had about 50 million friends, Kogan guesses, but not all were Americans. The contract he signed with SCL on June 4 was limited to Facebook users in only eleven states, so he handed over only 2 million profiles—names and demographic information on people and his predictions about their personal traits. "Then, later on, they came and they're like, *Hey you've got a lot of data, can we have the rest?*" he says. "We're like, *Sure* . . ." So millions more profiles went to SCL.

Kogan says that if he'd thought he was violating Facebook's terms, he would have stopped. "Look, in my world I have this incredible special relationship with Facebook—not a lot of academics have a relationship where Facebook shares their data with you. What kind of idiot would I have to be to do something that I think is gonna piss them off?"

Kogan does admit mistakes. "If I had a time machine and could go back, there are a couple of things I would do dramatically differently," he says. "One, I would do a lot more due diligence on *Who the hell is SCL*?"

KOGAN'S ARRANGEMENT WITH SCL infuriated Michal Kosinski. In his view, Kogan was duplicating work that he and Stillwell had pioneered and was now selling it for private gain. So Kosinski wrote a letter to John Rust, laying out his charges of unethical behavior.

Rust agreed it was a problem. He now says he never liked Kogan. He considered him "pushy." Researchers in the lab would give themselves nicknames, and Kogan dubbed himself "Beloved Commander" (a weird echo of young Mark Zuckerberg's self-description on the first version of Facebook). Worse, Kogan was publicly boasting about the database he had built. At a brown-bag lunch at the National University of Singapore on December 2, he promised to speak about "a sample of 50+ million individuals for whom we have the capacity to predict virtually any trait." The idea that Kogan was selling that to a political organization horrified Rust. "It's not what we do," he says. He told Kogan that duplicating the Kosinski-Stillwell work was wrong and he should stop it. *These are your colleagues,* he told him. *They've been working on this for years. Why don't you get on with your own work?*

Kogan disagreed. Rust suggested that they should take the matter to arbitration. The university would not pay the $4,000 fee, and Rust and Kogan agreed to split it. The arbitrator began investigating but Kogan suddenly withdrew, claiming that he'd signed an NDA that would prevent him from cooperating.

Rust wound up writing a letter to his dean on December 8.

> *I am becoming increasingly concerned about Alex's behaviour. All the hearsay from day one suggested that he completely ignored our letters and was continuing to operate his company within the University . . . Just to recap, the procedures he is using to build his database depends not just on obtaining predictions from the Facebook 'Likes' of 100,000 individuals, but on the fact that the Internet as currently set up allows him to obtain the same information on all the Facebook friends that are connected to them (none of whom have given any form of consent to this). As the average Facebook user has 150 friends, this makes a database of 15,000,000, and their intention is to extend this to the entire US population and use it within an election campaign.*

Actually, the database was many more than 15 million, maybe even more than the 50 million Kogan estimated. According to Facebook's calculations it could be as many as 87 million. But the world would not learn that for more than two years.

Kosinski was unhappy at the inaction. And he had a way to strike back. Some months earlier he had met a researcher, Harry Davies, who had been conducting interviews that would become source material for a theatrical play called *Privacy*. According to a press brochure about the play, "PRIVACY explores the ways in which governments and corporations collect and use our personal information, and what this means for us as individuals, and as a society."

That November—2014—Kosinski became a whistleblower. He told Davies, who had gotten a researcher job at *The Guardian*, about the Kogan–SCL connection, and handed over all the documents he had. Davies contacted SCL and asked about its relationship with Kogan, but got no answer. (Later, a Cambridge Analytica executive would explain that the team was headed to a party in Washington, DC, and the random person left in the office hung up on Davies.) He put the story aside.

But in the fall of 2015, Davies came across a *Politico* article that explained the relationship between SCL and Cambridge Analytica, the connection to Robert Mercer, and that the Ted Cruz presidential campaign was using the data. Davies dug back into the documents and in spare time from his researching duties, put together the story: how Kogan had gathered the data for a research project and then, violating Facebook's standards, sold it to Cambridge Analytica. The Cruz campaign insisted that all was kosher. "My understanding is all the information is acquired legally and ethically with the permission of the users when they sign up to Facebook," said a spokesperson.

Davies suspected otherwise. Before publishing, Davies emailed Kogan a summary of his upcoming story—basically charging Kogan with unethical behavior—and told him he had twelve hours to respond. Kogan freaked out. "That was certainly one of the most stressful moments of my life," he says. "I had never been in the press for anything negative." Kogan contacted the university PR office and worked on a response. He also gave a heads-up to his partner Joe Chancellor, who had left

Cambridge University and was now working for the Data Science team . . . at Facebook.

ON DECEMBER 11, 2015, Harry Davies's story appeared in *The Guardian*, reporting how stolen Facebook profiles were used in the Cruz campaign. The policy lords of Facebook were blindsided. No one in the DC office had ever heard of Kogan or Cambridge Analytica. But they *had* heard of Ted Cruz, and the idea that his campaign was directing ads using personal data mishandled by a Facebook developer was its nightmare of appearing to be a partisan force in the election. The team frantically tried to find out what it could. The person assigned to gather information was the lead official on the policy side of Developer Operations, Allison Hendrix.

It turns out that for months, the people in the Platform organization had been trying to deal with data misappropriation by political organizations, specifically Cambridge Analytica. Hendrix had been on the thread. On September 22, a political consulting firm in DC had asked Facebook if it could clarify the rules about using its data in campaigns. The request was spurred by competitors who seemed to be breaking those rules. "The largest and most aggressive [violator] being Cambridge Analytica, a sketchy (to say the least) data modeling company that has penetrated our market deeply," wrote the consultant, asking Facebook to investigate the company.

For the next few months, with not much urgency, various people in the Developer Operations organization gathered information. It didn't concentrate on Cambridge Analytica, but explored the practice of data scraping by political consultants in general. It lit on a right-wing site called ForAmerica, which was in the process of scooping up Likes from visitors to its popular Facebook page. After some initial confusion over whether these practices actually violated Facebook's policy, some of the employees on the chain confirmed that indeed it did. "I do suspect there is plenty of bad actor behavior going on," wrote an employee on October 21. But the investigation, if it could be called that, hardly went deep.

Then *The Guardian* story dropped, and suddenly learning about

Cambridge Analytica was a higher priority. In the frantic emailing inside the company, one employee unearthed an unsettling fact: "It looks like Facebook has worked with this 'Aleksandr Kogan' on research with the Protect and Care team."

"It was like the Wild West, with this guy having access, and we just didn't know what he was doing with it," says one of the Facebook people.

Facebook set up a call with Kogan, who recalls that Allison Hendrix instructed him to delete the data. He characterizes the conversation as friendly. Though he would have preferred to keep the data for research, he agreed. "Facebook to this point had been a really strong ally," he says. "Obviously I was not feeling great that they might be upset. Plus we had fifteen papers in the works with them!" Only a few weeks earlier, in fact, he had spent time on campus for his consulting deal with Facebook, helping the company with surveys.

Hendrix also contacted Cambridge Analytica/SCL, beginning an email thread with its director of data, Alexander Tayler, who at first professed that nothing was amiss. After a few exchanges, on January 18, 2016, he said he had deleted any Facebook data that Cambridge might have had. Hendrix thanked him. Hendrix, who previously had signed her emails with her proper first name, signed off this time with the diminutive "Ali."

Those loose promises were clearly insufficient. So Facebook began a process of negotiating binding agreements from the parties where they would vow they indeed deleted the data and were no longer using it. It assigned the task to outside counsel. It did not take steps to *confirm* if either actually deleted it, which would have been difficult in any case. How would they know if at some point Kogan put the data on a thumb drive and stuck it in his bag? While Kogan's app was bounced from the platform, neither Kogan nor Cambridge was banned. Kogan figured that everything would blow over and he would eventually be back in Facebook's good graces.

Throughout the process, the matter never seemed to reach Sheryl Sandberg or Mark Zuckerberg.

. . .

AS THE ELECTION season heated up in 2016, Cambridge Analytica was actively working for GOP candidates. After Ted Cruz dropped out, the company began working for the Trump campaign. Cambridge Analytica's vice president, Steve Bannon, became a top adviser to the candidate himself. Cambridge had contracted with a Canadian company called AggregateIQ—reportedly a Wylie connection—to implement a set of software services to exploit Cambridge's voter database, including the apparently undeleted profiles and personality summaries provided by Kogan.

What Facebook did not do for *more than a year after learning about the Cambridge Analytica data abuse* was get a formal affirmation that Cambridge had deleted the data. (Facebook's excuse: its outside law firm was negotiating.) While Kogan had not turned in his affirmation until that June, Cambridge did not do so at all during the entire election campaign, even as Nix had been boasting to his clients, current and prospective, about the huge database he had. Meanwhile, Facebook was a *partner* to Cambridge Analytica, which was a major political advertiser, enjoying support and advice from Facebook's Advertising team. At any time during the election, Facebook could have threatened to cut off access to its platform if Nix and company did not prove that they had deleted the ill-gotten personal information of 87 million Facebook users. Or Facebook could have demanded an audit. It did not. But it did collect millions of advertising dollars from Cambridge Analytica, without checking whether the money might be the fruit of the unauthorized profile data. In accepting advertising money, it accepted the company's claims that it wasn't, even while Cambridge had not yet signed an affirmation.

Cambridge Analytica did not formally affirm that it had deleted the data until Nix did so on April 3, 2017, after its candidate had been in the White House for months. Again, Facebook took its word and did not use the opportunity to conduct an audit to verify the claim. A year later, the UK Information Commission searched CA's computers and found that Cambridge Analytica might still have been using data models that benefited from the Facebook information. To this day, it isn't clear whether the company's election efforts used Facebook profiles, though *The New York Times* reported that it had seen the raw data in Cambridge Analytica's files, and former Cambridge Analytica executive Brittany Kaiser says that the data was indeed part of the election targeting.

And at no time during 2016 or 2017 did Facebook inform millions of users that their personal information had been operationalized—and their own News Feeds manipulated—for political purposes.

There is still a raging debate about whether Cambridge Analytica's data operation made any difference in the campaign's outcome. Before Trump was elected, the Cruz campaign had concluded that the data was not helpful. Brad Parscale would later tell *Frontline* that all but $1 million of the $6 million the Trump campaign paid Cambridge was for television; he says he used Cambridge employees as staff because of their talents, not their data. During the campaign, CEO Nix, however, boasted of his "secret sauce," and upon Trump's victory gloated that CA's "data-driven communication" played an "integral part" in the win. At Facebook itself, those experienced in the political scrum thought that Cambridge was something of a hype, one of countless wannabe consultants promising digital black magic. "They were sort of the Theranos of the campaign world," says one DC Facebook official. "Then after Trump won, there was this weird disconnect between people who saw them as evil geniuses and people in Washington who thought they were clowns."

Clown show or not, during the election Facebook lost track of the one fact that mattered: Cambridge had gotten hold of the private data of millions of Facebook users, and had yet to confirm that it deleted the data. And Facebook did not pursue the possibility that Kogan's database, passed on to SCL/Cambridge, was being used for the Trump campaign. CA's approach seemed a perfect way to exploit the vulnerabilities that Facebook had created in its drive to gain and retain users: use the data people shared to identify their hot buttons, and target them with manipulative ads—on Facebook—that pressed those buttons. That's what the Russians had figured out. As one Facebook policy person put it to me, "Could I guarantee that I could help you manipulate Facebook to win the election? The answer is no. But can you tap into people's fears, and people's worries, and people's concerns and people's bigotry, to activate and prime things? Absolutely."

When reporters followed up on the 2015 *Guardian* revelations in light of the election results, Facebook's responses were misleading. "Our investigation to date has not uncovered anything that suggests wrongdoing," a spokesperson told *The Intercept* in 2017, when clearly Facebook knew there *was* wrongdoing. That's why it had demanded that Kogan, SCL, and Wylie delete the profiles after 2015. Facebook also pointed reporters to a statement from Alexander Nix that Cambridge "does not obtain data from Facebook profiles or Facebook likes," even though it knew that Cambridge *did* license that data from Kogan. Citing that quote certainly seemed like misdirection, considering that CA hadn't certified that it deleted the data.

Wylie would later claim that he deleted the data in 2015 but was delayed in verifying it because Facebook's order didn't reach him until mid-2016. It had sent the forms by snail mail to his parents' home. "They just sent this letter saying, 'Can you confirm that you don't have the data?'" says Wylie. "It was like, you fill in a blank and then you sign it. It was like a blast from the past because I hadn't heard Kogan's name in a while." So much for urgency.

BY THEN WYLIE had a new digital pen pal, a reporter for *The Guardian/Observer* named Carole Cadwalladr. A feature writer and investigative journalist known for deep dives into her topics, often with a participatory twist (like working in an Amazon warehouse), Cadwalladr had become fascinated with what she perceived as the pernicious influence of big tech companies. In 2016, she began investigating Cambridge Analytica. She wrote a series of articles about the company—its involvement in Brexit, its methods, its ties to Robert Mercer and the ultraconservative movement that had backed Trump. And the Facebook data that Kogan had been called out for in December 2015.

She lit on Wylie as the key to the story. When she first contacted him in March 2017, he was wary, but eventually he handed over documents that helped inform her stories. But Cadwalladr wanted *him*. If Wylie cooperated fully, and told the Cambridge story from his point of view, it would be more compelling. "I sat on the documents for over a year," she says. "It's not enough to publish documents without the personal story."

Cadwalladr was a contract writer, paid by the article. But she turned down other assignments to keep working the Cambridge story. Eventually she convinced Wylie to go on the record.

But she had another concern beyond Wylie. In one of Cadwalladr's earlier stories, she had described how an intern had first suggested to Alexander Nix that SCL should get involved in data. This young woman, wrote Cadwalladr, was Sophie Schmidt, daughter of former Google CEO Eric Schmidt. According to Cadwalladr, *The Guardian* then heard from a top UK lawyer representing Schmidt. He did not deny the information but demanded that Schmidt's name be removed from the story because it was personal and of no public interest. "Our lawyers looked at it and said, *Yes, she can't win*," says Cadwalladr. "*But we might have to spend 20 or 30,000 pounds defending it.*" So the *Guardian/Observer* removed Schmidt's name from the story. "It really woke us both up to the problems of publishing this stuff in the UK," says Cadwalladr.

Her editor had an idea that might mitigate the problem: why not collaborate with a big US entity like *The New York Times*, which was less vulnerable to a bogus libel suit? Cadwalladr didn't like it—this was *her* story—but had no choice but to go along. The *Times* agreed to write its own story based on Cadwalladr's work and original reporting, and both would publish stories simultaneously, with Cadwalladr listed as a co-byline in the *Times*.

Cadwalladr's story portrayed the now-pink-haired, nose-ringed Wylie as a courageous whistleblower. This was something akin to Charles Manson blowing the whistle on Sharon Tate's murderers. Wylie had actively engineered the scandal. He had egged on SCL to create Cambridge Analytica, and he was responsible for enticing Kogan to unethically transfer Facebook user data to a political dark-ops consultancy. "To be a whistleblower, you have to be right at the dark heart of it," says Cadwalladr of the suspect hero of her narrative. Wylie would later characterize his transformation as a product of his disgust after the Trump election, a strange claim for someone who worked with Mercer and Bannon to form Cambridge Analytica. "I am incredibly remorseful for my role in setting it up, and I am the first person to say that I should have known better," he would later tell Parliament, "but what is done is done."

Early in the week before *The Guardian*'s scheduled publication that Saturday, Cadwalladr contacted Facebook. She always had problems getting responses from its communications people. She didn't know anyone in Menlo Park and had to filter her requests for comment though the UK office. A silence of several days was broken by Facebook's deputy general counsel, Paul Grewal. He took issue with her depicting the transfer of 50 million Facebook profiles from Facebook to Kogan and then to SCL as a "breach." Cadwalladr interpreted it as a threat to sue. (Facebook says this wasn't the case. Just a syntactic suggestion from a giant corporation's deputy counsel.)

While Facebook was technically correct about the term, it was an odd objection. A breach suggests carelessness, exploited by wrongdoing. In this case, Facebook had *given away* private data to Kogan without sufficient user permission. Giving social data to developers was in keeping with the rules that had been basically established with the 2007 Platform and continued with Open Graph, which enabled features like Instant Personalization. For years, as Facebook's user base expanded, those rules were seen as something that promoted growth, and they remained. Finally, in 2014, Facebook had acknowledged that those rules were flawed, and announced that it would close off that gaping privacy loophole—a year later. That extension allowed Kogan to build, and sell to Cambridge Analytica, his database of millions.

Facebook attempted to get ahead of the story, dropping a news post after the market closed on Friday. It explained that after the 2015 *Guardian* story, Facebook ordered Cambridge, Kogan, and Wylie to delete the data and they said they did. However, it continued, "Several days ago, we received reports that, contrary to the certifications we were given, not all data was deleted." So, in its ongoing crusade to "improve the safety and experience of everyone on Facebook" the company was banning the wrongdoers Cambridge Analytica, Kogan, and Wylie. Reading this without context, the move seemed to depict Facebook as a vigilant protector of user data. The announcement would be viewed in a different light soon afterward, when the *Times* and *Guardian/Observer* rushed their stories out.

Both stories worked the same explosive angle: Facebook had allowed the personal data of millions of its users to fall into the hands of Trump consultants during the campaign. Though much of the basic details had been revealed in December 2015, it seemed a lot more urgent—and shocking—now.

"For twelve hours it looked like we were taking proactive steps against CA, and then the bomb dropped," says one official in Facebook's DC office. "Whatever goodwill Facebook had earned in that period was just discarded."

Though Facebook had known the articles were coming for a week—and the larger story had been clear since the 2015 *Guardian* piece—the articles hit the company with the shock of a meteor. Perhaps it was because Facebook's chambered organizations had prevented the full Cambridge story from reaching Sandberg, and certainly Zuckerberg, who would consistently claim that before that week he had never heard of CA, Kogan, or the deleted data.

Facebook had been through meltdowns previously—News Feed, Beacon, the Consent Decree. Each time, though, Zuckerberg had speedily responded with a double-barreled message: First, apology. And then, plan of action.

But this time there was no plan.

"I'm not sure it would have worked if we had been like, *We're on it, we'll get back to you*," says Sheryl Sandberg in reconstructing those awful days where Facebook was, in a PR sense, burning, and its executives seemed to have been lost in the conflagration. "People would've been like, *They don't even know what happened!*" Which would have been the truth. "We were trying to make sure we understood the problem," says Sandberg. "We were trying to get real steps for a real problem, and we didn't have our arms around it. Looking back on it, that was not our best move. It was a bad move."

Zuckerberg would later agree. "I think I got this calculation wrong, where I should have just said something sooner even though I didn't have all the details and said, *Hey we're looking into this*, but instead my instinct is like, *I want to know what the reality is* before I go out and talk."

Rank-and-file Facebook workers were even hungrier than the public to hear the explanations. For months, Facebookers had been fending off queries from friends and relatives about what kind of company they were working for. Generally, the view from inside was that their employer was well intentioned but had made some mistakes. They could hold their heads up high. This was now in question. In addition, Facebook's stock—and the net worth of the workers—took a tumble when the market opened that Monday. They wanted to hear from their leaders.

Instead, the company sent its deputy general counsel Grewal—who had only days earlier menaced *The Guardian* with his letter—to explain Cambridge Analytica to the company. The absence of Sandberg and Zuckerberg was a morale breaker. "I was sympathetic to the employees," says Grewal. "No matter how well versed I was in the facts, the one thing that I could not do was suddenly transform myself into Mark or Sheryl."

After five days of executive lockdown—much of it arguing about PR options—Sandberg and Zuckerberg emerged and went on somewhat of an apology tour to selected media outlets. To some degree, they had figured out what went wrong in this particular case, and took responsibility: "We could have done this two and a half years ago," Sandberg said on the *Today* show. "We thought the data had been deleted and we should have checked." Just how much data, and what it was, they still weren't sure. Zuckerberg came closer to the cause when he told *Wired*, "I think the feedback that we've gotten from people—not only in this episode but for years—is that people value having less access to their data above having the ability to more easily bring social experiences with their friends' data to other places."

For the past twelve years, Zuckerberg had been ranking those values incorrectly.

In a larger sense Facebook's top leaders did not have "their arms around it," to use Sandberg's term. Cambridge Analytica was now a symbol of Facebook's bigger trust issue. The story had all the elements of Facebook's perceived flaws—a cavalier view toward user privacy, greedy manipulation, and the gut suspicion that the social network had helped elect Donald Trump. Every one of those flaws was the result of decisions made over the past decade to spur sharing, to extend Facebook's reach, and to step over competitors. To the public, Cambridge Analytica was now the lifted rock that exposed a hellish profusion of scurrying vermin.

For more than a decade, Facebook had skipped from one crisis to another without suffering serious consequences. It had *moved fast*, with little regard for what was overturned in its wake. Its motto may have changed. But Facebook was still breaking things. And Mark Zuckerberg was off to a very bad start to his year of rebuilding trust.

AFTER CAMBRIDGE ANALYTICA, Zuckerberg could no longer ignore Congress's cries that he submit to public hearings. Facebook's lobbyists and lawyers began negotiating. It was a sign of their lack of leverage that Facebook wound up agreeing to expose their CEO to two full days of interrogation, one before a joint hearing of two Senate subcommittees—Commerce and Justice—and a day later, to face the House Energy and Commerce Committee. They did manage to get one concession: Zuckerberg would not have to be sworn in. There would be no iconic image of Zuckerberg raising his hand before testimony, placing him in the category of tobacco executives and mafia chieftains. "As a practical matter, I will tell you that other than the visual, Mark was obligated to tell the truth, and any false statements he made would obviously have grave implications to him," says Grewal, who would quarterback the preparation efforts.

Zuckerberg's only previous experience in a public docket had been a year previous, testifying in the case involving Facebook's purchase of the virtual reality company Oculus. "Mark is not somebody who's often interrupted, not someone who's usually chastised, certainly in an open forum," says Grewal of the challenge of preparing for that trial. "He needed to understand what that would feel like, and understand that that's the lawyer doing his or her job. He took all that coaching and suggestion quite well." The elaborate plans for the Texas appearance had extended to the logistics—how could they get him in and out of the building without exposing him to cameras and press as if he were in a perp walk? "As a former federal judge, I remembered, there's always a second set of elevators in every courthouse, to take the prisoners up from lockup," says Grewal, who talked to the US Marshals Service to bend protocol a bit. "So Mark actually ended up taking the prisoner elevator up to the courts."

Facebook ran rehearsals for days, turning Zuckerberg's living room in Palo Alto into a virtual hearing room. Since the sweat-prone CEO would not be able to control the temperature, they aimed hot lights at him. Working from a set of "murder boards" a team of policy people took the roles of various legislators, firing every possible question at him.

On April 10, 2018, the hearing room in the Hart Office Building was packed. Protesters wearing cardboard cutouts of the Facebook CEO's face marched ghoulishly outside the room. In the gallery people held up signs with legends like, PROTECT OUR PRIVACY. Others wore goofy glasses that said STOP SPYING! Dozens of photographers surrounded the table where Zuckerberg would sit, while a cadre of his attorneys and policy apparatchiks would sit behind him. Finally, walking ramrod straight, Zuckerberg entered, wearing a dark suit and a loosely knotted sky-blue tie. (*The New York Times* would later devote an article to scrutinizing his non-hoodie apparel.) He read from his prepared statement.

Facebook is an idealistic and optimistic company. For most of our existence, we focused on all the good connecting people can bring . . . But it's clear now that we didn't do enough to prevent these tools from being used for harm as well. That goes for fake news, foreign interference in elections, and hate speech, as well as developers and data privacy. We didn't take a broad enough view of our responsibility, and that was a big mistake. It was my mistake, and I'm sorry. I started Facebook, I run it, and I'm responsible for what happens here.

The questioning began, and even Facebook's critics would become more annoyed with the preening, clueless questions from the legislators than with the stilted steadfastness of the billionaire former whiz kid in the public woodshed. Had the senators borne down and directed focused queries at him, they might have been more productive. But most chose to spend much of their five minutes lecturing Zuckerberg, or highlighting some mundane local technology issue. A number of legislators decided to spend their time on the fantasy that liberal techies at Facebook wrote algorithms to suppress conservative content. Even when zeroing in on privacy, they showboated. "What hotel did you stay in last night?" asked one senator, who beamed in triumph when Zuckerberg quite sensibly demurred. It was a rather weak analogy to the complex privacy issues exposed by Cambridge.

The hearings also exposed the technological ignorance of some of the senators. Generally the House representatives the next day were a bit sharper and more pointed, with some glaring exceptions. Utah senator Orrin Hatch expressed puzzlement that Facebook didn't charge its users. *How do you make your money?* he asked.

"Senator, we run ads," said Zuckerberg. Later, Facebookers would make T-shirts with that slogan.

Throughout, Zuckerberg responded humbly, if robotically, no matter how loony or hostile the questioner. In front of him was a detailed set of talking points, a fortune cookie–esque game plan for congressional rope-a-dope: *Breach of trust, sorry we let it happen . . . didn't think enough about abuse . . . made mistakes, working hard to fix them. . . .* If a legislator attacked him directly, he would revert to the scripted comment, *Respectfully, I reject that. It's not who we are.* And if he had any doubts whether he had the information at hand to answer a question clearly, he vowed that his team would supply the answer later. *Wired* counted that he did that forty-six times. "Please note," it wrote, "that this does not include the several occasions in which Zuckerberg claimed he didn't know an answer but promised no follow-up."

After two hours, Zuckerberg was offered a break and said, "Let's go on."

"That's when I knew we were all right," says Grewal.

Zuckerberg's comment on Cambridge Analytica: "When we heard back from Cambridge Analytica that they had told us they weren't using the data and had deleted it, we considered it a closed case. In retrospect, that was clearly a mistake."

When Zuckerberg returned to Menlo Park, he appeared at the all-hands meeting like one of the conquerors he had so admired as a young Latin student. But the upbeat mood was temporary. Investigations regarding Cambridge Analytica would continue for years. Within a few months, other venues would conduct their investigations, and be frustrated when Zuckerberg sent subordinates to appear in his stead. During the interrogations, officials would turn from Facebook's designated punching bag and address an empty chair reserved for the boss, directing pointed questions about every detail of Cambridge Analytica, and what it said about the company's practices.

It was almost anticlimactic when, in September 2019, documents unsealed from a class-action lawsuit against Facebook involving Cambridge Analytica indicated that the abuse of the Open Graph by Cambridge Analytica was far from a singular case. When Facebook looked into the matter, it discovered more than four hundred developers had similarly violated its rules. It suspended 69,000 apps, including 10,000 that may have misused Facebook user data. This astonishing figure got relatively little play—mainly because by that time, Facebook scandals were so widespread that this registered as just one more.

Those floodgates had opened with the odd concatenation of events that led to Cambridge Analytica. Aleksandr Kogan's happiness quiz turned out to be misery for Facebook. And it would not be the last time that Mark Zuckerberg would sit before a room of angry legislators.

17

The Ugly

ON MARCH 15, 2019, an Australian white supremacist entered the Al Noor Mosque in Christchurch, New Zealand, and murdered fifty-one worshippers. Brandishing multiple automatic weapons and playing military music on a portable speaker, he streamed the entire spree on Facebook Live. It is now a week later, and Monika Bickert is in Washington, DC, in a dimly lit cocktail lounge, wolfing down French fries and fighting back tears.

Bickert's job is setting the rules for content on Facebook. There has always been tension between Facebook's free-speech cheerleading and its need to keep the platform safe. But after the election, the scrutiny has been heightened. Bickert doesn't take it personally. Her mission isn't connecting the world. It's trying to stop Facebook from ruining it. Making the job harder was that, after Cambridge Analytica, the whole world was watching. And jumping on her each time her team failed at an impossible job.

BICKERT GREW UP in Southern California. She loved sports and excelled on the volleyball team while taking AP courses. Her high school history teacher was coach of the mock-trial team and urged her to join, and she loved it—the strategizing, the analysis, and especially performing in a virtual court. At Rice University, she studied economics and played volleyball. She had accumulated the credits required for graduation when an injury ended her varsity career after her third year. She went directly to Harvard Law, did a federal clerkship after graduating, and joined the US Attorney's Office, first in DC, and then in Chicago.

Bickert handled cases like the prosecution of forty-seven members of the Mickey Cobras, a street gang accused of selling heroin and fentanyl in the Dearborn Homes housing project. She put people in jail for government corruption and child pornography. She also fell in love with one of the top lawyers in the Chicago US Attorney's Office, Philip Guentert, a widower who had two young adopted girls. In 2007, to expose their Chinese-born daughters to life in Asia, they moved to Bangkok, still working for the DOJ. Bickert concentrated on sex-trafficking cases. Then Guentert was diagnosed with kidney cancer. They moved back to the States to address his medical problems. So when Bickert heard that Facebook was looking for someone with government and international experience, "I threw in a résumé and came to visit the campus, not really knowing what to expect," she says. No amount of prepping could have revealed what would eventually become the nature of her job there.

When she interviewed in 2012, Bickert was taken with the energy of the Facebookers, but more profoundly intrigued by the prospect of untangling some of the legal threads that came from running the biggest social network the world had ever seen. Nobody had ever dealt with those questions before.

Her first role at Facebook was responding to government requests for user data, where she got a sense of the power of the information people shared with Facebook. After about six months, the company tapped her as legal muscle to enforce its data policy with developers.

Because Bickert dealt with troublesome developers, she wound up in a room where Facebook's policy people were considering a video game that included what seemed like hate speech. The argument was whether it violated Facebook's rules. Bickert weighed in with an analysis that impressed Marne Levine, then the head of global policy. Levine realized Bickert might be the ideal match for the open position of arbiter of content policy. Bickert started in 2013.

Which is how Monika Bickert's supposedly low-key job at a tech company transformed into one of its most public and exposed roles: she became perhaps the world's most powerful arbiter of speech, operating in a fishbowl where every decision was subject to scorn and outrage, and implementing those decisions at a scale where failures were guaranteed.

Those failures could have consequences, particularly overseas, where the company had all too often rushed into countries without understanding the culture or setting up infrastructure to deal with sometimes dangerous abuses of the platform. Organized groups, sometimes actual governments, would post hate speech and inflammatory falsehoods that targeted dissidents or vulnerable minorities. Before such problems became public—and even sometimes afterward—Facebook would pay little attention, despite warnings.

During the Arab Spring movement in the Middle East, Facebook was celebrated as a force for freedom. Users running Facebook pages helped organize the 2010 Tunisian uprising. The Egyptian campaign to overthrow the government was hugely lubricated by Facebook; after a computer programmer was murdered by state police, a Facebook page called "We Are All Khaled Said" galvanized the protest movement that would overthrow the regime. "It felt like Facebook had extraordinary power, and power for good," former Facebook policy executive Tim Sparapani told *Frontline*. "I remember feeling elated to see people use this tool, a free tool, to do things that they could never have done before, to organize, to share their world, to show violence that was being foisted on them by people in their government who are trying to prevent this uprising. . . . It can't get any more real."

For years, the halo effect from empowering righteous activists would blind Facebook to the potential for abuses in other countries. From Menlo Park, it was hard to envision how the platform's political mojo that freed people could just as easily be used by those in power to divide and dominate them.

The Growth game plan was to spread Facebook around the world, and its operatives figured that some measure of good vibes and crowdsourced problem-solving would take care of unpleasant consequences of a mass free-speech platform popping up in regions unaccustomed to anything like it.

Maria Ressa, the Philippine journalist who had reported misinformation and hate campaigns to Facebook in 2016, was seeing this firsthand. After the Philippine strongman leader Rodrigo Duterte consolidated power, his followers kept using Facebook to demonize opponents, and then Ressa herself. She kept pushing Facebook for action. Ressa spoke to all the key policy people—Elliot Schrage, Monika Bickert, Alex Schultz, Sheryl Sandberg—and even was part of a small meeting with Zuckerberg at the annual F8 conference in May 2017, where the CEO met with global developers, telling them fake news was an issue, but it will take time to get it right. But the problem was now! To Ressa, none of those Facebook executives seemed to get it. "For the longest time I felt like not only were they in denial, I would get blank stares," she says. It wasn't until early 2018, she says, that Facebook responded emphatically.

Facebook's situation in Myanmar, formerly called Burma, was even worse. Myanmar was one of the countries where Facebook had rushed in without employing a single speaker of the language. Before things went bad, Chris Cox actually celebrated this approach to me in a 2013 conversation. "As the usage expands, [Facebook] is in every country, it's in places in the world and languages and cultures we don't understand," he said then. Facebook's solution at that time, he said, was not to hire dozens of people who knew the culture but to double down on algorithms that measured how much people were using it. The more engagement you had, the better it was working! Cox admitted that it was challenging to juggle the various factors around the world where people used the platform differently—like getting their news from it. He told me that a friend in "Burma" told him Facebook was a channel for news there. "It's like, *We have to get the news somewhere!*" (To his credit, Cox later became a force for more aggressive action in taking down offensive content. He would often clash with Zuckerberg on the issue.)

When Cox was boasting about Burma, Facebook was already being abused there. "What you've seen in the past five years is almost an entire country getting online at the same time," says Sarah Su, who works on content safety issues on the News Feed. "On the one hand, it's really incredible to have been a part of that. On the other hand, we realized that digital literacy is quite low. They don't have the antibodies to [fight] viral misinformation."

Both in journalistic accounts and in UN human rights reports, Myanmar's president and his supporters used Facebook as a weapon to incite violence against the Rohingya, a Muslim minority group. For instance, on June 1, 2012, the president's key spokesperson posted a call to action against the Rohingya in a Facebook post, warning of dangerous, armed "Rohingya terrorists," essentially garnering support for a government massacre that would indeed occur a week later. "It can be assumed that the troops are already destroying them [the Rohingya]," the post said. "We don't want to hear any humanitarian or human rights excuses. We don't want to hear your moral superiority, or so-called peace and loving kindness."

In November 2013, Aela Callan, an Australian journalist on a Stanford fellowship at the time, visited Facebook to alert the company to the situation, meeting with Elliot Schrage. She was told that Facebook had only a single content moderator, based in Dublin, who understood the Burmese language. She told *Wired* that she felt Facebook was "more excited about the connectivity opportunity" than the violence issues.

In 2014, a woman was paid to make a false report that Rohingya had raped her. The posts spread on Facebook and inspired a riot that led to the death of two people. Around that time Bickert hosted some civil society groups including one from Myanmar, who asked for help. Spurred in part by one of its policy people in Australia who took an interest in the problem, Facebook understood then that it needed more native speakers, but by its own admission was slow to hire enough content moderators who knew the language. An additional problem was the way the Burmese language is deployed online: sometimes it uses Unicode, which is the international standard; other times it uses a unique font that is tricky for

Facebook's system to read. It wasn't even until 2015 that Facebook translated its Community Standards manual to Burmese.

From the Philippines, Maria Ressa noticed that Facebook's struggles to grasp that problem resonated with its failure in her home country. "I think it took them a very long time to understand Myanmar," she says. "It was a combination of willful denial and lack of context. It's really, really a different world because they don't live in countries that are vulnerable. I just watched my life get torn apart by the stuff that Facebook allowed. They were looking at their problems within the context of Silicon Valley."

In June 2016, Facebook introduced Free Basics to Myanmar, making it even more challenging to police hate speech. When violence intensified, Facebook was unable to contain it. Bickert says one reason was that Facebook was hiring what it thought was a reasonable number of native speakers, but when violence breaks out, more people use Facebook and more provocative content gets posted. "We weren't well positioned," she says. "When the violence broke out, we didn't have good technical tools to find content. We were dealing with the font problem, with the report flow rendering incorrectly, and we didn't have sufficient language expertise."

Not until August 2018 did Facebook take major steps to take down content in Myanmar, removing eighteen Facebook accounts, one Instagram account, and fifty-two Facebook Pages. It also banned twenty people and organizations, including the commander-in-chief of the armed forces and a television network run by the military. Still, hate speech and provocations to violence continued. When Zuckerberg testified before Congress, Senator Patrick Leahy confronted him with an incident where a Facebook post called for the death of a journalist. Multiple pleas had gone to Facebook to remove the post before the company acted, said Leahy.

"Senator, what's happening in Myanmar is a terrible tragedy, and we need to do more," was Zuckerberg's response.

WhatsApp, which had become the dominant service in Myanmar,

presented special challenges because its content was encrypted, and Facebook could not know what was in the text exchanges unless a recipient of a message sent them the decoded information. The WhatsApp founders had decided to build encryption deep into the product, believing that this impenetrability was an unalloyed positive.

"There is no morality attached to technology, it's people that attach morality to technology," WhatsApp co-founder Brian Acton told me in 2018, looking back on the controversy. "It's not up to the technologists to be the ones to render judgment. I don't like being a nanny company. Insofar as people use a product in India or Myanmar or anywhere for hate crimes or terrorism or anything else, let's stop looking at the technology and start asking questions about the people."

Acton was expressing something that most of the company dared not speak out loud, but sometimes muttered in private conversation. Violence had persisted in many regions for centuries, well before there was a thing called Facebook. Naturally, the appearance of a communications platform like Facebook would be exploited by dark forces, just as previous technologies like radio, the telephone, and automobiles had been. In this view, Facebook was just the medium du jour.

But this was not a tenable argument for Facebook to make about a place like Myanmar, where people were using Facebook's unique properties of viral distribution to spread lies about a minority group, inciting readers to kill them. Facebook contracted with a firm called BSR to investigate its activity in Myanmar. It found that Facebook rushed into a country where digital illiteracy was rampant: most Internet users did not know how to open a browser or set up an email account or assess online content. Yet their phones were preinstalled with Facebook. The report said that the hate speech and misinformation on Facebook suppressed the expression of Myanmar's most vulnerable users. And worse: "Facebook has become a useful platform for those seeking to incite violence and cause offline harm." A UN report had reached similar conclusions.

When unveiling the BSR report in November 2018, Bickert announced in a press call, "We updated our policy so that we now remove misinformation that may contribute to imminent violence or physical harm. That change was made as a result of advice we got from groups in Myanmar and Sri Lanka."

Every reporter on the call probably had the same thought I did: *You mean, before 2018 it was okay?*

THE DOWNSIDE OF "moving fast" wasn't limited to Facebook's international expansion. There was also a carelessness regarding products that Facebook was launching. The real-time program Facebook Live was intended as a feel-good feature, but it misjudged the human capacity for mischief, self-destruction, and evil.

It started as a way to help the famous use Facebook to become more famous. Around 2014 a small team working on Mentions, a feature supporting celebrities, began developing a feature to stream video in real time. They convinced their manager, Fidji Simo, to back them. Simo was a hard worker—Facebookers recall with awe that when she was on bed rest in a troublesome pregnancy, she kept up her pace, having teams come to her house for meetings—and she decided to pivot her product team to focus on video.

By the time Facebook Live launched in August 2015, Twitter already had started a live-streaming product of its own called Periscope, and a start-up called Meerkat was also getting buzz. Facebook, unlike the others, not only streamed video live, but let it remain on the page after it was done, allowing users to continue posting comments about it. This allowed for the clips to spread virally over a period of hours or days, creating maximum engagement. Facebook initially limited Live to the certified celebrities that the Mentions team had worked with, but when Zuckerberg saw that people were watching—a Ricky Gervais stream got almost a million users' attention—he decided to open it to the world.

Facebook Live had huge impact from the start, in large part because the company tweaked the News Feed algorithm to favor the videos. Early on, a joyful thirty-seven-year-old Texas woman streamed herself in a Chewbacca mask and won herself more than 100 million views and a brief period of stardom. News and quasi-news outlets embraced it; when BuzzFeed, a publication built on the back of the News Feed algorithm, streamed the destruction of a watermelon in 2016, it became a national phenomenon, with 800,000 people tuned in live. Harmless stuff.

But there was also harmful stuff.

"We really didn't know how people would use it," says Allison Swope, who helped create the product. "We were like, *People can post videos right now of horrible things. Is this really that much different than a [prerecorded] video?* We tried to think of all the scenarios, but I still don't know why someone wants to commit suicide live on Facebook."

Ellen Silver, of Facebook's Trust and Protect team, insists there *was* some planning. "We definitely had the team think through potential new abuse factors that would occur, from a policy perspective and the enforcement perspective," she says. "And it was unfortunate that those behaviors did occur on Facebook Live."

Nonetheless, Facebook was unprepared. The Live team endured a three-month "lockdown" to deal with suicide videos soon after launch. "We saw just a rash of self-harm, self-injury videos go online," Neil Potts of Facebook's public policy team told *Motherboard*. "And we really recognized that we didn't have a responsive process in place that could handle those."

Suicide presented a tough case, but one the company had already been grappling with in an enlightened way, encouraging users to spot warning signs and using artificial intelligence to detect posts indicating an impending attempt. When one was flagged, the company would dispatch helpers, either Facebook friends or local authorities, or hotlines. (Later some critics would attack Facebook for doing exactly that, charging that by trying to identify impending suicides, the company was overstepping into the medical realm. Perfect example of a Facebook can't-win situation.) Video added another complication. The content was disturbing, but the video could alert people to action. Some people even charged that a suicide could be the *fault* of Facebook Live—that the temptation of a public exit lured people into the act.

There were murders, too, and Facebook had trouble dealing with those as well. For instance, in June 2016, a twenty-eight-year-old man named Antonio Perkins was streaming on Facebook Live when someone fatally shot him in the head and neck. Since the video did not show gore, Facebook said it did not violate policy and left it up. The murder happened only one day after a young man in France, who had just killed two police officers, ranted on Facebook Live for thirteen minutes. This triggered unease among Facebook employees. Andrew Bosworth addressed the issue with one of his notorious internal memos. He meant it as a conversation starter, but it wound up sounding too much like a manifesto. He titled it "The Ugly."

We talk about the good and the bad of our work often. I want to talk about the ugly.

We connect people.

That can be good if they make it positive. Maybe someone finds love. Maybe it even saves the life of someone on the brink of suicide.

So we connect more people.

That can be bad if they make it negative. Maybe it costs a life by exposing someone to bullies. Maybe someone dies in a terrorist attack coordinated on our tools.

And still we connect people.

*The ugly truth is that we believe in connecting people so deeply that anything that allows us to connect more people more often is *de facto* good. . . . It is literally just what we do. We connect people. Period.*

That's why all the work we do in growth is justified. All the questionable contact importing practices. All the subtle language that helps people stay searchable by friends. All of the work we do to bring more communication in . . .

I know a lot of people don't want to hear this. Most of us have the luxury of working in the warm glow of building products consumers love. But make no mistake, growth tactics are how we got here. . . .

We do have great products but we still wouldn't be half our size without pushing the envelope on growth.

"The Ugly" generated hundreds of comments from Facebook's employees, most of them appalled by the idea that fatalities might be collateral damage of Facebook's growth. But those objections were tame compared to the reaction when BuzzFeed leaked the memo in 2018. Zuckerberg had to issue a statement: "We've never believed that the ends justify the means," he wrote. Zuckerberg further disowned the Boz memo during congressional testimony, adding that controversial posts were part of Facebook's tradition of open internal debate.

Even *Boz* distanced himself from it. "I was putting a stake in the ground at what is the most succinct and extreme formulation about the philosophy we have towards growth," he says. He carelessly dashed it off to spur a conversation about growth, intentionally overstating the sentiment for the purposes of his thought experiment.

I suggested that maybe the reason his memo had created such a fuss was that it really *did* present the ugly truth. Wasn't Chamath Palihapitiya's obsessive drive to snare the entire Internet population of Earth really a huge risk for populations unprepared for a mass tsunami of sharing?

Bosworth rejected that conclusion, as did key people on the Growth team. But another Facebook executive provided a different perspective. "Mark realized this in 2007—with the first kidnapping, the first rape, the first suicide—that there were going to be consequences," the official says. "The world is full of bad people. No company in history has ever had to answer to the bad people in the world as much as Facebook has. It's mentioned in forty percent of divorces!" (It's not clear where he got that number, but a 2012 study found that Facebook was mentioned in a third of divorces.)

After the 2016 election, Facebook could not blow past those consequences, or minimize them by citing how tiny a percentage they were of the total content on the platform. It had to deal with the ugly. In 2017, Facebook created a group called Risk and Response, to try to get ahead of impending crises. "There had been a lot more interest and scrutiny of the way that Facebook was making decisions around content on the platform," says James Mitchell, who heads the group. "In that environment, one of the things you can do is say, *Well, let's do a better job internally of trying to find and identify these vulnerabilities.*"

If so, Facebook might have done better dealing with the mass murder in Christchurch, New Zealand. Facebook Live had been an integral part of the killer's social-media strategy. Using proven techniques of effective brand consultants, the terrorist promoted his deadly broadcast in advance of its premiere, using sites like 8chan, Reddit, and more obscure white-supremacist outposts. He knew that he didn't need many viewers initially, because he could count on hundreds of thousands of people, be they followers or trolls or voyeurs or simply curious, to repost his deadly selfie. The carnage stunned the world. And in a season of constant scandals, it was one more blow to a company whose reputation, it seemed, could not get lower.

Facebook's job—Bickert's job—was making sure as few of Facebook's users as possible watched the horrible video. Her job also required that she view the whole clip herself, no matter how much she would hate it.

She is telling me this after appearing on a panel in the nation's capital about free speech. On earlier trips to DC, Bickert had appeared several times before committees, often having to defend Facebook content by invoking rules that made sense only in conference rooms in Menlo Park or K Street.

Now we are in a cocktail lounge down the block from the conference and she is recounting Christchurch. The seventeen-minute video that depicted the massacre, including the assassin's hop from one mosque to a second, had only been viewed live by about two hundred people. Facebook heard about it twelve minutes after it ended, and took down the video. But then the video kept spreading on Facebook, even as the company used a digital fingerprint to thwart uploads. An elaborate cat-and-mouse game broke out where Facebook would block the video and persistent users would alter the file to slip past the censor. Within twenty-four hours, users attempted to upload a version of the video 1.5 million times, and Facebook blocked 1.2 million of those, meaning that 300,000 copies managed to appear on the platform. A week later, people were reporting that one could still find copies of the video.

Why thousands of people saw fit to upload that video was and is a mystery. Just another piece of evidence that connecting the world has a dark downside.

As she describes the experience of watching the video, her voice breaks, and the edges of her eyes begin to moisten. Even for Monika Bickert, former prosecutor of Bangkok sex trafficking, onetime Javert to drug-dealing Chicago street gangs, steadfast arbiter of speech for two and a half billion souls, and icy defender of Facebook against preening legislators, this was too much.

THE FRONT LINES of Bickert's efforts are the content moderators that Facebook began hiring around 2009, when it established its first international centers in Dublin. These were the successors to the customer-support people who in Facebook's early days blocked nude photos from parties, dealt with lactating activists, and frantically hired colleagues as the task became overwhelming. By now Facebook directed a cast of thousands—more than tripling post-election, to 15,000 by 2019. "A lot of that is because we feel like we under-invested earlier," says Bickert. The moderators work around the globe, sifting through millions of pieces of content that are either reported by users as improper or identified by artificial-intelligence systems as potential violations. And they quickly make decisions whether these posts are indeed violations of the Facebook rules.

Yet the vast majority of these moderators have little contact with the engineers, designers, and even the policy people who set their rules. Most aren't even employees. Since 2012, when Facebook started centers in Manila and India, it began outsourcing companies to hire and employ the workers. They can't attend all-hands meetings and they don't get Facebook swag.

Facebook is not the only company to use content moderators: Google, Twitter, and even dating apps like Tinder have the need to monitor what happens on their platforms. But Facebook uses the most.

While a global workforce of content moderators was slowly building to tens of thousands, it was at first a largely silent phenomenon. The first awareness came from academics. Sarah T. Roberts, then a graduate student, assumed, like most people, that artificial intelligence did the job, until a computer-science classmate explained how primitive AI was back then.

"That crystallized the problem for me," she says. "The only remedy had to be a massive, basically underclass of workers. In 2010, the firms were not admitting that they did this." Roberts and others clued in to the phenomenon identified a new kind of worker—not with the elite degrees and engineering background that the tech companies preferred, but still essential to their operations. They were also a reminder that the twenty-first-century Internet had veered from the idealism that marked its previous era. Though Mark Zuckerberg might have begun Facebook with expectations that a light hand would be required, his underlings in support roles figured out early that humans would be spending their days sifting through Facebook content to protect the masses from offensive and even illegal content. It was a natural evolution to put them in factories. They became the equivalent of digital janitors, cleaning up the News Feed like the shadow workforce that comes at night and sweeps the floors when the truly valued employees are home sleeping. Not a nice picture. And this kind of cleaning could be harrowing, with daily exposure to rapes, illegal surgery, and endless images of genitals. The presence of all that stomach-turning content was an uncomfortable fact for Facebook, which preferred to keep its armies of scrubbers out of sight.

A subgenre of journalism emerged, exposing the conditions of the moderation centers. Though Facebook says that the stories were exaggerated, some details were cross-confirmed by multiple articles and academic studies: The moderators are almost always employed by outsourcing companies like Accenture and Cognizant, and their wages are relatively low, generally in the $15 an hour range. They view an astounding amount of horrifying content at a brisk pace. The rules they use to determine what will stay up or be taken down are deceptively complicated. And the job messed with their minds. A series of stories by *The Verge*'s Casey Newton introduced some Dickensian elements: pubic hair and fingernails among the desks, lines to use the restrooms, and even a temptation to embrace the toxic conspiracy theories that were constantly posted on Facebook.

When I visited the Phoenix office, I spotted no pubic hair. I didn't have to get in line to use the bathroom. The space was clean, and a colorful mural greeted employees when they entered. There was no way the office matched the buzzy cacophony of an actual Facebook office, but it didn't have the dingy oppression of the boiler-room operation that some stories implied. The workstations consisted of display screens on long black tables; since moderators don't have assigned spaces, there were no personal items. That, along with the venue's status as a "paperless office," gave the unoccupied areas a sense of abandonment. At peak, I was told, four hundred moderators would be there; the office was staffed 24/7.

My guide was the Cognizant executive who set up the office. His expertise was outsourcing, not content policies. All the rules and their execution came from "the client," which is the way he referred to Facebook.

I met with a group of moderators who had volunteered to speak to me. About half had college training. They had all calculated that this particular job was superior to alternatives at this time in their lives. We went over the details of their work. Facebook expects moderators to make about 400 "jumps" a day, which means an "average handle time" of around 40 seconds for them to determine whether a questionable video or post must remain, be taken down, or in rare cases, escalated to a manager, who might send the most baffling decision to the policy-crats at Menlo Park. Facebook says that there is no set time limit on each decision, but reporting by journalists and several academics who have done deep dives on the process all indicate that pondering existential questions on each piece of content would put one's low-paying moderation career at risk. One of the moderators I spoke to seems to be lodging a personal war against this unwritten quota: his personal goal is 200 jumps a day. "When you do it too fast, you miss little details," he says, adding that he hopes that his high accuracy will provide him cover if his low average handle time is questioned.

How many errors are made? It's hard to say, but one indicator is the number of times that a user appeal of a decision was upheld. In the first three months of 2019, Facebook removed 19.4 million pieces of content. Users appealed 2.1 million of those decisions and were upheld—that is, the original decisions were wrong—a little under a fourth of the time. Another 668,000 removed posts were restored without an appeal. In other words, though most of the time the calls are correct, there are still millions of people affected by an inability of moderators to get things right in that stopwatch atmosphere.

Part of the challenge comes from trying to match questionable content to their playbook, the Community Standards successor to the one-page document that Paul Janzer referred to and Dave Willner began to expand in Facebook's early days. Moderators learn how to interpret the guide first in classroom sessions, then alongside a veteran before they're allowed to solo. Facebook made the guide public in 2018, after multiple partial leaks.

The Community Standards are a testament to the complexity of the task. The same set of rules applies all over the globe, despite variations in cultures as to what is deemed permissible. The standards apply to all of Facebook's properties, including the News Feed, Instagram, the Timeline on the profile page, and private messages on WhatsApp and Messenger.

The rules can venture into confounding, Jesuitical flights of logic. Some things are fairly straightforward. There are attempts to define levels of offensiveness in subjects like exposure to human viscera. Some exposure is okay. Other varieties require an "interstitial," a warning on the screen like the one before a television show that might show a glimpse of buttocks. Outright gore is banned. It takes a judgment call to fit a given bloodbath into the right box.

"If you rewind to Facebook's early, early days, I don't think many people would have realized that we'd have entire teams debating the nuances of how we define what is nudity, or what exactly is graphic violence," says Guy Rosen of the Integrity team. "Is it visible innards? Or is it charred bodies?"

To be sure, the twenty-seven-page document hardly covers every example. Facebook has created a vast number of non-public supplementary documents that drill on specific examples. These are the Talmudic commentaries shedding light on Facebook's Torah, the official Community Standards. A *New York Times* reporter said he had collected 1,400 pages of these interpretations. A cache of training documents leaked to *Motherboard* showed cringe-worthy images where anal sphincters were photoshopped into a picture of Taylor Swift, swapped for her eyes. The training slide says such defacement is permissible, because Swift is a celebrity. Doing this to someone in your high school class would be bullying, and disallowed. But the altered image of Kim Jung Un with his mouth swapped for an anus *with a sex toy inserted* is to be removed.

The toughest calls come with hate speech. Facebook doesn't allow it but has understandable difficulty in defining it crisply. "Hate speech is our most difficult policy to enforce, because of the lack of context," says Bickert. The same words used to joke with a friend are regarded much differently when directed at a vulnerable stranger or acquaintance. One case that found its way into the press was a post from a comedian that said "Men are scum." This got her a suspension. The rule says no blanket insults of a protected group. Genders are protected groups.

Monika Bickert and her team understand that it's not the same to say "men are scum" as it is to say "Jews are scum." But they feel it would introduce too much complication to distinguish between vulnerable groups and privileged ones. As it is, moderators have difficulty enough determining what hate speech is, according to Facebook.

Take the example of someone on Facebook describing a racist quote from a celebrity. If the user frames it as "Mr. Celebrity said this, isn't it shocking?" Facebook will allow that, she says. It is information that helps people assess the personality. If the user cited the same quote and said, "That's why I love this person!" Facebook would remove the post, because it affirms racism. "But what if I just give that racist quote and I say 'said by' and then I say the celebrity?" asks Bickert. "Am I saying he's great or am I saying it's bad? It's not clear."

Hate speech is so complicated that Facebook has laid it out in several tiers. Tier 1 includes calling men scum, as well as likening a group to bacteria, sexual predators, or "animals that are culturally perceived as intellectually or physically inferior." Tier 2 are insults that imply inferiority, like calling someone or a group mentally ill or just worthless. Tier 3 is a kind of political or cultural insult including calls for segregation, admissions of racism, or straight cursing. Penalties are commensurate with tiers.

Hate speech was one of the topics considered when I sat in on the Content Standards Forum meeting in 2018. In a building just down Route 84 from the Gehry structure, Bickert convenes this gathering every two weeks to consider changes to the rules. There are about twenty people in the room, and video connections to Dublin, DC, and other locations around the globe. They discuss either a "heads-up" issue identifying a potential problem and deciding whether to investigate, or a "recommendation" issue, where the team makes a decision on an investigation. Such inquiries, which usually get input from experts in the given field (civil rights, psychology, terrorism, domestic violence, etc.), involve weeks of data analysis, cultural studies, and feasibility consideration. In this meeting, a hate-speech issue similar to "men are scum" came up for discussion. The question was whether a hateful comment about a powerful group, like men or billionaires, should be treated as harshly as a slur against a protected group, identified by gender or race. The outcome was interesting: the report said that the best outcome would be making that distinction, and letting people vent on the group in power. But that course was rejected, because it would be asking the moderators to make overly complicated decisions.

The moderators themselves say they are ready for responsibility. Not surprisingly, these contractors hunger to ascend to employee status. Before I visited Phoenix's moderators, I spoke with Facebook's blessing to one who made the leap (I am permitted to refer to him only as "Justin"). He confirmed that it isn't easy, since the "skill sets" of moderation differ from those useful to create or market Facebook products.

Justin said that, somewhat unexpectedly, some of his more harrowing duties dealt with content that came into question not because of bad behavior but because the user had died. Facebook's algorithms often wind up surfacing dead people's accounts on the feeds of loved ones, and it can have the effect of a drowned corpse rising to the surface. Facebook now has elaborate "memorialization" protocols for dead members. "Memorialization was really stressful," he says.

But not the most stressful. "The worst video I ever saw was a man cutting his own penis off with a serrated knife," says Justin. "It was not a great time." By the time he saw that can't-unsee-it vignette, around 2016, Facebook was providing therapists. (When he'd started at the job, in 2015, there was no counseling provided.) He now goes once a week.

The moderators in Phoenix seem to regard exposure to disturbing images as an unpleasant but tolerable part of the job. Sometimes things trigger them, and they go to the therapist. One told me that a "hit me bad" post involved an animated video of animals and people having sex, being slaughtered, and defecating, among other things. "That had me for maybe a good two weeks," he says. But with support, he says, he got past it.

Facebook regularly reviews the decisions of its moderators, and when significant errors occur, it does postmortems for improvement. But the dynamics of time and money dictate that those decisions will be at a speed that ensures frequent errors. "If everyone reviewed one thing a day, we probably wouldn't miss things," the former moderator known as Justin had told me earlier.

It's Facebook's key dilemma: it keeps hiring moderators, but the volume of content they view is still so much that they have to move too fast to do their job right. People notice. When someone mistakenly has a picture removed, they go to social media to complain. When someone reports offensive content and it's not taken down, more complaints. The media notices: the journalistic equivalent of shooting fish in a barrel is following up on a report that Facebook blew a call that now looks terrible, but was probably the result of an overworked and perhaps traumatized moderator. Zuckerberg himself acknowledges this. "Nine out of ten issues that we have that are public are not actually because we have a policy that people broadly disagree with, it's because we've messed up in handling it," he says.

Despite the time pressure and the exposure to the worst of humanity, the people I spoke to said that, as jobs go, moderating for Facebook wasn't bad. They see themselves as unsung first responders, protecting the billions of Facebook users from harm. "I've had interactions with reviewers who have helped save somebody's life—someone who was trying to attempt suicide and they reported it to the law enforcement authorities," says Arun Chandra, whom Facebook hired in 2019 to lead the moderation effort. "The sense of satisfaction and pride in this work was a pleasant surprise."

Sarah T. Roberts told me that she had found in her work that moderators' best moments are often dampened by realizing that they are cogs in a machine where their employer—or the company that contracts their employer—isn't really listening to them. "If they are ever a part of the feedback loop, it's rare," she says. Once a moderator told her about a suicide threat that was resolved positively. "We never stopped to ask ourselves," said the moderator, "to what extent the crap that people see on our platform leads them to feel like they would want to self-harm."

The Phoenix moderators I interviewed had an unpleasant surprise a few months after I talked to them. In October 2019, Cognizant decided it no longer wanted to be involved in Facebook moderation. Facebook announced it would close the Phoenix office, and those digital first-responders would be out of a job.

. . .

FACEBOOK REALLY ISN'T pleased that it requires tens of thousands of people working in office parks to police its content at a rate of 400 posts a day each. But it has a long-term solution that will greatly improve its track record and mitigate the number of relatively low-paid workers who will require therapy to deal with the images posted by Facebook users. What if Facebook were to assess and remove all its troublesome content *before* people saw it, and not wait until someone reported it?

They believe the answer is artificial intelligence.

"Ultimately, the way we've been thinking about all of this space is how we move from a world where our approach to content is more reactive to one that's proactive," says Guy Rosen. "How do we keep building more and more AI systems that can proactively find more kinds of that content?"

That was the long-term solution to the seemingly intractable moderation issues. While Zuckerberg consistently warned that the content problems would never go away—angering those who felt that even a tiny percentage of missteps on Facebook's part meant hundreds of thousands of false or harmful posts left untouched—he fervently believed that salvation would come in the form of robots, perpetually patrolling the alleys of the News Feed like friendly local cops.

The company had been building its AI muscle for years, but not for that purpose. In the earliest days Facebook did hire some people adept in AI, and both the News Feed and the ad auction were fueled by learning algorithms. But beginning in the mid-2010s one particular approach known as machine learning began to accumulate amazing results, suddenly putting AI to use in a number of practical cases. This supercharged iteration on machine learning was called deep learning. It worked by training networks of artificial neurons—working somewhat like the actual neurons in the human brain—to rapidly identify things like objects in images, or spoken words.

Zuckerberg felt that this was another moment like mobile, where the winners would be those who had the best machine-learning engineers.

He wasn't thinking about content moderation then, but rather improvement in things like News Feed ranking, better targeting in ad auctions, and facial recognition to better identify your friends in photographs, so you'd engage more with those posts. But the competition to hire AI wizards was fierce.

The godfather of deep learning was a British computer scientist working in Toronto named Geoffrey Hinton. He was like the Batman of this new and irreverent form of AI, and his acolytes were a trio of brilliant Robins who individually were making their own huge contributions. One of the Robins, a Parisian named Yann LeCun, jokingly dubbed Hinton's movement "the Conspiracy." But the potential of deep learning was no joke to the big tech companies who saw it as a way to perform amazing tasks at scale, everything from facial recognition to instant translation from one language to another. Hiring a "conspirator" became a top priority.

Zuckerberg pursued Yann LeCun the same way he hunted and bagged Instagram and WhatsApp. In October 2013, he called LeCun. "We're just about to turn ten years old and we need to think about the next ten," he said. "We think AI is going to play a super-important role." He told LeCun that Facebook wanted to start a research lab—not something designed to get better ad placements, but to develop mind-blowing creations like virtual assistants that could understand the world. "Can you help us?" he asked.

LeCun presented a list of requirements that Facebook would have to meet if he were to set up a lab. It would have to be a separate organization, with no ties to product groups. It would have to be completely open—no restrictions on publishing. The results they came up with would have to be open-source so they would benefit everyone. Oh, and LeCun would retain his NYU post, working there part-time, and base the new lab in New York City.

No problem! said Zuckerberg, and the Facebook Artificial Intelligence Lab, or FAIR, now is centered in New York City, on the edge of NYU's Greenwich Village campus. It is the horizon-exploring partner to the company's Applied Machine Learning team, which directs its AI work to products.

LeCun says that the integration worked superbly. The applied group imbued the product with machine learning, and the research group worked on general advances in natural-language understanding and computer vision. It often worked out that those advances helped Facebook. "If you ask Schrep or Mark, like, how much of an impact FAIR has had on product, they will say it's much larger than they expected," says LeCun. "They told us, *Your mission is to really push the state of the art, the research. When things come out of it for a product impact, that's great, but be ambitious.*"

LeCun gave this rosy description of the relationship between FAIR and AML in late 2017. But only a few weeks later, Schroepfer created a new post, a vice president of artificial intelligence, who would lead both the research and applied branches of Facebook AI. The job went to Jérôme Pesenti, a French scientist who had worked for IBM. LeCun professed to be delighted at this move, which freed him from management tasks so he could concentrate more on actual science.

But after people turned on Facebook post-election, the company needed the whole field of AI to take a step forward, to produce algorithms and neural nets that would far exceed the capabilities of human beings to identify unsavory content, illegal content, hateful speech, and state-sponsored misinformation. The goal was that these could work proactively, finding rogue content before anyone reported it, maybe even before anyone saw it.

Pesenti says that AML now has a dedicated team in Integrity Solutions helping to address the company's issues with toxic content. However, the state of the art falls far short of what Zuckerberg is promising, and Facebook needs more from FAIR. Its scientists have to invent breakthroughs that might be as good as or better than humans in dealing with things like hate speech. But because LeCun set up FAIR as a research organization, Facebook can't order the scientists to focus their studies on specific domains. "One challenge we have is to map product problems to research," says Pesenti. "We haven't solved that, actually."

There have been some successes. It turns out that terrorist content was fairly easy for AI systems to identify, and Facebook would come to claim better than a 99 percent success rate in taking down such posts, even before users had the opportunity to view them. But the current state of AI can't really deal with complicated issues like hate speech. *Humans* can hardly deal with tackling the speech of 2 billion people with one set of rules that applies all over the globe, addressing wildly disparate cultures.

"A lot of work has gone into building and training these AI systems, understanding how they manifest across different languages," says Rosen. He cites a 2017 project that addresses Facebook's hate-speech system work in Burmese. It has helped increase the percentage of hate-speech posts Facebook blocks proactively—before anyone reports them—from 13 percent to 52 percent. Critics will note that means about half of the hate-speech posts in that dangerous region are still viewable.

Facebook also looks to AI to deal with another persistent problem: the mind-boggling number of fake accounts on the system. Not surprisingly, these are a huge source of fraud, hate speech, and misinformation. People were stunned when Facebook revealed that between January and March 2019, it blocked 2 billion attempts to open fake accounts—almost as many as actual users on the system. Overwhelmingly, these are clumsy, though persistent, attempts to create phony Facebook identities in bulk. As Alex Schultz told *The New York Times*, "the vast majority of accounts we take down are from extremely naïve adversaries." But not all are so naïve. Despite AI, or anything else that Facebook could throw at the problem, the company concedes that around 5 percent of active accounts are fake. That's well over 100 million.

That's Facebook's dilemma: its scale is so massive that even when it makes improvements, the scope of what's left is staggering. And those motivated to make the posts will learn to adjust to Facebook's tactics. In 2018, for example, Facebook proudly announced that its AI teams had learned to read the content of messages embedded into graphics. Previously, its systems could only read words when they were stored as text. That shortcoming had allowed Russian operatives to slip their inflammatory ads on immigration, racism, and Hillary Clinton's identity as Satan past Facebook's digital monitors.

In other words, Facebook had figured out a defense for a war that it had lost the last time around. Who knows what tactics its foes will adopt in the future?

Meanwhile, it's left to the 15,000 or so content moderators to actually determine what stuff crosses the line, forty seconds at a time. In Phoenix I asked the moderators I was interviewing whether they felt that artificial intelligence could ever do their jobs. The room burst out in laughter.

THE MOST DIFFICULT calls that Facebook has to make are the ones where following the rule book creates an outcome that seems just plain wrong. For some of these, moderators "elevate" the situation to full-time employees, and sometimes off-site to the people who sit in the Content Moderation meetings. The toughest ones are sometimes elevated to Everest, to the worktables of Sandberg and Zuckerberg. Even then, the decisions are hard to make. There are times when the rules of offensive content come in conflict with what is going to look best for Facebook. They can involve essentially politically charged decisions, with powerful supporters on each side of the argument. No matter what Facebook decides on these, it loses.

As with many of its problems, Facebook didn't really confront this until the election year of 2016. That September, a Norwegian writer named Tom Egeland posted on Facebook a story he had written about six photos that "changed the history of war." One of them was an iconic image that would have been familiar to anyone who lived in America during its tragic fiasco in Vietnam. The picture had won the 1972 Pulitzer Prize for photography. It was known as "Terror of War" or "Napalm Girl." It showed a group of children running down a road screaming in pain because of napalm burns. Behind them were American soldiers in uniform. The child framed in the center of the photo, Kim Phúc, was naked.

For Facebook's moderators—especially since the case seems to have been handled outside the United States, where the photo wasn't familiar—this was a no-brainer. The rulebook clearly bans nude images of children past infancy, and so Facebook quickly removed the image. Egeland was annoyed, and tried to repost the photo. Facebook suspended his account. By then, the issue had reached Menlo Park. Bickert's team now understood that Facebook was censoring a photo of historic value, but supported the takedown. If you make exceptions for one naked kid, where do you stop?

Then the story broke wide open. Egeland had been writing for the most popular newspaper in Norway. His furious editor wrote a front-page editorial, with huge letters saying DEAR MR. ZUCKERBERG. . . . It claimed that Facebook—"the world's most powerful editor"—was acting as a censor. Norway's prime minister reposted the photo, only to have Facebook take it down again. Other news outlets picked up the story. The comms people got flooded with queries.

This caused a crisis in Facebook's policy world. For years, Facebook had been dodging complaints of what it left up—a year earlier, it had left up Donald Trump's anti-Muslim posts. Now it was under fire for what it took down. Could naked children also qualify for an exception? Plenty of people felt that, Pulitzer or no, there was no room for Kim Phúc's terror on Facebook.

Chaos broke out in Facebook's policy world. "We were all in this together trying to fix it, but we just don't know how to fix it," says someone involved in the discussion. What made it a major decision for Facebook was not the choice itself, but the outrage generated by Facebook's adherence to its own rules. Interpretations that seemed logical when the rule book was written often could look outrageous when exposed to public scrutiny.

"That photo got posted all the time," says Dave Willner, who had taken a job at Airbnb doing similar work by then. "If you do not know that it is a nonconsensual nude image of a child who has had a war crime committed against her—if it were not a Pulitzer Prize–winning photo—everyone would lose their damn minds had Facebook not censored it." Another keep-it-down advocate was Andrew Bosworth. "I would have said, *Hey if you want that picture up, change the laws in your country. Like, listen, buddy, I think that is a tremendously important photograph, historically, but I can't have it on the site, not legally.* Change the laws!"

But that's not what Zuckerberg thought. Ultimately he and Sandberg had to sign off on the decision. From that point on "newsworthiness" was a factor in determining exceptions to the general rules. Napalm Girl, in all her shocking nakedness, was back on Facebook.

Facebook's heads of policy, Elliot Schrage and Joel Kaplan, saw the incident as a watershed. "That was—internally—the clearest example that our impact and influence in America had changed," says Schrage. "Facebook was no longer about sharing interesting and relevant information; we shaped larger cultural conversations too."

Monika Bickert puts it another way. "We learned that it is okay to make exceptions to the letter of the policy to maintain the spirit of the policy," she says.

From that point, the pageant of exposure, pressure, and correction would play out on a regular basis. The most striking examples came in criticism of Facebook's handling of fringe right-wing content that seemed to violate Facebook's rules. Nearly every time a Facebook representative would testify before Congress, GOP legislators would rant about the conspiracy of Menlo Park liberals to suppress conservative speech. Their complaint was not only that in some cases Facebook took down posts from extremists—the Republicans believed that Facebook cooked up algorithms that favored liberal content. The data didn't prove it, and it wasn't even clear if the legislators actually believed it or were just trying to game the ref.

As a result, Facebook had a torturous time with trash-talking provocateurs from the right. When the white-nationalist conspiracy monger Alex Jones repeatedly posted comments that seemed to violate Facebook's hate-speech rules, the company was loath to ban him. Complicating matters was that Jones was an individual, and his Facebook page, InfoWars, was an operation staffed by several people.

The situation was radioactive. Fringe as he was, Jones had a huge following, including the president, who had been a guest on the InfoWars radio show. Did Jones's newsworthiness make him a figure like the president, worthy of a hate-speech free pass? During the summer of 2018, the controversy raged, as reporters kept citing hateful posts. Ultimately, it was the pressure. Within hours after Apple took down his podcast, Zuckerberg himself pulled the plug on InfoWars. Jones was suspended for thirty days, and ultimately Facebook would ban him as "dangerous." It did this in tandem with the expulsion of the fierce-tongued Nation of Islam leader Louis Farrakhan, in what appeared to be an unmistakable play for balance.

When I pressed Zuckerberg in early 2018 about Facebook's delicacy in handling GOP complaints, he bent over so far backward in respecting their point of view that I worried his chair would hit the floor. "If you have a company which is ninety percent liberal—that's probably the makeup of the Bay Area—I do think you have some responsibility to make sure that you go out of your way and build systems to make sure that you're not unintentionally building bias in," he told me. Then, ever balancing, he mentioned that Facebook should monitor whether its ad systems discriminated against minorities. Indeed, Facebook would commission studies of each of those areas.

Part of Zuckerberg's discomfort arises from his preference for less oversight. Even while acknowledging that content on Facebook can be harmful or even deadly, he believes that free speech is liberating. "It is the founding ideal of the company," he says. "If you give people a voice, they will be able to share their experiences, creating more transparency in the world. Giving people the personal liberty to share their experiences will end up being positive over time."

Still, it was clear that Zuckerberg did not want the responsibility of policing the speech of more than 2 billion people. He wanted a way out, so he wouldn't have to make decisions on Alex Jones and hate speech, or judge whether vaccines caused autism. "I have a vision around why we built these products to help people connect," he said. "I do not view myself or our company as the authorities on defining what acceptable speech is. Now that we can proactively look at stuff, who gets to define what hate speech is?" He hastened to say that he wasn't shirking this responsibility, and Facebook would continue policing its content. "But I do think that it may make more sense for there to be more societal debate and at some point even rules that are put in place around what society wants on these platforms and doesn't."

As it turns out, Zuckerberg was already formulating a plan to take some of the heat off Facebook for those decisions. It involved an outside oversight board to make the momentous calls that were even above Mark Zuckerberg's galactic pay grade. It would be like a Supreme Court of Facebook, and Zuckerberg would have to abide by the decisions of his governance board.

Setting up such a body was tricky. If Facebook did it completely on its own, the new institution would be thought of as a puppet constrained by its creator. So it solicited outside advice, gathering a few hundred domain experts in Singapore, Berlin, and New York City for workshops. After listening to all these great minds, Facebook would take the parts of the recommendations it saw fit to create a board with the right amounts of autonomy and power.

I was one of 150 or so workshop participants at the NoMad Hotel gathering in New York City's Flatiron district. Sitting at tables in a basement ballroom were lawyers, lobbyists, human rights advocates, and even a couple of us journalists. For much of the two-day session we dug into a pair of individual cases, second-guessing the calls. One of them was the "men are scum" case that had been covered a few times in the press.

A funny thing happened. As we got deeper into the tensions of free expression and harmful speech, there was a point where we lost track of the criteria that determined where the line should be drawn. The Community Standards that strictly determined what stood and what would be taken down was not some Magna Carta of online speech rights but a meandering document evolved from the scribbled notes of customer support people barely out of college.

The proposed board would be able to overrule something in that playbook for the individual cases it considered, but Facebook provided no North Star to help us draw the line—just a vague standard touting the values of Safety, Voice, and Equity. What were Facebook's *values*? Were they determined by morality or dictated by its business needs?

Privately, some of the Facebook policy people confessed to me that they had profound doubts about the project.

I could see why. For one thing, the members of this proposed body—there will be forty members, chosen by two people appointed by Facebook—can take on only a tiny fraction of Facebook's controversial judgment calls. (In the first quarter of 2019, about 2 million people appealed Facebook content decisions.) Facebook would have to abide by the decisions on individual cases, but it would be up to Facebook to determine whether the board's decisions would be regarded as precedent, or simply limited to the individual pieces of content ruled on, because of expedience or because they were lousy calls.

One thing seems inevitable: an unpopular decision by a Facebook Supreme Court would be regarded just as harshly as one made by Zuckerberg himself. Content moderation may be outsourced, but Facebook can't outsource responsibility for what happens on its own platform. Zuckerberg is right when he says that he or his company should not be the world's arbiter of speech. But by connecting the world, he built something that put him in that uncomfortable position.

He owns it. Christchurch and all.

18

Integrity

FACEBOOK'S "M TEAM" consists of around forty of its top leaders, the people who make its biggest decisions and are responsible for executing them. They gather a few times a year in a large room on the Classic campus. The July 2018 meeting was one of the first of these held after Cambridge Analytica.

It started out as usual. In M-team meetings, the executives all do a brief check-in, saying what's on their minds both in business and in life. It can get pretty emotional: *my kid's sick . . . my marriage ended . . .* When it was his turn to talk—he always goes last—Zuckerberg made a startling announcement.

He had been reading a book by venture capitalist Ben Horowitz, the partner of Facebook board member Marc Andreessen. Horowitz defined two kinds of CEOs: wartime and peacetime. A good CEO, he wrote, has to interpret circumstances in a given time and decide which to be. "In wartime, a company is fending off an imminent existential threat," he writes. Wartime CEOs must be ruthless in confronting those threats.

Since Facebook had been under siege for the past two years, this made a big impression on Zuckerberg. In earlier times, Zuckerberg told the group, he had the luxury of being a peacetime CEO. (This was a debatable self-definition for the Cicero-quoting leader who went into lockdown mode to thwart perceived challenges from Google, Snapchat, and Twitter.) He told the group to hereafter consider him a wartime CEO.

He emphasized one shift in particular. Horowitz put it this way: "Peacetime CEO works to minimize conflict. Wartime CEO neither indulges consensus building nor tolerates disagreements." Zuckerberg told his management team that as a wartime CEO he was just going to have to tell people what to do.

Some in the room thought that he was saying from that point their role was to shut up and obey his directives. Zuckerberg resisted that characterization when I later brought it up to him. "I basically said to people this is the mode that I think we're in," he says of the declaration. "We have to move quickly to make the decisions without the process of bringing everyone along as much as you would typically expect or like. I believe that this is how it needs to be to make the progress that we need right now."

I wondered whether he found the role of wartime CEO more stressful or more fun.

A Zuck silence. Eye of Sauron.

"You've known me for a long time," he finally said. "I don't optimize for fun."

Zuckerberg's internal announcement reflected the huge amount of thought he'd been devoting to how Facebook might negotiate its woes now that he and Sandberg had turned on the apology fire hose. Zuckerberg felt that the company had backed up its mea culpas with a flood of new products and systems that addressed the flaws and vulnerabilities that had provided opportunities for misinformation, election tampering, and data privacy. Facebook had made changes in time for the 2017 French election and had avoided some of the worst effects that the US and the Philippines had seen earlier. It was embarking on a labor-intensive strategy to get through the 2018 midterms in America.

But though Zuckerberg had declared that his primary job was now fixing Facebook, he wasn't going to curb his ambitions in a unilateral surrender to competitors. Facebook had to move forward as well.

On the eve of the 2018 F8 that May, I spoke with Zuckerberg about both what new products he'd be announcing and the thought process behind announcing *anything*. He knew he'd be obligated to show penance and talk about fixing things. But Facebook also had to keep introducing new products. "On the one hand, the responsibilities around keeping people safe—the election integrity, fake news, data privacy, and all those issues—are just really key," he said. "And then on the other hand, we also have a responsibility to our community to keep building the experiences that people expect from us."

With an engineer's logic his keynote speech for the conference would begin with fifteen minutes of trust-building, and then an equivalent amount of time on new products. The "going forward" part.

Facebook was proceeding with caution on some fronts. It had developed a product called Portal, a display screen with a camera and microphone to enable video connections with friends and family. But Facebook's cooler heads realized that introducing what could be seen as a home surveillance device might not be a smart move only weeks after the Cambridge Analytica disaster. Zuckerberg did have another product to announce at the event, though: called Dating, it would create an entirely new, and very personal, dossier about Facebook's users.

I asked him at the time if he was sure that it was a good idea to introduce something involving such intimate information at what appeared to be the nadir of people's trust in the company.

He replied that Facebook focused on meaningful relationships—what could be more meaningful than people you date? He also ran through the privacy protections in the new feature. The conversation drifted elsewhere, but then he abruptly returned to the concern I'd expressed about the Dating product. "Obviously you're asking this question," he said, "but do you think that this is a bad time to be talking about this?"

Well, yes, I said.

"This is the threading of the needle that we talked about up front," he said. "I'm curious if you think that moving forward on new products will feel like we're not taking the other stuff as seriously as we need to. Because my top priority is making sure that we convey that we are taking these things seriously."

He wasn't going to kid himself. Winning back people's trust was a long process. It would probably take three years. But he felt that the rebuild had started.

Despite bad press, Facebook Dating rolled out in several smaller markets that year and hit the United States in September 2019.

And Facebook began selling its Portal before the end of 2018. Reviewers thought it was a good product but advised against buying it because, they said, no one could trust Facebook.

ZUCKERBERG WAS RIGHT to worry about the reaction from users and developers. In 2018, the year he determined to win back trust, Facebook's trustworthiness had tanked. Even as Facebook worked to improve its products, a relentless stream of headlines kept dragging down its reputation. First came the revelations that Facebook's cutback on data-gathering—the one that supposedly ended after the one-year grace period that started in 2014—had not been uniformly employed. Some major companies, like Airbnb, Netflix, and Lyft were white-listed, allowing them to continue accessing information. (Also on the white list was Hot or Not, the inspiration for Zuckerberg's 2003 folly, Facemash.) Especially embarrassing: some of these revelations came out in a lawsuit from a company, Six4Three, that actually *was* blocked from receiving user data. Facebook had quite sensibly denied access to its app, Pikinis, which allowed users to find posts of friends in bathing suits and other forms of undress. Facebook's reward was litigation that would wind up revealing a trove of damaging emails.

That was only one example of how things were spilling out. Every day dozens of reporters, working for top-tier newspapers or philanthropically funded investigative units, woke up in the morning and began digging up dirt on Facebook. It wasn't hard.

Sometimes exposing Facebook could be as simple as using its ad product to discover a shocking flaw, like targeting an ad toward "Jew haters." That was one of many questionable categories algorithmically generated by Facebook's self-service ad product when one typed in the word "Jew" to start the process. Investigative reporters at ProPublica found 2,274 potential users identified by Facebook as fitting that category, one of more than 26,000 provided by Facebook, which apparently never vetted the list itself. "I know exactly how this happened," says Antonio García Martínez, a former ad product manager who helped launch this feature. "Facebook feeds a bunch of user data into what pages you've liked, profile data, whatever. I used to call it Project Chorizo, like the sausage maker, because that's what it was. You put in all this data, and it would pop out topics." Essentially, to work at scale, Facebook had built a system where an AI with little understanding of what was offensive to humans was empowered to create those categories like . . . Jew haters. Facebook later removed the categories. "We know we have more work to do," said Rob Leathern, a Facebook executive working on ad integrity.

Other scandals stemmed from Facebook's own panicky attempts to address its precarious reputation. In November 2018, *The New York Times* revealed that Facebook's policy team had hired a firm called Definers Public Affairs to impugn its competitors—and even to cast aspersions on financier George Soros, who had criticized Facebook in a Davos speech. Adding pungency to the claim was that Soros was also a favorite target of anti-Semitic hate speech, including attacks on Facebook. (It is exceedingly odd that a company headed by Jewish executives frequently found itself in situations that involved alleged anti-Semitism.)

Policy head Elliot Schrage, whose own family suffered losses in the Holocaust, publicly claimed responsibility. Observers saw this as Schrage taking the fall for his boss, Sheryl Sandberg, who insisted she knew nothing about it. Then emails emerged showing that Sandberg might have known something about it after all. This particular episode turned out to be overblown—companies often would hire outside mouthpieces to impugn competitors, and Definers didn't hide its connection to Facebook—but no one was giving Facebook the benefit of the doubt in 2018.

Other wounds were totally self-inflicted. One might think that the last thing Facebook's head of global policy might want to do was drag the company into the middle of the utterly radioactive controversy regarding Supreme Court nominee Brett Kavanaugh, whose angry response to a charge of a youthful sexual assault polarized the nation. But right there on television, sitting behind the nominee, was Joel Kaplan, taking a day off to support his friend from the Federalist Society. The outrage at Facebook was widespread. Kaplan was forced to publicly apologize at an all-hands a week later. Not for helping his friend, but for the failure to give Facebook a heads-up. Someone at the meeting later told *Wired* that Kaplan looked to be in shock, like "someone had just shot his dog in the face." The apology's sincerity came under question only a day later, when Kaplan threw a celebratory party for Kavanaugh after the Senate voted the justice to the Supreme Court.

Within weeks of *that* disaster, Facebook announced that it had discovered that hackers had exploited flaws in its infrastructure to get access to the information of 50 million users, including that of Sandberg and Zuckerberg. Unlike Cambridge Analytica, this was a literal breach. The intruders had exploited a vulnerability that had been exposed for more than a year. A few months earlier, as part of his post-Cambridge apology tour, Zuckerberg had planted a stake in the ground: "We have a responsibility to protect your data," he wrote, "and if we can't then we don't deserve to serve you." It's hard to imagine a metric that wouldn't indicate that his promise was broken. In a rare audio press conference, he was twice asked if he thought that his own resignation was appropriate. The answer was, both times, no.

SHERYL SANDBERG WAS fighting too, not just for Facebook, but for the personal brand she had built over a lifetime. In addition to her leadership role in the company's rehabilitation, she still took time to support the extensive Lean In organization that had arisen from her first book. She believed that she had helped women and was proud of it. Facebook's woes had affected the movement—it must have hurt Sandberg when Michelle Obama, speaking at Brooklyn's Barclays Center, said, "It's not enough to lean in, because that shit doesn't work all the time." But Sandberg kept her head down, still striving for the A+.

Every now and then, Sandberg would do a Facebook Live session in Building 20, in talk-show format. It was in her role as the leader of the Lean In movement rather than her day job. She would sit across the table from her guest, often someone promoting a newly released inspiring tome of her own, each sipping from coffee mugs with the Lean In logo, and conduct a gentle interview. You could tell it was the best part of her day. Behind them, visible through the glass walls of her conference room, you could see Facebookers bustling past, some surely creating new pixels to dampen whatever crisis was blowing up in that minute. Inside her conference room, dubbed Only Good News, Sandberg would be asking her subject the final question she always posed: *What would you do if you weren't afraid?*

What if *you* were asked that question, I said to Sandberg in 2019. It was our final interview, a two-hour extravaganza that I begged for after our previous meetings had ended just when things were getting good. "What I would do if I wasn't afraid is try to be the Facebook COO and grow this business and say I'm a feminist," she says. People forget, she adds, but when she wrote *Lean In*, it was a risky and unpopular thing for a female corporate leader to declare.

Sandberg had recently been before Congress. She had been on a panel with Jack Dorsey; the committee wanted Google CEO Sundar Pichai to attend as well, but he declined. A piqued Senate committee left an open seat before an empty space on the table with his name on the folded paper.

Sandberg had prepared for the ordeal with her usual zeal, blocking off entire days for rehearsal. Every detail was considered, down to her interaction with her fellow panelist—she decided that she and Dorsey should not engage in their usual friendly hug, as it might suggest collaboration. She also determined not to bash Pichai, as it would look bad to attack him in his absence.

Sandberg's testimony was successful. Nobody asked her, as they had done with Zuckerberg, what hotel she was staying at. Her former government work had taught her how to devote proper deference to showboating politicians, some of whom she had visited in days previous to pitch her case personally.

It also helped that, besides Dorsey's hipster demeanor, other distractions drew attention from any gaps in her replies. Flitting around the hearing room was the conspiracy artist Alex Jones, newly banned from Facebook. "Beep-beep-beep-beep—I am a Russian bot," he yelled, giving credibility to any social network that chose to shut down his divisive meme machine.

During our long session, Sandberg denied what virtually everyone assumed was her career goal: public elected office. Definitely not, she said. But she *would* have taken an appointed office. Circumstances thwarted her. One logical exit point might have been after the IPO in 2012. She had promised Zuckerberg five years, and 2013 would mark the spot. But the company had taken more than a year to struggle back to its opening price. It was a bad time to leave.

Her husband's death was in a class of disaster by itself, she says, "a cataclysmic moment." After that, she wasn't taking anything on, except being with her kids and working her way back to Facebook. And then came 2016. "I was hitting a decade [at Facebook]," she says. "But I knew after the election and Russia started happening, and the fake news, we were in for a rough ride. And now I feel tremendous responsibility to stay and make sure this gets to a better place. Mark and I are the most likely people to fix what needs to be fixed."

I ask her about what I thought might have been the roughest of all recent moments: her meeting with the Congressional Black Caucus in October 2017. Sandberg had gone to meet with them with Facebook's global diversity officer, Maxine Williams. Basically, the caucus schooled her. They were outraged that Facebook had hosted Russian propaganda that fueled white prejudice against black people. And there was more. Facebook had been hit with civil rights violations because it allowed advertisers to discriminate against African Americans. Its workforce had too few people of color, and it had no black directors on its board. The members also zeroed in on Williams—why was she the head of diversity and not a C-level chief diversity officer? (Facebook later addressed each of these, giving Williams the title change and appointing former American Express CEO Kenneth Chenault to its board.) As each member vented, Sandberg kept repeating, almost like a mantra, "We will do better," promising answers to their queries. After the meeting, Representative Donald Payne of New Jersey expressed his dissatisfaction, telling *The New York Times* that he once had an uncle who hated when people said they were "gonna" do this or that to fix their messes. "He used to say, *Don't be a gonna*," said Payne. "And that's what I said to her—*Don't be a gonna.*"

"It was one of the hardest meetings ever," Sandberg says. "I listened through the whole thing. I took really careful notes. I walked out of there saying we, and I, have a lot of work to do. Over the next couple months I called every single member who was there and even others, and now I am personally leading our civil rights work."

What made it especially hard is that those were *her people*—Democrats, human-rights supporters, fighters for justice. "I'm very progressive, I am a big donor, I am a big funder. Look, people were upset we missed it. Like, *we're* upset we missed it."

Sandberg was getting very emotional while reliving that meeting and needed a few seconds to pull herself together. And it then all came out with the tears. The frustration, the *people are not getting it* frustration, the pain of the past two years. She'd said earlier, quite wisely, that after losing her husband, how bad could dealing with Facebook's troubles be? But she has been battered by how people have turned on Facebook, seen her reputation questioned, and here in her conference room she's now addressing it without notes or talking points.

"I mean, look, this is a big deal," she says, emotion still saturating her voice. "This company got built by Mark, by me, by all of us because *we really believe*. I believe *so much* in people having a voice. I started my career working on leprosy in India and I have been to villages and homes where there is no electricity. And I watched the head of leprosy of India step over a patient as if she was nothing. I know what it is to not be connected.

"And so for me, coming to Facebook, we connected people and gave people voice. And that voice is everyone from Hillary Clinton to Donald Trump to . . . to people all over. And so the fact that those same tools . . ."

She catches her breath. The unfinished part of the sentence seems to be *those same tools that we use at Facebook are used for evil.*

"I remember when the Arab Spring happened, we thought that was incredible," she continues. "We didn't do it—it was just a tool. But people were connecting. And for *this* to happen. Our election was and is a super big deal. And it's not just that other people were upset. I think maybe because we're not good at expressing it or maybe because I'm not this honest and open and this is probably the most I've ever shared or maybe I haven't had the forum, but *I'm upset.* I don't need the board to be . . . Like, I'm upset. Erskine [Bowles] and I are close; he's upset, I'm upset, we're all upset."

It's getting close in the room, and the PR person staffing the meeting—a recently hired reinforcement to replace a recent escapee—has stopped the incessant typing that PR nannies usually do throughout these meetings, and her eyes are so wide they look like those googly stickers on Messenger. "I knew this was going to be a really hard time at Facebook and I was hitting ten years," says Sandberg. "But I am here to do the hard work."

In these doldrums, it seemed almost as if weeping were part of the job description. In an interview for *The New York Times*, CTO Mike Schroepfer teared up while describing the inability of Facebook's artificial intelligence to stop the spread of the murder video from Christchurch, New Zealand. I heard secondhand that one Facebooker remarked—I could never confirm this—that on some days all stalls in the women's bathrooms would be occupied by crying employees; those coming in to weep had to queue up to wait their turn.

"That's a terrible story," says Sandberg when I shared it with her. "I mean, *I* cry. Cry right at your desk!"

. . .

ONCE FACEBOOK WAS Silicon Valley's prime raider of talent. Now its competitors were preying on the Facebook workforce. Employees found it a good time to leave for start-ups. A computer-science teacher at one of the big AI schools told me that Facebook used to be the top employment choice. Now he guesses that about 30 percent of his students won't consider it, for moral reasons.

Inside Facebook there were doubts too. Facebook regularly surveys its employees, and *The Wall Street Journal* managed to see the results of a poll taken by 29,000 Facebookers in October of 2018. (Even the fact that the poll leaked was a sign of trouble in the ranks.) Only a little more than half of the workforce expressed optimism about the company, a drop of 32 points from the previous year. Thousands of workers in the previous year no longer believed that Facebook was good for the world, a sentiment that also won a bare 53 percent majority.

The doubts had even reached the highest levels of the company. One executive described to me a meeting of the very top leaders, known as the small group, around mid-2018. "I don't think everyone believes the stuff we're working on," Zuckerberg told them. He asked them to write down Facebook's big efforts on a piece of paper, and rate them, one to ten.

The results were disheartening.

"Basically, everyone was like, *Everything we're doing is not good*," says that executive. "*Why are we trying to compete with Google with Search? Why are we doing Watch? Why Oculus?*" (Another person in the room contends that it wasn't *all* so negative, but confirms the episode.) Zuckerberg was unfazed. He told his team that all big efforts face skepticism at first. He had always done well by blowing past the doubters.

Up until then, Zuckerberg's magic had always been the ability to make the right call. Sam Lessin, the Harvard classmate who later worked as a Facebook executive and remains a close friend, says that multiple times he would be in a room where Zuckerberg made a decision that conflicted with the opinion of everyone else. His view would prevail, and he would be right. Time and again. After a while, people came to accept it.

Now some of those decisions didn't look so good. Maybe even a wartime CEO might take objections more seriously. "It's within every leader's right to make edicts," says someone in the room for many Zuckerberg decisions. "But leaders fail when they convince themselves that everyone disagreeing with them is a signal for them being right."

Zuckerberg's loyalists stuck with him. On his thirty-third birthday, he posted a celebratory photo with twenty of his closest work friends, who presented him with a cake depicting different cuts of meat. Among those gathered around their leader with huge smiles were people now defined by their ties to Zuckerberg: Sandberg, Bosworth, and his empath alter ego, Chris Cox.

But there was a growing number of disaffected former loyalists. Roger McNamee, the investor who wrote Zuckerberg and Sandberg about fake news in 2016, had been only the first of a vocal covey of apostates who paused in their glamorous lives to publicly condemn the company that made them wealthy. In a public interview at the National Constitution Center in Philadelphia, Sean Parker attacked Facebook's addictiveness. "The thought process that went into building these applications, Facebook being the first of them . . . was all about: *How do we consume as much of your time and conscious attention as possible?*" he said. "The inventors, creators—it's me, it's Mark, it's Kevin Systrom on Instagram, it's all of these people—understood this consciously. And we did it anyway." Justin Rosenstein, who co-invented the Like button, now was condemning the "bright dings of pseudo-pleasure" that come from clicking on the thumbs-up sign.

Perhaps the most bruising critique came from Chamath Palihapitiya. Speaking at the Stanford Graduate Business School in December 2017, the man who drove Facebook's growth said, "I think we have created tools that are ripping apart the social fabric of how society works." He cited an incident in India where the WhatsApp rumor mill spread news about a kidnapping that never happened—and seven people were lynched in the outrage that followed. "It's just a really, really bad state of affairs," he said. Though some good does come from Facebook, he added, he personally avoided the service and his children "aren't allowed to use that shit."

That could not stand. Sheryl Sandberg got in touch with him. Neither party will disclose what was said, but Palihapitiya publicly clawed back his statement.

DESPITE THE PUMMELING Facebook was taking, its business had never been better. Its core advertising strategy—which merged the voluminous data it gathered with outside information to help advertisers reach the most promising audiences—was proving unbeatable. After years of developing its techniques and calculating metrics that proved its value, Facebook was the undisputed leader in what was known as PII, or personally identifiable information.

Marc Pritchard, P&G's chief brand officer, remembers a conversation that he had years earlier with Sandberg about cookies, the little data markers that websites plant on your computer when you visit. "I remember very clearly," he says. "Sheryl said, *Cookies are going to die and the future is going to be PII data.* The difference is with PII data, you got a lot more to manage than cookie data. Cookie data is anonymous. The future being PII data was right."

As journalists and regulators began to expose how much Facebook knew about its users and how skillfully it was packaging the information to deliver ads, the company made some mild concessions to transparency. That hardly slowed down momentum. For one thing, the ad system was so complicated that even Zuckerberg didn't understand all the intricacies. When he went before Congress, he deferred a number of questions about Facebook's ad practices. It wasn't part of his preparation. "I was expecting the congressional testimony to be primarily about Cambridge Analytica and maybe to some extent about Russian interference," he told me soon after testifying. "I figured that the other product questions that came up, I'd basically be able to answer because I built our product." He punted the questions, and on his plane ride home vowed to look into it himself. "I actually felt like I didn't understand all the details around how we were using external data on our ad system, and I wasn't okay with that," he said.

What Zuckerberg found was a system so fortified with information that even seemingly significant changes would essentially make no difference. Just before Zuckerberg testified, Facebook had ended one of its most controversial practices, called Partner Categories. Until then, Facebook matched its own information with the extensive files that data brokers (including the giants like Equifax and Experian) compiled on consumers, so advertisers could target individuals more accurately. If, for instance, a publication wanted to reach its own subscribers on Facebook—or those subscribing to a competitor—it could use the combined data to hit them directly.

When I asked an ad executive a few months later if this change had any effect on the business, the Facebooker laughed. *None!* was the answer. While Facebook stopped buying data from brokers, its policy stated clearly, "Businesses may continue, on their own, to work with data providers." Facebook made it easy for advertisers to plug that purchased data into their system, just as before. The only difference was that the advertisers were now paying the data brokers directly.

While the Europeans imposed relatively strict privacy regulations, everywhere else Facebook was free to participate in the ongoing bonanza of Internet tracking, a widespread practice where every website people visited and every search term they used was routinely logged and used to sell things to them. The US legislators kept talking about privacy laws that might roll that back but never seemed to come up with any. No one took more advantage of this than Facebook, because its own invisible pixel was on millions of sites. If you lit on a page for a brand of sneakers, or checked out a car, or, heaven forbid, vetted an over-the-counter drug, you could reliably count on an ad for what you had just perused popping up in your News Feed. The creepiness of it made people shudder.

The phenomenon gave rise to a widely shared suspicion that Facebook was somehow listening to everyone's conversations. Senator Gary Peters spoke for many Americans when he asked Zuckerberg about it during testimony. "I hear it all the time, including from my own staff," he said. "Yes or no: Does Facebook use audio obtained from mobile devices to enrich personal information about its users?"

"No," said Zuckerberg.

The truth was that Facebook didn't need to spy on people's audio. It already had all the PII data it needed to help advertisers hit the mark not just on the kinds of audiences it wanted to reach, but on the exact individuals who would be in that audience.

As a result, Facebook was a must-buy for advertisers, as digital advertising headed toward the majority of all ad spending in the United States, a mark it would actually reach in 2019. Its only serious competition, particularly in the dominant mobile field, was Google; the two companies combined had around 60 percent of all digital advertising, and more than two-thirds of the mobile market.

All during Facebook's tumble from grace, no matter which story about the company's missteps was leading the news, one could expect its earnings call to tell a different story: either Sandberg or its chief financial officer, David Wehner, would say, "We had a very good quarter," more often than not reporting record revenues. The company that Mark Zuckerberg started with a thousand dollars from a classmate was now raking in more than $50 billion a year, and its Wall Street valuation was more than $500 billion.

One call, however, did not go so well, and that was to report second-quarter earnings in July 2018. As always, after the stock market closed, Zuckerberg, Sandberg, and Wehner trudged to a conference room on campus to report the results and take questions from analysts. This time they had bad news.

For months Zuckerberg had been promising to hire thousands more people to work safety and security, and some of that was now affecting profits. That wasn't new. "As I've said on past calls," Zuckerberg read from his notes, "we're investing so much in security that it will significantly impact our profitability." What did make an impact was that Facebook indicated that its momentum from its current ad model had slowed: sponsored stories on the News Feed may no longer be the future. But

Facebook did have a replacement in mind: ads placed among the strip of clips known as Stories, which started on Instagram and had now moved to Facebook, WhatsApp, and Messenger. Facebook had yet to figure out how to make those equally profitable, and its advertisers were still learning how to use those ads. Facebook was confident, though, that all of this would happen. Just not right away. And for multiple quarters, the gap would affect revenue.

It was like someone had yelled *Fire!* in a jam-packed nightclub. Investors panicked, selling shares in the after-hours market. When Zuckerberg and his team left the conference room, Facebook's stock price had fallen 20 percent, losing $120 billion in value. Zuckerberg himself had lost $17 billion in the hour that the call took.

"I think we had the biggest stock drop in the history of the world or something," Zuckerberg told me later. "That was a very big correction based on trying to reset expectations about how we were going to run the company."

But even that setback was temporary. Facebook's users weren't going anywhere. And neither were its revenues. "What is clear is that people are still using Facebook and Instagram," says Pritchard. "And people are still advertising on Facebook and Instagram."

FACEBOOK *WAS* DOING hard work, led by the Integrity team morphed out of the Growth organization. According to Guy Rosen, the thinking was that since Growth had built Facebook to more than 2 billion users, it would be best suited to fixing safety and security at scale. "And the Growth team is very analytical in how the work is approached and how things are measured," he says.

The Integrity team had a motto for this process: "Remove, Reduce, Inform." And it seemed to be having an impact. Three independent reports studying Facebook between 2016 and 2018 concluded that the company was making progress on fake news. A study by researchers at the University of Michigan estimated that what it called "iffy content" had been reduced by half.

Citing such statistics didn't move the public much, since the headlines were all about the repercussions of Facebook's previous sins, which regulators were pursuing with zeal. None less than the Federal Trade Commission. It now appeared that Facebook had not lived up to the promises it made in the 2011 Consent Decree. One was that Facebook must inform users in advance if their data were to be handed over to other companies. Since that's exactly what happened for at least 50 million users in the Cambridge case, Facebook had the very difficult task of explaining why it failed to notify those people, and did nothing while CA pelted Facebook with ads that may have used targeting based on Aleksandr Kogan's personality profiles.

Compliance with the order took two paths. One was with the FTC itself: when Facebook launched something new, it would brief staffers on the commission, noting the privacy protections of the product or feature. Sometimes they would even take guidance on tweaking the product to protect users even more than in the original design. Facebook also retained, as the order demanded, an outside auditor, in this case PwC (formerly PricewaterhouseCoopers), one of the "Big Four." (At the time Facebook hired PwC, Joel Kaplan's wife had been its partner in chief of public policy, a post she kept until 2016.) Periodically, a PwC team would hear from Facebook lawyers and policy people about how the company was complying with the order, and then shuffle back to their offices to prepare a report for the FTC. Apparently the auditors did not flag the issue that Facebook had failed to inform 50 million users that a developer had violated its terms of service and handed their data over to political consultants funded by the far right. People learned about this from reporters, not Facebook.

The FTC felt rightfully burned by Facebook's behavior. A new investigation found the company in violation of its 2011 agreement. Its sins included "deceptive privacy settings, failure to maintain safeguards over third-party access to data, serving ads by using phone numbers provided to Facebook for the purpose of account security, and lying to certain users [when it said] its facial recognition technology was off by default, when in fact it was on." The complaint was a devastatingly detailed description of a company that seemed to have earned the sobriquet of "digital gangsters" that the UK Parliament had bestowed on it in February 2019. What made things worse was that all of the deception and trickery occurred while the company was supposedly on good behavior after its previous perfidies.

The finding kicked off a protracted settlement negotiation with Facebook, a complicated game of chicken where the commission tried to eke out the most punishment it could without Facebook rejecting the deal and taking the issue to a trial, whose uncertain outcome would not be determined for years. One of the key points of contention was the personal responsibility of Zuckerberg and Sandberg. Many observers expected they would be named, as it was up to them to uphold the previous settlement, and they'd failed spectacularly.

The FTC blinked first. On July 24, it announced its settlement, without naming Zuckerberg or Sandberg. They had not even been deposed, as is common in such investigations. As expected, it hit Facebook with a $5 billion fine, by far the largest the agency had ever levied. (The previous high had been $100 million.) Even so, two of the five commissioners dissented, feeling that the settlement went too easy on Facebook. Bolstering their claim was that Facebook's stock hardly budged at the announcement. In an earnings call soon after the settlement, Facebook reported $17 billion in revenue that quarter. The term used most often in the reports of the settlement was "slap on the wrist."

IN JUNE 2018, longtime VP of communications and public policy Elliot Schrage resigned (though he would remain at the company in an advisory role). After a long search, Sandberg began pursuing former UK politician Nicholas Clegg, who had once been deputy prime minister before suffering two humiliating defeats, losing his cabinet post and then his seat. Since then he had been looking at what was happening in the tech world. "The more I looked at the rhetoric and language of backlash against technology, in particular social media, the more alarmed I became that the backlash would kind of throw the baby out with the bathwater," he says. This clearly resonated with Facebook.

Clegg was reluctant to take another public beating by representing the fattest piñata of the tech world, but Sandberg convinced him to fly to California and meet with Zuckerberg and Chan. "When Sheryl has a target in mind, she is pretty implacable and pretty remorseless," says Clegg. He warned Sandberg that he would be blunt. Indeed, on meeting the CEO, he told him, "Your fundamental problem is that people think you're too powerful and you don't care."

"Yes, totally understandable, I get that," said Zuckerberg. Clegg later would say the reply surprised him, but Zuckerberg had been absorbing criticism for two years, with no tears whatsoever dripping from his unblinking eyes. Clegg got the job.

Clegg's arrival came at a time when internal tension at Facebook had been so high that it welcomed anything that seemed like a breather. Whether it was simply fatigue or a feeling of genuine improvement, morale was stabilizing at Facebook, and Clegg contributed to that. Some months earlier, Zuckerberg had handed down several edicts that fit with his wartime CEO stance. No longer would anyone be given a C-level title. (Zuckerberg says that this was not a "broad company effort," but a decision that came from the fact that high-ranking executives like Olivan were C-deprived, while others with no more power got that perk.) It was a surprisingly easy order to enforce because most people besides Sandberg who had a C—the chief security officer, the chief marketing officer, and the CEOs of Instagram, WhatsApp, and Oculus—were already gone or headed out the door. Another was that Facebook would not cooperate with any media profiles of its executives. Clegg's view was looser. He allowed a long, thoughtful *Vanity Fair* story on content moderation that focused on Monika Bickert's role.

Late in the year Clegg was the final speaker at an all-hands meeting, following Zuckerberg's assurances that the company was making progress, and Guy Rosen's presentation on how it was proceeding on Integrity issues. Clegg's straightforward, encouraging style resonated. "I said that [while] some of the coverage might be unfair, you can't con your way out of what is true," he told me afterward. The good news, said Clegg, was that while Facebook had screwed up, it was now on the road to redemption. "It's just true that this company is now trying to retrofit onto its extraordinary creations and inventions a bunch of stabilizers, seat belts, and the like." One person at the meeting—someone who only months earlier had made scathing remarks about the company—seemed reassured. "The coverage had gotten so over the top, to the point of caricature," says the employee. "Even people internally who had their doubts about the company felt like, *Hang on a second, this is not right—we are better than this.* That all-hands was one of the best I've seen them do. I heard from a lot of people how good it made them feel."

The public view of Facebook was still brutal. But the company was making changes and the new PR regime promised at least not to make things worse.

"It's a shared view that we've turned the corner and that we now have confidence that we can not only address problems that come up but we can systematically get ahead of them in the future," Andrew Bosworth told me in late 2018.

Not all of them, it turned out.

While Facebook's critics were abundant, one in particular seemed to get under Zuckerberg's skin: Apple's CEO, Tim Cook. As Facebook's problems became more public after the election, Cook began to voice reservations about social media, and Facebook in particular. Apple's business model, Cook noted at every opportunity, was based on a straightforward exchange: you pay for the product and use it. The Facebook business model, Cook would note, provides a service that seems free but actually isn't, as you are paying with your personal information and constant exposure to ads. With a touch of his native Alabama in his tone, Cook would say, "If you're not the customer, you're the product." He implied that Apple's model was morally superior.

Years after the death of Apple's fabled CEO and co-founder Steve Jobs, the company still had the aura of the elite operation in Silicon

Valley. Zuckerberg had gotten along well with Jobs, and seemed to have been a willing mentee. Jobs recognized Zuckerberg's intelligence and seemed to get a kick out of his brash approach. They would often go on walks together, with the older executive sharing his pointed insights.

Cook and Zuckerberg's relationship was chillier. Cook disagreed with Zuckerberg's comments about privacy, and did not use Facebook personally. Basically, Cook didn't seem to trust Zuckerberg as a partner, and didn't go out of his way to hide it. Complicating matters was the dramatic pivot of the press and government, and to some degree the public, against the giant tech companies that suddenly seemed to dominate everyday life. Insiders referred to it as the "Techlash." Of the West Coast behemoths being lashed against, Facebook was the biggest source of scorn and concern, with Zuckerberg seen as the guy who helped lose the halo that once hovered over the tech world.

Just as leaders of great national powers would summit despite their hostilities, Zuckerberg and Cook would generally set aside time to talk at the annual Herb Allen summer gathering. In 2017, Zuckerberg had been upset at a remark Cook had made at a commencement speech; the Apple CEO had told the graduates not to measure their worth with Likes, and Zuckerberg had taken that personally.

Tim Cook wasn't about to run his speeches past Zuckerberg. By then Cook was promoting privacy as a pillar of Apple's deal with its customers. The targets of his jibes were Google and Facebook, but only Google was a direct competitor, and Zuckerberg felt sideswiped. After Cambridge Analytica, Cook was asked what he would do if he were in Zuckerberg's place. "I wouldn't be in that situation," he said. In an interview soon afterward, Zuckerberg called the comment "extremely glib."

In mid-2018, Zuckerberg arranged a CEO sit-down at Apple Park, the company's exotic spaceship-like headquarters. Once again, Zuckerberg complained about Cook's remarks. And once again, Cook brushed him off.

Zuckerberg says he can't get into Cook's head but is disappointed that he hasn't convinced Apple's leader that Facebook's business model is as valid as Apple's is. "It's widely understood that a lot of information businesses or media businesses are supported by ads, to make sure that the content can reach as many people as possible to deliver the most value," he says. "And there is a certain bargain there, which is, you're going to be able to use this service for free and there will be a cost, which you'll pay with your attention, and advertisers will want to target ads to the type of people who are using whatever that service is."

On January 30, 2019, the tension between Apple and Facebook exploded into a hot war. The escalation began when Apple looked into an app called Onavo Protect. It was the successor to the application created by the Israeli spyware company Onavo, purchased by Facebook in 2013. The application followed Onavo's original plan of providing a free service to consumers and sucking the hell out of the data to do business analysis. The app promised users a secure network connection and used the Facebook name to signal trustworthiness. Once users installed it, it protected their information from everyone but Facebook, which aggregated all the data from Protect users to figure out what people did with their phones.

This approach violated Apple's terms of service. Onavo Protect, Apple concluded, was a surveillance tool marketing itself as a secure VPN, and harmful to users. It told Facebook to withdraw the app, or Apple would ban it.

Facebook did withdraw it, in August 2018. But it was not ready to give up its data. In fact, it already had a tool that used similar VPN technology to monitor users' activity, called Facebook Research. Facebook paid subjects to use it and was transparent that it would gather data. That still put it in violation of Apple's terms, but in this case, Facebook had a plan in mind to bypass the rules. Since it paid the app users a paltry sum, but still a sum, Facebook now considered them contractors. (Those users also included thousands of teenagers, a practice that possibly ran afoul of laws that protected the privacy of minors.) This enabled Facebook to include the app in Apple's "enterprise" program. Since apps in the enterprise program weren't available to the public—most often they were used for pre-release prototypes or utilities limited to employees only—they didn't have to go through the usual Apple certification.

Then Apple discovered the repackaged app and decided that it was abusing the enterprise program. So Apple decided to turn off Facebook's access to the entire program. Without warning. In terms of internal applications, this was like cutting off a company's electricity. Not only was the Onavo app rendered dysfunctional, but all the test versions of programs in development stopped working. In addition, a set of helpful services for people who worked at Facebook, like the one that listed the menus in various campus cafés, also suddenly stopped working. Facebook's employee shuttles, which people widely use to get around the sprawling headquarters complex, also relied on an internal app, which went down as well.

The cutoff coincided with a quarterly earnings call. Zuckerberg, Sandberg, and CFO Dave Wehner entered the conference room to do the call. They had good news. Last year, 2018, had been the company's best yet. "Full-year 2018 revenue grew 37 percent to $56 billion, and we generated over $15 billion of free cash flow," said CFO Wehner.

Zuckerberg boasted about how much Facebook had risen to his trust challenge. "We fundamentally changed how we run this company," he said. "We've changed how we build services to focus more on preventing harm. We've invested billions of dollars in security, which has affected our profitability. We've taken steps that reduced engagement in WhatsApp to stop misinformation, and reduced viral videos in Facebook by more than 50 million hours a day to improve well-being. . . . I feel like we've come out of 2018 not only making real progress on important issues but having a clearer sense of what we believe are the right ways to move forward."

As he spoke, people on Facebook's campus could not test new products and were canceling meetings because they could not get the shuttle.

It was Facebook's split-screen moment, symbolizing the disconnect between the erosion of its reputation and the robustness of its business.

Facebook's campus had come to standstill, a direct result of its dicey privacy practices. But the money kept pouring in.

THE DISCONNECT EXEMPLIFIED Facebook's difficult 2018. Its leaders felt that the company was making progress, but in the harder-to-measure market of reputation, its stock had bottomed out. People will remember Facebook's 2018 for Cambridge Analytica, a huge data breach that was actually a breach, and maybe a hundred other mistakes and violations. But Facebook would prefer that people recall its Election War Room.

This was a conference room set aside to prepare for various plebiscites in the summer and fall of 2018, prime among them the American midterm congressional elections. I visited it twice, and once, after much prodding, I was allowed to drop in for a few minutes on the day that voters went to the polls. But tours of the War Room were common in the weeks leading up to the election. In fact, the pride in the War Room aroused suspicion that the whole thing was a charade, as phony as the Facebook pages set up by the Russians during the 2016 election.

The twenty-four people staffing the room, a spokesperson told me, were backed up by the 20,000 people now working in safety and security at Facebook. (A year later, Facebook would be reporting a higher number: 35,000!) Inside, the room was like a security Ginza, with hundreds of screens displaying dashboards reporting real-time results. Others conferenced in Integrity workers from around the globe, including Brazil, where an election was also being held. Even with the AI, the War Room was an expensive, labor-intensive solution. But whether or not the physical facility was actually required, or was essentially a showroom for the press, Facebook got through the 2018 midterms without searing consequences. Facebook's head of civic engagement says the system actually did block tampering, citing an example of a false-information attack from Pakistan (or Macedonia, he doesn't remember which) directed at Wyoming voters.

The fact that an election was held without Facebook having a role in screwing up the outcome was now considered a victory for the social network.

"I'd like to have our track record of the 2018 election for 2016, for sure," Sandberg told me. "In 2016 we had never thought of this form of interference, we didn't know what it was, no one in the government knew what it was, no one in any administration told us anything about it, before or after, at all." (Actually, Maria Ressa told them.)

Still, Facebook was making progress. But reading newspapers (if anyone still did that) or checking the online news, you would never know this. Scandals kept popping up. It became a joke among journalists, a parody of the sign in factories listing how many days since the last industrial accident, with the daily number ideally reaching triple figures, at the least. With Facebook, the number seldom reached double figures, and often reset at one or two. The reporters kept digging (or wrote up what fell into their laps), the regulators kept investigating, the courts kept deposing, and the public was still thinking about whether it should #deletefacebook.

Enough with those tweaks, people seemed to be saying. The question was whether the next crisis might be big enough to ruin the company—and if Mark Zuckerberg was really plotting change, *fundamental* change.

And it turns out he was.

19

The Next Facebook

ON WEDNESDAY, MARCH 6, 2019, Mark Zuckerberg tossed down another bolt from his fishbowl Olympus. After all the incremental tweaks since the election, and all the ongoing public concern and scrutiny, he was finally making his big move. The name of the post was "A Privacy-Focused Vision for Social Networking." The king of sharing was now recalibrating to emphasize the feature that set Thefacebook.com apart in its very first appearance: privacy.

He made the decision in his wartime CEO mode: by declaration. "I basically just found I could've invested years in talking to our team internally and still not gotten everyone to the same place," he says. "At some point we just needed to make a decision."

His observations of Internet behavior had convinced him that people were gravitating toward services without the drawbacks of the News Feed. "Private messaging, ephemeral stories, and small groups are by far the fastest growing areas of online communication," he wrote. "Many people prefer the intimacy of communicating one-on-one or with just a few friends. People are more cautious of having a permanent record of what they've shared."

Since the Zuckerberg Way is to see upcoming paradigm shifts as opportunities, he was now shifting Facebook to take advantage. "As I think about the future of the Internet," he wrote, "I believe a privacy-focused communications platform will become even more important than today's open platforms."

Since the premier open platform was the Facebook Blue franchise, that might have presented a problem for Zuckerberg. Fortunately (for him) he owned three other communications platforms besides Blue. He also owned what he believed would become the dominant platform in the century's third decade, the VR company Oculus. And was bringing all of them under tighter control.

Zuckerberg had actually been preparing for this shift for the last year. While all the headlines had been about scandals and elections and earnings, he had been quietly changing the company.

Beginning with the dispatch of the founders of all the key companies he had bought.

Elaborating on his vision at the 2019 F8 conference a few weeks later, he took pains to say that the shift wouldn't mark the demise of legacy Facebook. It would have a place as what he called "the town square," a public forum where well-meaning people and just plain awful people could participate in a two-billion-headed conversation. But people increasingly wanted protected spaces where they could converse privately. This would be akin to "the living room," where conversations were not public. Facebook was now betting that the living room would be more popular than the town square. It would present fewer problems for Facebook than the current free-for-all where misinformation, hate speech, and stupid distractions thrive in plain sight. With strong encryption, no one would see the content that now causes such outrages.

When I talked to him after his keynote, I noted that he did not once mention the News Feed in his speech.

Short silence. "Yeah, that's maybe true," he finally said. "But it's still important."

Just not the future of the Internet.

What enabled Zuckerberg to be so confident about his company's future in the post–News Feed world was the astounding success of the franchises he'd bought and created between 2012 and 2014. One of them,

Instagram, was turning into an epochal win. Its growth rate far surpassed Facebook's and, after a deliberate start, ad revenues began pouring in. Though it too had been used as a tool for Russian misinformation, the stigma never seemed to settle on the photo-based social-media site. Instagram was beloved. Its CEO, Kevin Systrom, was viewed as a Silicon Valley guru, one who, in contrast to his boss, was known for probity, design rigor, and empathy. Instagram was so successful that, in the words of one Facebook executive, "The sideshow became the show." Which in itself would become a problem for Mark Zuckerberg.

When I visited Instagram's new headquarters in early 2017, though, everything seemed copacetic. In contrast to Building 20's warehouse chic, Instagram's new palace had a minimalistic aesthetic and big windows that shooed in natural light. Just like an Instagram post. Five years after selling to Facebook, Kevin Systrom was still firmly in charge.

"When we joined [Facebook], I think our big question was, Were we going to have independence?" he told me. "Because when you start something, it's your baby and you want to take care of it and you want to nurture it. There's something that makes Instagram special, and it's the community, and I did not want Instagram to become a feature of something larger."

He and his co-founder, Mike Krieger, essentially ran Instagram as a separate franchise, but reaped the benefit of Facebook's infrastructure, marketing, and even its artificial-intelligence research. For example, Instagram had recently used Facebook's machine-learning skills to change its feed from a chronological stream to a ranked one. Though Systrom reported to CTO Mike Schroepfer (that would soon change to Chris Cox), he still had direct contact with Zuckerberg. They would meet for dinner once a month or so, and the interaction was more like a business meal between peers than a meeting with the boss.

Systrom insisted to me that, as of that moment, Zuckerberg did not meddle. Systrom cited the recent redesign of the Instagram logo, which had been a fairly realistic image of a '60s Polaroid-ish camera, with a rainbow-colored patch at the top-left corner. Though a redesign seems trivial, it was actually a massive deal. Times had changed; Instagram had become a global business. And the app was not just a way to share in a fun and vintage way but an important part of the way people expressed themselves. So it deserved a logo that discarded the realistic view, known in software as a skeuomorphic image, and went to something more abstract, a glyph of rectangles and circles that suggested a camera. Replacing the rainbow were warm color gradients that gave the logo a shimmery look. Because the change was so dramatic, Systrom was a little concerned when he showed it to Zuckerberg over one of their dinners. "By the way, I forgot to mention we're designing our logo," he said late in the meal, bracing himself for a long discussion. But Zuckerberg just looked at it and said it was nice, though maybe he didn't care for the rainbow gradient so much.

Systrom had even reached a good compromise with a potential flash point: inserting ads in the stream of Instagram posts. Systrom thought the News Feed has suffered with too many ads, and did not want Instagram degraded in the same way. He insisted, at least at first, on a limit to how many ads would be in the feed, and the right to approve the ads that did run. If that meant that Instagram would forgo the revenues that came from opening up its feed to the millions of businesses who used the self-serve process of the Blue app to reach targeted customers, so be it. Systrom personally signed off on each ad that appeared, making sure that he, and not some algorithm, determined whether it met the aesthetic standard to look good in the Instagram flow.

"Everything we do in product here is very independent," Systrom told me. "We're all kind of marching towards a similar destination, but we have very different ways of doing it."

Neither of us knew it at the time, but that 2017 conversation might have been the last time Systrom would be touting his freedom. In the months that followed, Instagram would be fettered by an increasingly short leash, looped around the wrist of the guy in the gray T-shirt.

One other thing would change. In 2018, Systrom would become a father. He'd take the generous parental leave that all Facebook employees are encouraged to enjoy. And just before that leave was over, he would quit.

THE MARRIAGE OF Facebook and Oculus was bound to be rougher. Zuckerberg was excited about what Oculus would be in a decade, when virtual reality would presumably become as dominant as mobile proved to be after the smartphone. In the meantime, Oculus was basically a game company selling hardware, an alien business to Facebook.

Survival as a game company would require Facebook to toss in billions of dollars and compete in an industry it didn't care much about. Oculus had a chicken-and-egg problem. Ideally there would be a wide library of great software that ran on its flagship product, Oculus Rift. But it was expensive: the $500 price tag did not include the supercharged computer required to run the software, which brought the total to $1,500, more than most people could afford. Because the user base remained tiny, the big game developers did not see the value of spending the million-plus dollars it would take to create a first-class title.

So Facebook paid them. It set up a division called Oculus Studios that handed out grants for companies to produce content that would run on the Rift. Meanwhile, software wizard John Carmack, a pragmatist, worked on cheaper, mobile-based products with broader appeal. Oculus did a deal with Samsung to create Gear, a $100 headset where your own smartphone would provide the visuals. It was VR on the cheap. Gear sold a lot more units than Rift, but it was an inferior experience.

In Zuckerberg's keynote for the Oculus developers conference in 2017 he set a goal of a billion people using virtual reality. This was news to Oculus's top executives, who were unaware of the prediction until they heard it in rehearsal.

While Zuckerberg was fixated on virtual reality as a social technology, the gamers at Oculus felt this was something for the future. "Social is probably number four on my personal list for what I think is important for VR," John Carmack says. (He adds that this may be due to being "an antisocial person with hermit tendencies.")

It was symbolic that Facebook's social VR team was not even part of the Oculus organization, but an engineering group reporting through Zuckerberg's chain of command. To satisfy his desires to get social VR now, they began building a product called Facebook Spaces (note: not Oculus) where people could interact in VR with the Rift. Since the Rift's relatively tiny user base consisted mainly of hard-core gamers—who had zero interest in capturing baby's first steps, especially since grandma had no headset to view it—Spaces had difficulty finding an audience. While the demos seemed cool—Zuckerberg did an elaborate one himself, with his own family as cartoon avatars superimposed into his actual living room—there was something visually unsettling about them.

Facebook's demo for the 2017 F8 developer conference was not as charming. The company had just donated to the Red Cross's effort to aid hurricane-ravaged Puerto Rico, and to explain it, Zuckerberg and his social VR head Rachel Franklin took a VR tour of the damaged island. The video showed their cartoon avatars viewing the wreckage with inappropriate giddiness, including a high-five celebrating Facebook's generosity. Zuckerberg's subsequent apology, if one could call it that, was of the "sorry to anyone I offended" variety. The episode did not help Facebook's virtual-reality efforts.

Zuckerberg's hope lay in Oculus's research facility in Seattle, which had hired top scientists to help devise low-cost goggles that would solve long-term VR problems and deliver an "augmented reality" experience, superimposing computer graphics as an overlay on the real world. He exercised patience, confident that the lab's head, Michael Abrash, was gathering the best scientists to advance the field. Oculus would need them, because Apple, Microsoft, and other companies were also devoting their resources to such a product.

That patience did not extend to the performance of the Rift, which he described in an earnings call as "disappointing."

And then there was the Palmer Luckey problem. While the Oculus founder had participated in products like the hand controls for Rift, his time was increasingly spent as Facebook's ambassador of virtual reality.

He gave celebrity demos of the "Toy Box," where two people could share a virtual space and do things like play Ping-Pong. In *Time* magazine's big 2015 story on virtual reality, the cover subject was Luckey in a headset, superimposed on a tropical beach background. "Palmer focused on being the face of VR, doing press and evangelizing," says Brendan Iribe, Oculus's co-founder and CEO. "And it was great until the Nimble America situation."

Luckey was a political conservative, supporting the right wing with the same enthusiasm he devoted to fast food, cosplay photos with his girlfriend, and soldering artisanal computer peripherals. He was a huge admirer of the military. Iribe remembers that once he got a call saying Luckey had driven a tank on the Facebook campus. The police had been called. The vehicle was Luckey's Humvee, repurposed from military service with toy machine guns attached to the postings. To Facebook's workers, though, it might as well have been a nuclear bomb. Luckey defused the situation and wound up posing for pictures with the cops, but the incident was a black mark on his record. "Here at Facebook, you can't drive Humvees with guns—military vehicles—onto the lot and have the police show up," says Iribe. "That's not what we're focused on here."

In the summer of 2016, Luckey had come across a group of seemingly like-minded Trumpers in a sub-Reddit called The_Donald, devoting themselves to "shitposting in realtime." They called themselves Nimble America, and he anonymously donated $10,000 to allow the group to fund a billboard outside of Pittsburgh with a cruel cartoon face of Hillary Clinton captioned, in uppercase, "TOO BIG TO JAIL." When Luckey later confirmed his contribution to a *Daily Beast* reporter, he thought the exchange was off the record. The reporter thought otherwise. The story, "The Facebook Near-Billionaire Secretly Funding Trump's Meme Machine," ran on September 22, 2016.

This ruined Luckey at Facebook. The press eviscerated him. Luckey insisted he was misunderstood. His donation to Nimble America, he said, was only to buy billboards and maybe print some T-shirts. He had no role in Internet trolling, meme creation, or posting racist comments.

Still, the overwhelmingly liberal Facebooker workforce was appalled, and some called for his resignation. Ironically, this was happening at the same time as the top policy people at Facebook were intentionally doing nothing about the actual Trump meme machine that was running rampant on their platform. Worse, some Oculus developers said that they were abandoning the platform because of Luckey's actions.

Luckey drafted a letter to his colleagues to explain the situation, but Facebook insisted he sign a different letter. Facebook's deputy general counsel, Paul Grewal, emailed him to say, "I need to tell you Mark himself drafted this and the details are critical." Luckey was shocked to see that the post going out under his name claimed that he was supporting third-party candidate Gary Johnson for president. Overall, the letter reeked of the inauthenticity of a hostage dispatch, and failed to satisfy anyone.

While Facebook did not fire Luckey—yet—it put him on ice, instructing him to stay off campus and not communicate with coworkers or on social media until after the election. His customary appearance at the annual Oculus developers conference was canceled.

Luckey's treatment contrasted with that of another public Trump supporter, board member and original investor Peter Thiel. When Thiel said he would donate $1.25 million to the candidate, Facebook employees called for his ouster. Zuckerberg defended him in an internal post, literally as he was banning Luckey from the Facebook campus. "We care deeply about diversity," he wrote. "That's easy to say when it means standing up for ideas you agree with. It's a lot harder when it means standing up for the rights of people with different viewpoints to say what they care about."

Luckey assumed that once the election was over, the controversy would blow over. But when the unthinkable happened—Trump won—his return became impossible. Still, Facebook had good reason not to fire him in 2016. In January 2017, Luckey was scheduled to testify at the intellectual property trial springing from Oculus's initial sale, and the company needed him as part of its defense. He dutifully prepared, and delivered his testimony, hoping that he would then be able to return. Oculus was his life.

But Facebook would not let the founder of Oculus return to Oculus. "I can say internally we looked very hard at what role he could have after this situation happened," says Iribe. Each technology division head at Oculus was asked if he could use Palmer Luckey, the guy whose invention had created the division. And not one said that they had a place for him. For all practical purposes, Palmer Luckey was fired.

A few days later, Facebook hired an outsider, former Googler Hugo Barra, to become the new head of Oculus. Iribe was demoted, and would leave in 2018, after Zuckerberg figured out the person who would pull together Oculus, the social VR team, and all of Facebook's other hardware efforts, like the Portal display. Meet the new VP of Hardware: Andrew "Boz" Bosworth. Zuckerberg tapped his go-to bro because he still thought VR was destined to become the next big thing. When he talked about it, his voice would rise a notch.

But his near-term goal of a billion users in virtual reality seemed unrealistic, considering the poor performance and huge losses of Oculus so far. "There's no doubt that we thought we would be further in at this point," says John Carmack. "We've squandered a lot of resources. We have been lavishly funded and there have been so many projects that are spun up and abandoned, internal directions that got staffed way up, lots of money spent and then we decide not to do them for good or bad reasons, lots of projects that have not been managed particularly well."

Bosworth inherited what was to be Oculus's breakthrough product, a standalone headset that promised to deliver almost all of the "wow" experience of the Rift. At $400—and no need to buy a specialized computer—it promised to be a breakthrough in the field. Oculus successfully launched the product, which even won the respect of a gaming press that had grown cynical.

Quest was not something that people would use on a persistent basis, and it would not fulfill Zuckerberg's dreams of virtual or augmented reality being the platform for social interaction. That could only be done by ditching those cumbersome headsets and creating the technology that would allow people to become a form of cyborg—part human, part Facebook. That would happen, he hoped, by the efforts of Oculus Research, the lab in Seattle working on long-range projects. It was making progress on its wear-all-the-time Augmented Reality eyeglasses. Beyond that, Facebook was exploring how to get its products literally in people's heads. It hired a team of neuroscientists to create typing-free interfaces between thought and action. And in 2019, Facebook bought a company called CTRL-Labs, which picked up brain signals from one's wrist so one could control apps just by thinking. Every time the project got a mention in the press people would joke, *Oh, Facebook now wants to get inside your brain.* But that was actually true.

WHILE 2016 MARKED an election-year fiasco for the News Feed on the Blue app, Instagram introduced its most successful feature that year, one that would change Facebook forever. It would come from, of all places, Snapchat.

A few months after turning down Zuckerberg for good in 2013, Evan Spiegel realized that Snapchat had a piece missing. Sometimes people would snap a picture or a video and want to send it to a bunch of their friends. To do that on Snapchat's one-to-one service, they'd have to do it serially, starting over again for each friend. How could Snapchat let people tell their daily stories to all their friends while maintaining the ephemeral spirit of the app?

"We really felt like that needed to be done in a respectful way," says Spiegel. Meaning: not like Facebook, which he felt encouraged people to be inauthentic versions of themselves, a red-carpet distortion of their true, fun, goofy personalities. Even worse, the content was presented in a feed that unfolded in reverse chronological order. Unless you were Harold Pinter, who once wrote a play where the end came first, this was no way to tell your story. People knew this instinctively: when you come home and tell your family about your day, you don't start at the end. You don't tell the story of your birthday in reverse!

Spiegel's answer was a feature that allowed its users to pictographically share the fun stories of their day, beginning to end. The defining feature of Snapchat—its impermanence—was even more valuable when users were sharing something to a group of friends as opposed to just one buddy. "There's something really optimistic and inspiring about waking up every day as if it's a new day, and not being defined in terms of who you were yesterday," says Spiegel.

The name of the feature was obvious: Stories.

Stories would be the anti–News Feed.

Spiegel gathered a small team in the Blu House, one of several facilities his company now was working out of in downtown Venice. The product they created allowed you to post a series of photos or short videos and festoon them with the weird and silly stickers and virtual masks that the regular product offered. Users could swipe to the Stories page and see a strip of the stories available to them. After twenty-four hours, the stories would be gone. Spiegel thought it was brilliant.

But no one used it. "Like, literally no one had any idea what it was for," he says. "*What is this Stories thing?*"

Spiegel did not panic. When Snapchat itself first launched, it had been a flop. "That's always the challenge with new ideas," he says. "It takes time for people to change their behavior." That is what happened with Stories. After a few months, the graph that showed the rate of pickup roused itself from the basement and began to take the satisfying trajectory of an S-curve.

Facebook noticed. But this time it was not Mark Zuckerberg who would be trying to copy a Snapchat product. It would be Kevin Systrom. And that would be very bad news for Evan Spiegel.

Systrom has never denied that his Instagram Stories feature is essentially the same idea as the original Snapchat product. But he resists the idea that his team simply swiped someone else's concept and slapped it onto Instagram. "You can view it one of two ways," he says. The first, he says, is that while Instagram was growing, someone else changed the world with a competing product and the company needed to react by copying that product. The other view, which he embraces, is that Instagram's own success was so vast and dramatic that it overstepped itself, and created a natural gap that needed filling. By Stories.

Instagram had started as a way for people to visually share highlights of their lives. But as it got bigger, the larger scale of the network made it less personal. More people were using it, but for some, it was no longer the first online place they would go in the morning. "The world needs a place where people can share funny, goofy things with their closest friends where they are not going to be judged," says Systrom, sounding a lot like Evan Spiegel.

Systrom concedes that Snapchat filled that gap first, but now Instagram needed to fill it, too. "It was part of our ecosystem that we just left open," says Systrom. "We wanted to allow people to share [and] highlight epic moments of their life, and if we wanted them to [also] share the silly moments throughout their day, people would embrace that."

Instagram treated the project with the highest priority, and soon had its version of Snapchat's idea. Then there was the question of what to call the feature. Everyone thought of it as just "stories." Which was what Snapchat already was calling its product. "We started realizing that there's no reason to call it anything different," says Kevin Weil, Instagram's head of engineering at the time. "Let's just embrace it. This is going to be a common format for lots of apps and services, not just Snapchat and Instagram. So we'll call it a story. The same way they have."

Instagram was so confident—or perhaps needed the product to work so badly—that it went all in on it, in a way that had been seldom seen at Facebook in the past few years, where innovations were gingerly integrated, or in some cases even released as separate experimental apps. New features rolled out gradually, usually after painstaking testing on a small percentage of users in some remote country that no one normally paid attention to. Not Stories. Instagram released it to almost the entire world, all at once, hitting users like a sudden thunderstorm. Thumbnails for the Stories galleries were sited on top of the screen, indicating prominence over the feed that had been the core of IG's product since the Burbn days.

Systrom had braced for a slow takeoff as people acclimated to the format. But the user base wolfed down Stories as if they were cheeseburgers dropped onto a desert island. "I didn't realize how much we had created this vacuum to be filled," says Systrom. (Or maybe Snapchat had trained them.)

In a way, Instagram had been failing users by evolving into a showcase for celebrities and influencers. The world of Instagram had belonged to its stars, and the rest of the crowd was just living in it. All of a sudden a new use for the service had popped up where you shared casual moments with your friends, with no pressure—in twenty-four hours, the story was gone. It was almost as if Instagram had suddenly achieved the fun and intimacy of . . . Thefacebook.com of college days, where silliness was the rule, and the concept of FOMO had yet to stir the persistent anxiety that gripped people in the new age of social media.

What's more, Stories did not seem to be cannibalizing the Instagram feed. "People still love showing off their epic vacation in that one photo, but they also really like taking fifteen vacation photos that they don't want to show people forever," Systrom told me in 2017.

Snap CEO Evan Spiegel refused to comment on the blatant appropriation of his idea. (Snapchat had shortened its name in 2016.) His subordinates, though, were apoplectic. "It was like a bombshell going off," says a Snap executive at the time. Spiegel would not comment for some time, even inside the company. Spiegel's future wife, Australian supermodel Miranda Kerr, was not so circumspect. "I cannot stand Facebook," she told London's *Telegraph*: "When you directly copy someone, that's not innovation—that's a disgrace. . . . How do they sleep at night?"

Very well, apparently. Stories, Zuckerberg boasted in an earnings call, was on track to be bigger than the News Feed. But if Systrom and Krieger felt their success would be rewarded by Zuckerberg, they were wrong.

IN ZUCKERBERG'S MOVE to bring his properties under control, the toughest challenge was WhatsApp. WhatsApp had a cloistered culture, its workforce bounded within the walls of the unmarked office in

Mountain View. They eschewed traditional metrics of success. Its mission was not merely to connect people but to bestow them with freedom to connect without constraints—from mobile services and even from governments.

So it was entirely consistent with that mission that WhatsApp pursue a scheme to encrypt all its messages by default. Co-founder Brian Acton especially felt WhatsApp's users should be able to communicate in a way where government eavesdroppers could never access the secrets they shared with friends, family, and business associates.

In the summer of 2013, Acton began working on an end-to-end encryption model for WhatsApp. Creating a cryptosystem to protect the communications of more than a billion people, and withstand the attacks of everyone from wizardly hackers to sophisticated state intelligence agencies, is the ultimate don't-try-this-at-home enterprise. It was a blessing when Acton connected with Moxie Marlinspike, a crypto-activist and master cryptographer who believed that encryption was core to freedom in the digital age.

Marlinspike was being funded by donors to create a mass-market easy-to-use encryption tool called TextSecure. Acton convinced him to help build the TextSecure technology into WhatsApp. Though senders and recipients would not necessarily realize it, every single message sent would be as protected as a missive from one spy to another. Snoops, spooks, hackers, and divorce lawyers might be able to intercept a message, but they never would be able to read it, because from the time the Send button was pressed to the moment it was read, the content would be scrambled. Even Facebook would not be able to read it.

There was considerable risk. The FBI and the NSA had been warning about a scenario called "Going Dark," where security and safety would be at risk if they were not able to retrieve the content of messages. Facebook could be slapped with fines, or, if it turned out that one of those encrypted messages involved the planning of a murderous attack, the company could be shunned, or worse.

The purchase deal still had not closed when Acton informed Zuckerberg—he pointedly did not ask permission—that WhatsApp was pursuing end-to-end, and the CEO took it in with his typical inscrutable form of assent. "We were like, *Mark, we are building end-to-end encryption,*" says Acton. "He's like, *Okay, okay fine, you guys go ahead and do that, I don't care.*"

Actually, Zuckerberg had done a considerable amount of thinking on the subject. He had been outraged in 2014 when Facebook learned, via Edward Snowden's leaks, that the US government was snatching its communications from Facebook's data centers. Zuckerberg also had an emotional bias toward encryption. If his own early communications—the IMs and emails regarding ConnectU when he was at Harvard—had been encrypted, he might have been spared embarrassment.

When Zuckerberg did express reservations about encryption, his issues were not about addressing the concerns of law enforcement, but about Facebook's bottom line. In mid-2017, Facebook was implementing a new financial strategy in its messaging apps, opening what had previously been a person-to-person communications service to one where businesses and customers could connect. Messenger was already well along on that path. According to Acton, "The question Mark kept raising was, *If we have end-to-end encryption, are we leaving money on the table?*" The problem wasn't that the actual messages between businesses and customers would be hampered, but that Facebook itself would not be able to scan the messages to see what was in them and use the information to make a better user experience, or even to serve users better ads or add-on services. "People were questioning end-to-end encryption in terms of its business value," says Acton.

WhatsApp kept its encryption. But the conflicts about making money from WhatsApp became increasingly heated.

Not long after the deal closed, discussions began between the founders and Zuckerberg on whether WhatsApp should drop its yearly fee of a dollar. The revenue was minuscule in light of Facebook's much bigger economy. Acton was against it; he felt that the fee was like an insurance policy. "Mark was like, *Kill it, kill it, kill it*," says Acton. "[If] the boss says kill it and the minion says don't kill it, you kind of lose the argument."

The next compromise was bloodier. Koum and Acton had designed WhatsApp to be the opposite of Facebook in terms of gathering information; all it knew about users was their phone numbers, by design. Their assumption—one they communicated to their users via the WhatsApp blog when they sold to Facebook in 2014—was that they would be allowed to continue along those lines.

> *Respect for your privacy is coded into our DNA, and we built WhatsApp around the goal of knowing as little about you as possible. You don't have to give us your name and we don't ask for your email address. We don't know your birthday. We don't know your home address. We don't know where you work. We don't know your likes, what you search for on the Internet or collect your GPS location. None of that data has ever been collected and stored by WhatsApp, and we really have no plans to change that.*

But Zuckerberg was not going to pay more than $20 billion for a service that would run counter to his own core business model. In mid-2016, he made another argument that, as CEO with total voting control, he could not lose: Facebook should be able to use some WhatsApp data and merge it with its other services. One move in particular stood out: integrating the phone numbers of WhatsApp customers into Facebook's databases. This would allow Facebook to link this most valuable of personal identifiers to millions of users of the Blue app who had previously withheld that information.

Doing this would require rewriting the terms of service contract between WhatsApp and its users. Users, of course, rarely read those lengthy and impenetrable TOS contracts. But regulators do, especially in the privacy-conscious European Union, which parses those legalistic word salads as one of the few reliable clues to what companies are actually doing with user data.

Complicating matters was that Facebook had specifically promised when it bought the company that it would not integrate WhatsApp data with its own. The pledge had been necessary to win approval of the persnickety European bureaucrats, and in the United States, the FTC had also obtained that promise.

The TOS change seemed to run afoul of this agreement. Particularly egregious was the fact that the change would be opt-out rather than opt-in. That meant that if the user did nothing, the data would be shared. Only the savviest and most motivated users knew about the change and figured out how to stop their WhatsApp data from being merged with the giant repository of Facebook data. Acton later told *Forbes*'s Parmy Olsen, "I think everyone was gambling because they thought that the EU might have forgotten because enough time had passed."

The post on the WhatsApp blog of August 25, 2016, which Koum and Acton signed onto even as they despised the move, put the change in a positive light. "By connecting your phone number with Facebook's systems, Facebook can offer better friend suggestions and show you more relevant ads if you have an account with them," it said.

The EU wasn't fooled. Because the change violated the promise Facebook made when it submitted the acquisition for review, it fined the company 100 million euros (about $122 million) for the turnaround. Facebook claimed, "The errors we made in our 2014 filings were not intentional."

Zuckerberg kept pushing. In early 2017, he insisted that WhatsApp move to the Menlo Park campus. The move was as harmful to WhatsApp's culture as Acton and Koum feared. The WhatsApp people were accustomed to a different atmosphere from the boisterous, close-quartered dorm-room spirit permeating Facebook's offices. Porting WhatsApp's more heads-down vibe to Menlo Park created friction. To Zuckerberg's credit, he allowed WhatsApp employees to keep their larger desks, and even had the bathrooms remodeled to accommodate them—the privacy-obsessed WhatsApp folk wanted stall doors that reached the floor. But according to a *Wall Street Journal* article, other Facebook folk resented the idea that these newcomers were special. Some Facebook veterans, offended that WhatsApp people had the temerity to hand out posters requesting that visitors "please keep noise to a minimum," would catcall the newcomers. "Welcome to WhatsApp—Shut up!" they'd chant, wrote *The Wall Street Journal.*

Acton didn't have much of a relationship with Zuckerberg. When the two of them met, Acton would try to start a conversation about their kids—they both had young children and even had used the same obstetrician to deliver their babies. Acton felt that at those points Zuckerberg would always change the subject. "He got really good at keeping people away," says Acton. "And the dude lives about a mile from where I live!"

Acton tried to bring up his problems with Sandberg as well, with poor results. He saw her as a political animal, and felt she did not regard him as a peer. Once he was at a meeting with her and in the middle of it she spotted a visiting acquaintance—"some high-profile guy from ESPN or something" he recalls—and interrupted the meeting to talk to the visitor.

Sandberg would frequently argue that WhatsApp should carry ads, comparing it with Instagram, which embraced advertising and was making a fortune for Facebook. Acton told Sandberg he disagreed with the "monetization initiatives" Facebook was suggesting, and even invoked the clause in his contract that said he didn't have to stick around if Facebook drew revenue from WhatsApp in a manner he disagreed with. Sandberg told him that all of that was above her pay grade.

The spring of 2017 and into the summer, Acton kept going to Jan Koum. *Dude, I can't do this anymore,* he'd say. He knew Koum wanted out as well, and suggested they leave together.

But Koum's plan was to stick it out and negotiate his exit in stages: first a departure from the board, then an extended period of time where he would be technically employed at Facebook. And then he would collect the vast majority of the money coming to him—about three-quarters of the $2 billion remaining.

Acton could not wait. He rashly told Zuckerberg he was leaving before invoking the monetization clause. "I didn't hit him with any hard-hitting punches like, *Oh, this is a shitty place, this ad stuff is wearing me out,*" he says. "I kind of regret it, because I feel I was a little disingenuous to Mark. But I didn't ever feel I had the rapport that I could even share that with him."

Acton believed that because of the clause in his contract about monetization, he was entitled to speed up his vesting, even though he was leaving the company. But he hadn't mentioned it. There was about a billion dollars at stake. About two weeks after the meeting he wrote an email to invoke that clause. They met again, Acton, Zuckerberg, and Facebook's deputy general counsel, Paul Grewal. Zuckerberg told him it was probably the last time they would ever be talking. Acton said that he wanted to make it clear that this was about the monetization. "That was one of the last things I said to him," says Acton. "I just didn't want to put ads in the product."

Both sides tried to hammer out a settlement, but ultimately Acton didn't feel good about it and just walked away, leaving almost a billion dollars behind. His departure went public in September 2017.

Eight months later, on April 30, 2018, Facebook announced Koum's departure. In a blog post Koum wrote, "I'm taking some time off to do things I enjoy outside of technology, such as collecting rare air-cooled Porsches, working on my cars and playing ultimate Frisbee." With Koum's kitty now estimated at $9 billion, the Porsche market just got more interesting. His last day was in August 2018.

As bitter as Acton's exit was, he did have the salve of walking away with $3 billion. "That's a good way to look at it," he says when I point this out to him at the end of a long interview we had in downtown Palo Alto. ("Anything you want, I'll tell you," he promised.) He used $50 million of that stake to create the Signal Foundation, which uses Moxie Marlinspike's crypto tools for an easy-to-use and impossible-to-break communications service for the masses. He sees it as penance for yielding his business, if not his soul, to Mark Zuckerberg.

"I espoused a certain set of principles, even publicly, to my users and I said, *Look, we are not gonna sell your data, we are not gonna sell you ads*, and I turned around and sold my company," he says. "That is my crime and my penance [is] that I have to pay to that crime. I live that every day. Signal is my hope to pay back on that."

Acton took one last shot at the company that made him a self-loathing billionaire. On March 20, 2018, on the heels of the Cambridge Analytica news, he invoked a hashtag that had been trending for a while on Twitter:

It is time. #deletefacebook.

The most popular response to the tweet came from Elon Musk. "What's Facebook?" he tweeted, not even softening the blow with an emoji.

Acton's former colleague David Marcus, himself in charge of Facebook's message app, posted an indignant response on his public Facebook feed. "I find attacking the people and company that made you a billionaire, and went to an unprecedented extent to shield and accommodate you for years, low-class," wrote Marcus. "It's actually a whole new standard of low-class."

Marcus's post won a lot of Likes from Facebook's executives. But Brian Acton's critique—both in the *Forbes* interview and in the one he had with me a few weeks earlier—was directed as much toward himself as toward Facebook and Zuckerberg. "I am the first to declare," he told me, "that I am a sellout."

Replacing Koum as the head of WhatsApp was a longtime Facebooker named Chris Daniels, who had paid his dues as head of the troubled Internet.org initiative. He struggled to win over his subordinates, and some loyalists left, but he began to put the service on a path to include some of the things the founders had long fought against. In November 2018, ads began appearing on WhatsApp.

Not long after he took the job, Daniels reported the progress he had made at a small group meeting. "I want to say one thing," said

Zuckerberg, signaling a rare meta-comment. "There are some good things that Jan brought to the table, but I also realize how much he was standing in our way." Then he mentioned how it made him reflect about a few other areas in the company where he should be thinking the same thing. It was an odd thing to say when one of the Instagram founders was in the room.

In March 2019, after Zuckerberg announced that he was combining all the services, including WhatsApp, Daniels left as well. Zuckerberg replaced him with another executive who'd been in the birthday photo, Will Cathcart.

STORIES, WHICH INSTAGRAM had taken from Snapchat, were such a success that Zuckerberg initiated a heist of his own—Zuckerberg announced internally that Facebook, the Blue app, was going to add its own version of Stories.

Facebook's adoption of its younger sister's breakaway feature came at an interesting moment. The growth of Facebook's Blue app had slowed: in North America it was actually declining. Meanwhile, Instagram had topped a billion users, reaching that level even faster than Facebook had. What's more, it was now beloved in a way that Facebook no longer was. While Facebook was increasingly viewed like taxes—something unpleasant that was part of your life whether you liked it or not—people enjoyed Instagram. Young people in particular embraced Instagram and barely ever checked Facebook. When I spoke to a few high school classes in 2018, I would ask the room how many people used Facebook, and perhaps one or two hands would be raised. But almost everyone held their hands up when I asked about Instagram.

Zuckerberg could rightfully take pride in his purchase. But to some inside the company it seemed he wanted to take . . . credit. When speaking about the success of Instagram he made it a point to note that while the co-founders did a good job, their success was equally due to Facebook's support. He pointed this out in the earnings call when he announced Instagram's cross into the billion-user realm, and it came up one afternoon when he and I were taking a long walk on the faux savannah on Building 20's rooftop. The conversation started by revisiting his refusal to sell to Yahoo! in 2006, and how happy he was to have made the difficult choice. He told me he now advises young entrepreneurs not to give in to pressure and sell their companies if they feel that those companies have potential to succeed independently.

I couldn't help but make the connection to a pair of co-founders of a rising company who accepted an identical offer. "Does that mean that Kevin and Mikey made a mistake by selling to you?" I asked him.

He paused for a minute, as if he were a chess grandmaster startled by a move from an inferior player who suddenly shifted the board to his disadvantage. He didn't want to denigrate two executives who had done a fantastic job for him. But he did anyway. "On the one hand, I think they would've done a good job; they're very talented and could have built the business to be worth more than a billion dollars," he told me. "On the other hand, I really do think that without the work that Facebook has done, and I think we're the best in the world at this, I don't even think that they'd be half as big as they are today."

By late 2017, it was clear that Zuckerberg's relationship with Instagram had changed. *How can we help you?* had changed to *How are you helping us?* And then, in keeping with his wartime CEO speech, *Maybe you should just listen to me.*

At first, Zuckerberg seemed to simply want to tap Instagram for more revenues. The number of ads on the system was always a concern of Systrom's, and previously Zuckerberg would look favorably on his arguments to take a long view and not clutter the feed, or more recently, Stories, with a lot of ads. Now Zuckerberg was ordering an increase. It was almost as if Zuckerberg wanted to increase Instagram ads so he could lower the volume on Facebook Blue, thus making Blue more attractive.

As Instagram hit a billion users in early 2018, it seemed to those at Instagram that his nos to their requests for resources became even more frequent. Zuckerberg instructed Growth leader Javier Olivan to list all the product benefits Facebook provided to Instagram, basically so he could trim them back.

One conflict came from Systrom's plans for its internal messaging service, Instagram Direct. Systrom and Krieger wanted to build it out to a separate app, like Facebook had done with Messenger. It would be a competitive answer to Snap. As with Snap's service, the messages would disappear a day after the recipient viewed them. Since none of Facebook's messaging services had captivated the youth market as much as Snap did, it might have been Facebook's best chance to win that valuable demographic.

At one point, this would have been the kind of development that Systrom and Krieger would have mentioned to Zuckerberg as a fait accompli. But once Zuckerberg began ruminating on a vision for a unified Facebook, those days of independence were over. For its 2018 budget, Instagram allocated a certain number of hires to further develop Instagram Direct. Zuckerberg rejected the request. It was part of a pattern. Though Instagram was the fastest-growing property at Facebook in 2017, Zuckerberg cut back on its hiring in the 2018 budget.

Still, Instagram began testing the new app in several countries. It was received well enough to extend the test. But Zuckerberg put a halt to it, saying he wanted to gauge its impact on other properties. Then he formally issued an order to stop the progress. Months later, Facebook would announce that in the future, all messaging on Instagram would be handled by the Messenger team.

Another benefit from Facebook to Instagram on Olivan's list was cross-promotion. When a user shared an Instagram photo on the News Feed there would be a notation of that, a little way to expose Instagram to Facebook users.

Gone.

Even more serious for Instagram, Zuckerberg was reconsidering cutting back on Instagram's use of the Facebook friend graph. This was one of Instagram's most valuable growth tools: when new users signed up they could instantly connect with all their Facebook contacts, making the service valuable from that moment on. Instagram's leaders could live with more ads and less promotion, but not losing the friend graph. Zuckerberg had promised the Instagram team he would never do that. A few months later, Facebook decreased Instagram's access to the graph. Soon after that, Zuckerberg began experimenting with a total cutoff.

The pattern was unmistakable: Zuckerberg was changing Instagram's direction from something that could be bigger than Facebook Blue to Blue's giant satellite. Those close to Systrom noticed his frustration, as he endured what seemed like constant micro-humiliations. Despite his senior service, Systrom was not part of Facebook's de facto ruling cabal. He was not usually included in the emergency crisis meetings dealing with things like the Russian invasion or Cambridge Analytica. Zuckerberg hadn't even made a visit to Instagram's headquarters. Ever.

The last big Instagram event with Kevin Systrom presiding was the launch of a service called Instagram TV, or IGTV. The idea was to leverage Instagram's popularity with celebrities and influencers to take on YouTube, which was the go-to platform for people whose talents largely lay in brilliantly sharing—or faking—authenticity so that millions of people would connect with them. But first Instagram had to convince a skeptical Zuckerberg that the product would not draw eyeballs from the video efforts on the Blue app. Under the leadership of Fidji Simo, Facebook was spending billions of dollars on a service called Facebook Watch, even producing its own programs. Zuckerberg finally let the project go forward, but Instagram was directed to agree that the IGTV videos would be posted to Facebook by default before it could launch the product.

The launch, which was to take place in Menlo Park with a live connection to journalists and influencers gathered in Facebook's East Village New York City office, was a disaster. Instagram had contracted with a top-end events organizer, which produced an elaborate set with a revolving stage. It didn't work, and the presentation bombed. By the time a newly improvised presentation was ready, many of the journalists were gone.

And then Systrom left for his paternity leave.

. . .

BY MAY 2018, when Zuckerberg reorganized his executive team, it was universally understood that the alpha dog among them was Chris Cox, who until then was the head of products. He had been the voice welcoming thousands of Facebookers as they joined the company, and people would regularly tell him how they still drew inspiration from his speech. If Zuckerberg had been wiped out during his drive in a NASCAR vehicle on his national tour, the smart money for his successor would have been Cox.

So when Zuckerberg rearranged the musical chairs in the executive suite—promoting Javier Olivan, giving Growth to Alex Schultz, extending Schrep's domain—he could trust Cox in a new role as the person in charge of what was becoming known as the Family, the collection of apps that, in total, were bigger than Facebook itself. In its financial results, Facebook had changed the way it reported its Monthly Average Users to a total of all people using at least one of the services. This had the advantage of demonstrating momentum despite the mediocre growth of the Blue app. The total number was on a trajectory to reach a mind-blowing 3 billion by 2020.

Not long after he moved into the role, Cox tried to explain it to me. "I'm going to be focused on making sure we can keep these unique cultures and values of these different products, but build a really strong, solid infrastructure across [the apps]." The key part of that, he said, was making sure that the safety and security measures that Facebook was developing for the Blue app were built into the messaging apps that Facebook owned.

But that didn't match what Zuckerberg had been thinking. Maybe at one point it would have been appropriate for the cultures of the individual companies to thrive. But now, it was time for those properties—because that's what they were, properties—to be recognized as cogs in the Facebook machine. For a while, this was something Zuckerberg communicated internally, at first obliquely. He would be making moves that, if you took a step back, you could figure out—he was drawing those franchises closer. Like getting rid of each service's name in its employees' email addresses: no more instagram.com, or whatsapp.com or even oculus .com. Everybody would use the mother ship domain of fb.com. Even the names of the services themselves would be adjusted. No more just plain Instagram. It would be Instagram by Facebook. (At least he wasn't exhuming "A Mark Zuckerberg Production.")

Cox was in the difficult position of middleman for many of the disagreements between founders and Zuckerberg. Of particular concern was the dissatisfaction of Systrom and Krieger. It was clear they were unhappy, but Zuckerberg continued on his path.

So it could not have been a surprise to Zuckerberg when Systrom and Krieger quit. They broke the news to Cox, their immediate boss. There was no need to physically meet with Zuckerberg to give him a chance to change their minds.

Adam Mosseri, who had become Instagram's chief operating officer earlier that year, would now head the franchise. He had also been in the birthday picture.

After their departure I asked Zuckerberg directly about what I had been hearing from the Instagram team: Were you in any way jealous of Instagram?

"Jealous . . ." he repeated.

Yes, I said. And that you would prefer growth of Facebook's Blue app to Instagram's?

He said no and explained to me how he thought about it. Early on, Facebook was the main product and Facebook, Instagram, and Messenger were just starting. It made sense to leave the founders alone and let them build their best products. "That was incredibly successful," he says. "And it made sense for the first five years. But now we're at a point where all these products are big and important. I don't want to just build multiple versions of the same product. We should have a more coherent and integrated company strategy."

And if that meant losing the founders, so be it. "I can understand if you're an entrepreneur who built one of those things and had awesome success, you'd wake up and say, 'Okay, I'm proud of what I did but this isn't for me going forward.' That's how I see it, and we're going in the right direction."

Those close to Kevin Systrom, though, believe that had Zuckerberg not asserted control, he would have remained at Instagram for twenty more years.

Systrom and Krieger did not trash-talk their employer on the way out. No #deletefacebook hashtags. Their farewell post was gracious. Normally, a departure of this significance would have generated a blizzard of questions at the weekly all-hands. But that was also the week that Joel Kaplan thumbed his nose at his liberal colleagues and showed public allegiance to Brett Kavanaugh. Also that week was the discovery of the security breach that exposed the personal information of 50 million Facebook users, the biggest information-security disaster in the company's history. The exit of Instagram's founders was downranked to outrage number three that week.

Systrom said nothing publicly until he appeared at a *Wired* conference in November. He revealed that he'd just gotten his flying license and was excited about that. He was spending time with his infant daughter. He didn't want to lay out the details of his departure. But he didn't pretend that it was a happy split. "You don't leave a job because everything's awesome," he said.

WHEN ZUCKERBERG MADE his announcement about a new Facebook, where all the franchises would be integrated into one giant infrastructure, it seemed like a great opportunity for Cox, whose role would be to quarterback the integration. But Cox had no appetite for the job. He disagreed with the whole Privacy-Focused Vision. In particular, he had concerns about Zuckerberg's insistence that products would be protected by strong encryption. In part, Zuckerberg was doing this as a reaction to his own experience: if some of his early communications had been encrypted—or vanished in the way Stories went away—his early IMs and emails would never have been exposed. And, of course, making privacy a centerpiece of the Next Facebook was a firm answer to critics charging Facebook with being an Orwellian snoop.

Cox saw the other side. Besides posing a technical challenge, encrypting the contents of all the messaging services so that even Facebook couldn't read the posts would hamstring the company's efforts to fight hate speech and misinformation.

A week after the Instagram duo left, Cox quit Facebook. He still loved the company, and had been giving his usual welcome to new employees even as he prepared his resignation. He simply disagreed with the strategy.

"As Mark has outlined, we are turning a new page in our product direction, focused on an encrypted, interoperable messaging network," he wrote in a News Feed post, with a photo of him and Zuckerberg, both with big smiles, the best of friends. "This will be a big project and we will need leaders who are excited to see the new direction through."

Meaning . . . *not me.* Cox was thirty-six years old and had spent thirteen of those years at Facebook. He was not going to spend the next two years working on an integration he did not believe in.

"For years, the company has been so rotated towards these town-square products that when you say, now we're going to lead with the living room, it creates conflict," Zuckerberg tells me. "Some of the best people at the company looked at this and said, 'I'm not here for this.' This is a deep cultural evolution, and I don't have the answer or even understand the complexities of how this will play out. But it's going to be a multi-year thing." At least now with members of his inner circle installed in the services he wanted to integrate into Facebook, he could work through the process of melding operations without the impediment of possessive founders.

But a more immediate obstacle had arisen. Critics and regulators were questioning just why Zuckerberg had been permitted to collect those properties in the first place. They invoked a word destined to grow dramatically in Facebook's word cloud over the next few years: antitrust.

When Senator Lindsey Graham had asked Zuckerberg in 2018 who his competitors were, the Facebook CEO stumbled a bit before saying there were eight large social apps. Zuckerberg didn't mention that he owned four of them.

For months, anti-Facebook activists had been urging the FTC, the Department of Justice, and state attorneys general to take antitrust action against Facebook. Some of those opponents surmised that Zuckerberg's integration strategy was a way to bind the properties so closely together in a technical sense that Facebook could thwart an order to sell one or more by credibly claiming that it was impossible to sever them. Tim Wu, a Columbia University law professor and antitrust expert who had once consulted with the FTC, had joined with Scott Hemphill of the NYU School of Law to create a thirty-nine-slide PowerPoint presentation charging that Facebook purchasing nascent competitors "in order to maintain its dominance in the provision of social media services" violated the Sherman Act and the Clayton Act, both laws against unfair competition. They took the presentation on the road, showing it to federal and state agencies and prosecutors.

Wu and Hemphill's effort got a boost in May 2019 when a third partner joined them: Chris Hughes. As one of the original co-founders of Facebook, his was the sharpest *et tu?* of all the apostates to date. Hughes, who had headed a small social-justice nonprofit, now regretted his role in building a company he had come to consider unhealthy for the world. (Though his remorse apparently did not go as far as returning any of the $500 million he had gotten in Facebook stock.) *The New York Times* published his long op-ed, entitled "It's Time to Break Up Facebook." While affirming that Zuckerberg was "a good, kind person," he shared inside stories that made his former classmate look like social media's Al Capone. Hughes begged legislators and regulators to strip Instagram and WhatsApp from Facebook. For good measure, the *Times* created a five-minute mini-documentary on Hughes and devoted its lead editorial to a call for antitrust action.

Hughes's pleas were hardly necessary. By mid-2019, Congress, state, and federal agencies were actively pursuing antitrust investigations against Facebook, as well as Apple and Amazon. But Facebook seemed the biggest target. By October, forty-six states and the District of Columbia had joined the investigations, while the FTC and DOJ were both gearing up for their own inquiries. The House of Representatives issued a sweeping subpoena asking basically for every piece of documentation, including personal emails, relating to Facebook's pursuit of its start-up targets. Meanwhile, presidential candidates were inveighing against the company. Elizabeth Warren, one of the front-runners in the Democratic field, had a plan to split Instagram and WhatsApp from Facebook.

Facebook's defense involved warning that if it were weakened, giant tech companies from China would rush in and fill the gap. None of its detractors seemed to buy that. And the clouds of regulation continued to loom.

Mark Zuckerberg was never one to pull back when people were challenging him. If he had, Facebook might have never connected almost half of humanity. Publicly, he would say that breaking up Facebook was a bad idea. A Balkanized Facebook would be less able to police its content, he'd say, and once again evoke the threat of Chinese companies invading the social graph. Privately, he was channeling Cato the Elder again. At an all-hands meeting in 2019—an employee leaked the transcript, a sign of eroded loyalty within the company—Zuckerberg said that if Warren won the White House and persisted in her views, Facebook would "go to the mat" to keep its properties.

But even more important than playing defense was moving forward. Facebook needed new initiatives so it could own the future as much as it owned the present. So in mid-2019, in the midst of fighting for his company's life, Mark Zuckerberg was about to unveil what might have been the company's most audacious project since the News Feed itself.

It was going to create "the Internet of money."

Facebook had been trying for more than a decade to build commerce into its products, dating back to the Facebook currency struggles it had with Zynga. And now that Zuckerberg was binding all his franchises together, he envisioned businesses accessing multiple Zuckerberg services to enhance their commercial activities. But managing payments among those services, particularly in the developing world where people had no bank accounts or credit cards, was problematic.

The solution—and an opportunity to grab a space in the net paradigm shift—came in an email to Zuckerberg from one of his favorite executives, David Marcus, who headed the Messenger team. On his 2017 Christmas break, Marcus had been on a family vacation in the Dominican Republic. He was musing about cryptocurrencies, which was not such a stretch for a former PayPal executive. A technology called blockchain had the potential to lock down the security of digital currencies, but so far the electronic currencies in circulation were more objects of speculation than exchange. Marcus felt that Facebook could change this.

What if Facebook created a global digital currency? Marcus had ideas on how it could do this and dashed them off to Zuckerberg.

With Zuckerberg's passion for cryptography in full gorge, he welcomed the idea. It would especially be helpful after the merger of Facebook's properties. Now they would be freed of the difficulties of navigating the hundreds of different national currencies in the world. By creating a universally accepted global coin, Facebook could monetize everything it owned, everywhere.

Marcus quickly turned Messenger over to his lieutenant and began building a team. Two of his top engineers were refugees from Instagram. The team grew over the next year. The engineers tackled some of the tricky problems of scaling a digital currency to handle millions of transactions. And the policy team worked on the values and messaging of this ambitious plan, which would be unveiled in a white paper that summer. Sounding very much like the marketing promise of Internet.org, the white paper presented the project as directed at underserved people in poorer regions of the globe. While Internet.org tried to provide broadband to these people, this new mission promised to provide economic power to the 1.7 billion adults in the world who have no access to a bank.

Facebook named the currency Libra. The word refers to three things—an ancient Roman unit of measure, the astrological sign whose symbol is the scale of justice, and a phonetic resemblance to the French "*libre*," meaning "free." "Money, justice, and freedom." A "Libra" would be worth around a dollar, or a euro.

Libra had a convoluted plan for administration, mainly to address the skepticism people would bring to a global currency established by a company that was now among the least trustworthy on the planet. The company would turn over administration of this currency to an outside body, the Libra Association. It would have one hundred partners—each one a "node" on the blockchain, able to make direct transactions. Facebook would be only one of those nodes, with but a single vote. An outside director would run the association. Facebook would also make the Libra code available via open-source software. No secrets.

Ceding control actually would make Libra *more* valuable to Facebook, because not being owned by Facebook would make Libra attractive to those suspicious of Facebook, which was just about everyone.

Of course, Facebook would have the unique status of having created the Libra technology. And before the Libra Association even met, formalized its charter (Facebook had helpfully created a draft), or hired a director, Facebook had invented the first implementation of the currency, which it called Calibra. When the company announced the Libra project, it also revealed screen shots of the Calibra "wallet," which was stuffed with a currency that did not exist yet.

Facebook announced the plans in July, including twenty-seven partners who raised their corporate hand to join the nascent Libra Association. Impressively, it included payment giants Visa, Mastercard, and PayPal. Notably missing were other tech giants, who might have removed the stigma of a plan so closely associated with one company.

Still, Facebook's Libra project was genuinely worth discussion, as the company had tackled hard problems and come up with some innovative approaches to the serious issue of the rising trend of cryptocurrencies. But before those issues could be seriously discussed, one had to deal with the giant woolly mammoth in the room. This was Facebook doing this. Facebook!

When Marcus first revealed the Libra plan to me in May 2019—I was the first journalist briefed on the plan—he acknowledged the challenge of "trying . . . to create a public utility from a place of a great deal of skepticism because it's coming from Facebook." But there was no sense of how extreme the reaction would be after its unveiling. Regulators, legislators, and Facebook's many, many critics pounced on the idea, suggesting that instead of "Libra" the unit of currency should be called "Zuck Bucks."

Zuckerberg was unfazed. As always, he had been the one in the room to approve it. Perhaps as it was with the News Feed, so would it be with Libra. Once people tried it, they would love it. He clearly felt that way as well about his Privacy-Focused Vision for Social Networks, unpopular as it was in his own company, so much so that his most valued employee had left in order to avoid it.

In July 2019, Marcus testified before a skeptical Senate Banking Committee. His testimony failed to change minds. Over the next few weeks, several partners, including Visa, Mastercard, and PayPal, pulled out of the Libra Association. Trying to stop the tides, Facebook promised that it would not implement its plans without regulatory approval. On October 23, 2019, Zuckerberg himself came to Washington to answer questions on Libra before the House Financial Services Committee. Only a week before, he had been in the area, sharing his views on free speech in a lecture at Georgetown University; he was attempting to defend Facebook's recently announced policy of not fact-checking political ads. Facebook's position was that would do nothing to restrain outright lies in sponsored posts circulating on Facebook. For a company that had been devoting considerable energy to purging or minimizing toxic content on its platform, this was an odd stance.

The hearing began with a sour note for Zuckerberg. In her opening remarks the chairwoman, Representative Maxine Waters, who had asked for a moratorium on Libra, said that Zuckerberg's proposal was so egregious that it had opened the door in her mind to breaking up Facebook. "It appears you are aggressively increasing the size of your company and are willing to step on or over anyone, including your competitors, women, people of color, your own users, and even our democracy to get what you want," she told him.

A few legislators—mainly Republicans, apparently sated by the company's consistent attention to their bogus complaints about bias—seemed to defend Zuckerberg, expressing concerns about stifling innovation. But mostly Zuckerberg, only three years ago the poster child for American ingenuity, was greeted with hostility. A recurrent theme was the company's corporate rap sheet. For instance, Representative Nydia Velázquez of New York invoked Cambridge Analytica and Facebook's broken promise not to merge WhatsApp data with its other databases. Zuckerberg, who appeared for much of the six-hour session as in need of concussion protocol, said it was an important question. "We certainly have work to do to build trust," he admitted.

"Have you learned you should not lie?" asked Velázquez.

Low moments abounded. "You have ruined the lives of Americans," said Representative Joyce Beatty, taking him to task for Facebook's civil rights disappointments. Representative Alexandria Ocasio-Cortez pummeled him on his political ad stance.

Four hours into the session, Zuckerberg asked for a pause so he could go to the bathroom, waving his water bottle to indicate his distress. The chairwoman wanted one more round of questioning before a vote and dictated that he should submit to another questioner before he could relieve himself. That next interlocutor, Representative Katie Porter, started by making fun of his haircut and ended with a request that he commit to spending one day a week as a content moderator.

After the hearing, the chairwoman had a private conversation with Zuckerberg and then took questions. I asked her if she had heard anything to make her better disposed toward Libra. She said no.

The basic problem with Libra was this: Facebook, with its exquisitely talented engineers, its unparalleled drive, and its canny sense of product, may well have come up with the best implementation of digital money, topping dozens of less proficient previous stabs. But ultimately the quality was secondary to who made it. The determination would be made by assessing the legacy of Mark Elliot Zuckerberg, the man who set out to connect a world that was perhaps not ready to be connected, and did it anyway.

After Beacon, after Cambridge Analytica, after News Feed–fueled violence in multiple countries, after fines for civil rights violations, privacy misrepresentations, and security breaches from the FTC, SEC, EU, and UK Parliament . . . after *everything*, people wanted to know:

Why would anyone trust Facebook with their money?

Epilogue

THE FINAL TASK on Mark Zuckerberg's calendar before a July 4 vacation trip is our last interview for this book. As I walk the path through the well-landscaped front yard of the Zuckerberg house in old Palo Alto—a pleasing, century-old Craftsman-style house to be sure, but not the imposing mansion that one of the world's richest people might occupy—I am thinking of Andrew Weinreich.

Weinreich, you'll recall, was the lawyer and entrepreneur who first concocted the idea of an online social network as we know it today, and envisioned that the entire world might be bound in a single network. I wonder what it would have been like if Weinreich, and not Zuckerberg, had fulfilled that vision. Certainly my talk with the founder of the defunct sixdegrees would have been conducted in a vast headquarters building. As it was, I'd met with Weinreich in a small conference room he'd reserved in a noisy WeWork office. As he described how sixdegrees was ahead of its time, through glass walls we could see a bustling pageant of Millennials and Gen Zs striving to be the next Mark Zuckerberg.

Weinreich, fit at fifty, had a quick answer for my question as to whether he's haunted by what someone else accomplished with his idea: No. He also had no hesitation when I ask him, in light of Facebook's experiences, whether he still thought connecting the world is a good thing. He does. If he had been in Zuckerberg's place, though, he is sure he would have been quicker to pick up the warning signs when things were going wrong.

Mark Zuckerberg would undoubtedly tell Weinreich it's not so easy. He greets me with his sheepdog, Beast, whose exuberance leads Zuckerberg to consign the animal to a windowed parlor.

We've been talking for this book for three years at this point, and by now our interviews are conducted with as much candor as I think he's ever going to exercise. He has come a long way from the reluctant communicator I first met in 2006: He now sees interviews as not only opportunities to make his points but also a means to learn more about how he is being perceived. In this meeting, and in a similarly frank interview we'd had a few weeks earlier, he's willing to take on the question of culpability, dancing between contrition and defiance. (Since those two sessions were like one extended interview, I'm including comments from both here.)

Yes, he does take responsibility for Facebook's inattentiveness as the platform became a host of fake news, misinformation, and hate. But his mea culpas are tempered by a refrain that the emergence of those problems, and Facebook's inattention as they proliferated, were results of over-optimism rather than complacency or greed.

"The big lesson from the last few years is we were too idealistic and optimistic about the ways that people would use technology for good and didn't think enough about the ways that people would abuse it," he says.

And he acknowledges that his decision to basically delegate key parts of the company to others compounded the problem.

"Maybe it just takes someone far better than me to do this," he says. "Starting at nineteen, having not had the life experiences in all these areas, I think it would be, at least for me, impossible to internalize all of the different parts of what running a company could be. And Sheryl was so good at doing this stuff that maybe it was easier to turn that over." By necessity, he's more dialed-in now. "Part of my personal journey over the last fifteen years has been taking more responsibility for each of those pieces."

There is merit to the point that a bunch of college students would be unprepared for the unprecedented consequences of opening up an unfettered global platform for speech and commerce. Who could have anticipated what would come of connecting people on such a massive scale? How could someone be faulted for pursuing the idealistic goal of binding humanity?

But the naïvety/idealism defense only goes so far. Didn't Facebook ignore the problems that arose from its relentless pursuit of growth? There is also the fact that the company's business model led it to become a supremely polished machine, running on the high-octane fuel of an unmatched trove of personal data. And while Facebook did start in a dorm room, within a year Zuckerberg was getting advice from Silicon Valley's most experienced entrepreneurs and investors, as well as respected CEOs like Don Graham. Furthermore, its problems emerged while one of the best executives in the industry was its chief operating officer.

I do think that Zuckerberg is sincere when he says he has continued to believe in the values of sharing and free speech he embraced a decade earlier. But the decisions he has made over the course of the past fifteen years have reflected a secondary set of goals—growth, competitive supremacy, and seeking massive profits. Because executing these sub-goals would help Facebook in its quest to connect the world, they became hopelessly intermingled with the mission itself, often leading Zuckerberg to make decisions that, in isolation, seemed anything but idealistic.

When I share these thoughts, he pushes back.

"I think you can either look at [our problems] as a result of idealism or you can look at it and say that it was a result of cynicism," he says. "And I think people who know me think that it's not [cynicism]. I've never been running the company saying, *Well I'm just trying to optimize this to make as much money as possible so let's run ahead.* I just think we did not pay enough attention to the way that people would abuse things, because we were too idealistic about how the technology could enable a lot of good."

That *idealism* thing again.

"It's true that the people who know you don't say you are a cynic," I say to him. "But they do say you are hypercompetitive."

He gives me one of his trademark zone-outs for a few seconds. "I think that's right, yeah," he finally replies.

Zuckerberg has other concessions. When I suggest that the Twitterization of the News Feed opened the door to unintended consequences such as misinformation and dopamine distractions, he agrees. "In retrospect, we shouldn't have pushed so far," he says, noting that as part of Facebook's recent cleansing it has reversed course to some degree. "But we learned important things," he says.

He has a similar upbeat view on privacy. As we spoke Facebook was just weeks away from the FTC settlement that would cost it $5 billion and force a set of oversight rules on the company. But he feels that despite mistakes, Facebook's current role as a villain is overstated. "If you ask people what their perception of Facebook and privacy is, it's generally true today that we don't have a good reputation," he says. "People think that we've eroded [privacy] or contributed to eroding it. I would actually argue we have done privacy innovations, which have given people new types of private or semi-private spaces in which they can come together and express themselves."

I ask him about growth, suggesting that he had empowered a team whose mission was solely to add and retain users, subsuming Facebook's putative mission to connect the world and make it a better place. Didn't this make growth in itself the North Star of the company? "I agree with a bunch of what you're saying but not all of it," he says. "I think you can look at this from a cynical perspective that we were trying to grow because growth went in its own direction. [But] the reason people use social products is to interact with other people. The most valuable thing we could do for people was make sure that the people they cared about were on the service."

Maybe it was a mistake to distance himself from the policy issues that would cause Facebook so much trouble. Maybe he pushed the Twitterization of the News Feed too far. Maybe he was so busy growing Facebook that he was late to realize the importance of monitoring content. But a worse sin, he believes, would have been timidity.

"I think a lot of people would be more conservative and say, *Okay, this is what I believe should happen but I'm not going to mess with it because I'm too afraid of breaking something. I am more afraid of not doing the best thing we can than I am of breaking the thing that we currently have.* I just think I take more chances and that means I get more things wrong. So in retrospect, we have certainly made a bunch of mistakes in strategy, in execution. If you're not making mistakes, you're probably not living up to your potential, right? That's how you grow."

He says this even while conceding that some of those mistakes have had terrible consequences. "Some of the bad stuff is very bad and people are understandably very upset about it—if you have nations trying to interfere in elections, if you have the Burmese military trying to spread hate to aid their genocide, how can this be a positive thing? But just as in the previous industrial revolution or other changes in society that were very disruptive, it's difficult to internalize that, as painful as some of these things are, the positive over the long term can still dramatically outweigh the negative. If you can handle the negative as well as you can.

"Through this whole thing I haven't lost faith in that. I believe we are one part of the Internet that's part of a broader arc of history. But we do definitely have a responsibility to make sure we address these negative uses that we probably didn't focus on enough until recently."

Fishing for rosebuds is a futile pursuit with Mark Zuckerberg. He is who he is. Facebook may have to change, but Zuckerberg doesn't believe he has to.

"I couldn't run this company and not do things that I thought were going to help push the world forward," says the man who some think has done as much destruction to that world as anyone in the business realm. And as he looks me in the eye, it's clear he believes it.

IT'S TIME TO leave. Zuckerberg frees Beast from the sunroom. He walks me to the door. Just as I go, standing on the top of the steps outside his house, Zuckerberg mentions his notebook. Early in the interview, I'd told him I had pages from the Book of Change he wrote in 2006, and he'd told me he wished he still had it. It would be cool to see it, he now says. I happen to have a scan of it on my phone, and I open the file and hand it to him.

Zuckerberg gazes at the cover page—with his name and address and the promise of a $1000 reward to anyone locating it—and his face lights up. "Yes, that's my handwriting!" he confirms.

I realize that I am, in a sense, doing what twenty-two-year-old Mark Zuckerberg asked in his offer of a reward: restoring a precious treasure that seemed irretrievably lost.

As he swipes the pages, a rhapsodic smile spreads across his face. It's as if he has been suddenly reunited with his earlier self: the fresh-faced founder, unacquainted with regulators, haters, and bodyguards, blissfully parsing his visions to a team that would alchemize them into software, and then change the world in the very best way.

He seems almost reluctant to break the trance and hand me back the phone, but of course he does, turning back to his house and closing the door.

Acknowledgments

A company takes a leap of faith when it grants access to a journalist, and I appreciate Facebook's decision to grant the invaluable time and energy of their employees in interviews with me. I am especially grateful that the company kept that faith after it became clear that my book would be taking a turn that neither Facebook nor I expected. Elliot Schrage and Caryn Marooney were key in granting me that access, forwarding my pitch to Mark Zuckerberg and Sheryl Sandberg. My guides to Facebook, Bertie Thomson and Derick Mains, skillfully managed to support the project while serving their employer well. My thanks also go to the rest of Facebook's communications staff, who did their best to provide me with information and interviews, including a late arrival, John Pinette. The promises of access, especially to Mark and Sheryl, were met and exceeded, and in the final months, both of those leaders made a special effort to engage me, with particularly constructive (I think!) sessions on our final meetings. I thank them both, as well as the hundreds of Facebookers, past and present, who spoke with me.

Facebook has been the subject of deep attention since its origin, and I appreciate the work that all my fellow journalists and authors have done. In particular, David Kirkpatrick's *The Facebook Effect* is an invaluable look at the company's first half decade. I owe a special thanks to Jessi Hempel, who shared some of her interviews with me, as well as her wisdom and comments. An associate computer-science professor named

Michael Zimmer has created an amazing resource called the Zuckerberg Files, an attempt at a comprehensive archive of every single interview of its namesake; I feasted on it. Casey Newton's newsletter, *The Interface*, helped me keep current with the daily torrent of Facebook news.

I spent a lot of time in California working over the past three years, and I owe a lot to those who provided me with shelter and support. Lynnea Johnson and Caroline Rose's bungalow was my bivouac for much of the time, until I imposed on John Markoff and Leslie Terizan. Katie Hafner and Bob Wachter were splendid hosts in San Francisco. Leslie Berlin generously lent me her son's car before he rightfully claimed it for his campus use. Thanks also to friends on two coasts: Bradley Horowitz, Irene Au, Brad Stone, Kevin Kelly, Megan Quinn, M. G. Siegler, and Steve and Michelle Stoneburn.

Back on the East Coast, I had two other writing homes. The Writers Room in Manhattan is a wonderful retreat. And the Otis Library in Massachusetts provided precious connectivity—and convivial hospitality—in broadband-starved western Mass. (Fiber-optic is coming, though!)

Lindsay Muscato was a meticulous and omnivorous researcher. She also did a masterful pivot to the captain of my fact-checking team, which included the talented and tireless Rosemarie Ho and Rima Parikh. (All errors are mine, of course, but many errors have been avoided due to their heroic labors.) Lu Zhao provided superb help with the notes. Transcriber par excellence Abby Royle spent many hours with the voices of Facebookers in her ear.

Serena Cho, a journalism student who looked into Zuckerberg's Exeter years, helped me with background on the school.

Thanks also to my colleagues at *Backchannel* and then *Wired*—which merged halfway through the book. Nick Thompson patiently waited out my book leave, and my editors Sandra Upson and Vera Titinuk understood the balancing act. Issie Lapowsky, who was on the Facebook beat at *Wired*, was more than generous with thoughts and connections.

My agent, Flip Brophy, as always, was as solid an advocate and adviser as any writer could hope for. Thanks also to Nell Pierce at Literistic.

Thanks to all the Dutton/Blue Rider team. John Parsley patiently waited for the manuscript and then handled it well. Cassidy Sachs skillfully managed logistics. I appreciated Rachelle Mandik's sharp copyediting eye. Alice Dalrymple skillfully captained a speedy production process. *Facebook: The Inside Story* was originally signed by David Rosenthal of Blue Rider Press—hey, David, it's done!

As always, the most thanks go to family, some of whom—sister, in-laws, niece—I keep up with via private Facebook groups unaffected by fake news and Russian misinformation campaigns. I couldn't do this without Andrew Max Levy and Teresa Carpenter's support and love.

Notes

Facebook was drawn mostly from more than three hundred interviews of company employees past and present as well as knowledgeable outsiders who had direct interaction with the people and events in the book. (Unless otherwise noted, direct quotes in the book come from those interviews.) I benefited from the work of David Kirkpatrick, who wrote the definitive history of the company's early years: *The Facebook Effect: The Inside Story of the Company That Is Connecting the World* (Simon & Schuster, 2010). An invaluable resource was the Zuckerberg Files (zuckerbergfiles.org), a comprehensive source of more than one thousand interviews and videos, administered by Michael Zimmer. And Casey Newton's daily newsletter, *The Interface*, always kept me apprised of the latest twists in the Facebook story.

INTRODUCTION

8 **"Tech's most popular CEO":** Hillary Brueck, "Facebook Boss Still Tech's Most Popular CEO," *Fortune*, February 26, 2016.

9 **people were shaken:** This characterization of the Facebook office after the 2016 US election comes from interviews with Sheryl Sandberg.

10 **"I've seen some of the stories":** David Kirkpatrick, "In Conversation with Mark Zuckerberg," Techonomy.com, November 17, 2016.

15 **undercooked goat:** Brian Hiatt, "Twitter CEO Jack Dorsey: The Rolling Stone Interview," *Rolling Stone*, January 19, 2019.

CHAPTER ONE: ZuckNet

19 **Weinreich:** Information about sixdegrees from personal interviews and Julia Angwin, *Stealing MySpace: The Battle to Control the Most Popular Website in America* (Penguin, 2009). The video of the launch event is on YouTube.

19 **six connections away:** The discussion of the "six degrees" problem is drawn from Duncan Watts's influential book, *Six Degrees: The Science of a Connected Age* (W. W. Norton, 2003).

19 **"Chain-Links":** The 1929 short story is out of print in English, but a translation of Karinthy's "Chain-Links" by Adam Makkai is available on the website https://djjr-courses.wikidot.com.

20 **published two years later:** Jeffrey Travers and Stanley Milgram, "An Experimental Study of the Small World Problem," *Sociometry* 32, no. 4 (December 1969), 425–443.

22 **the pending patent:** Teresa Riordan, "Idea for Online Networking Brings Two Entrepreneurs Together," *New York Times*, December 1, 2003.

22 **Mark Elliot Zuckerberg was born:** There are a number of good accounts of Zuckerberg's early life that I drew from besides personal interviews, including his parents. Particularly valuable was Matthew Shaer, "The Zuckerbergs of Dobbs Ferry," *New York*, May 4, 2012. In 2011, Ed Zuckerberg spoke in detail in a radio interview with a town supervisor on local station WVOX, reported in an Associated Press story: Beth J. Harpaz, "Dr. Zuckerberg Talks about His Son's Upbringing," Associated Press, February 4, 2011. Other useful accounts of early Mark Zuckerberg include two *New Yorker* profiles: Jose Antonio Vargas, "The Face of Facebook," *The New Yorker,* September 13, 2010; and Evan Osnos, "Can Mark Zuckerberg Fix Facebook Before It Breaks Democracy?" *The New Yorker,* September 10, 2018. Also, Lev Grossman, *The Connector* (TIME, 2010), ebook of *Time* magazine's 2010 Person of the Year; and Kirkpatrick, *The Facebook Effect.*

23 **he did not object:** Shaer, "The Zuckerbergs of Dobbs Ferry."

24 **"My wife was a superwoman":** Ed Zuckerberg, WVOX radio interview.

24 **"Good Jewish mother":** Mark Zuckerberg at Y Combinator Startup School, 2011, *Zuckerberg Transcripts*, 76.

24 **When a magazine writer visited:** Shaer, "The Zuckerbergs of Dobbs Ferry."

25 **"If you were going to say *no*":** Ibid.

25 **"strong-willed and relentless":** Lev Grossman, *The Connector,* 98.

26 **"I'd go to school":** Bill Moggridge, "Designing Media: Mark Zuckerberg Interview" (MIT Press, 2010), *Zuckerberg Videos,* Video 36.

26 **"it reached a point":** Interview with James Breyer at Stanford University, October 26, 2005, *Zuckerberg Transcripts*, 116.

26 **Teachers would later:** Matt Bultman, "Facebook IPO to Make Dobbs Ferry's Mark Zuckerberg a $24 Billion Man," *Greenburgh Daily Voice*, March 12, 2012.

27 **"princely":** Jessica Vascellaro, "Facebook CEO in No Rush to 'Friend' Wall Street," *Wall Street Journal*, March 3, 2010.

27 **Zuckerberg's digital version:** Michael M. Grynbaum, "Mark E. Zuckerberg '06: The Whiz Behind Thefacebook.com," *Harvard Crimson*, June 10, 2004.

27 **his creations were terrible:** Mark Zuckerberg, Menlo Park Town Hall, May 14, 2015, Accessed via Facebook Watch.

27 **"Everything was tech":** Randi Zuckerberg made her comments on *The Human Code with Laurie Segall* podcast, February 2, 2018.

28 **more than his peers:** *Masters of Scale* podcast, September 2018.

31 **Harkness method:** Phillips Exeter Academy explained on one of its web pages "The Exeter Difference."

31 **Alex Demas later told:** "A Greek Schoolmate Uncovers Zuckerberg's Face(book) and Its Roots," Greek Reporter, May 14, 2009.

31 **"affable and game":** Petrain shared his Zuckerberg recollections with me via email.

31 **Petrain would later write:** Petrain shared his Zuckerberg recollections with me via email.

32 **"knows no boundaries":** Vargas, "The Face of Facebook."

32 **Marty Gottesfeld:** David Kushner, "The Hacker Who Cared Too Much," *Rolling Stone*, June 29, 2017.

33 **collected his dollar:** Todd Perry, "SharkInjury 1.32," *Medium* posting, April 4, 2017.

33 **push-ups:** The story is recounted by Todd Perry in Alexandra Wolfe, *Valley of the Gods* (Simon & Schuster, 2017), 109–10.

33 ***There's really no reason:*** Grynbaum, "Mark E. Zuckerberg '06: The Whiz Behind Thefacebook.com."

34 **Photo Address Book:** Screenshots of the book, as well as Tillery's online version, are included in Steffan Antonas, "Did Mark Zuckerberg's Inspiration for Facebook Come Before Harvard?" *ReadWrite*, May 10, 2009.

CHAPTER TWO: Ad-Boarded

Zuckerberg's Harvard years are voluminously documented, though often different agendas are at play. Besides personal interviews, some of the more consistently useful sources were *The Facebook Effect*, the published excerpts from depositions in the ConnectU case, the excellent coverage in the *Harvard Crimson*, and many comments from Zuckerberg himself in various interviews in the Zuckerberg Files.

37 **"I am not even sure":** *The Human Code with Laurie Segall* podcast, February 4, 2019.

38 **His formal acceptance:** Zuckerberg shared the video on Facebook, May 18, 2017.

39 **Paul Ceglia:** The *Ceglia v. Zuckerberg* court filing provides the information about the $1,000 fee. After the suit was dismissed because Ceglia allegedly forged the document that was his claim to owning Facebook, Ceglia fled to Ecuador to avoid prosecution. As of June 2019, the United States has been unable to extradite him. Bob Van Voris, "Facebook Fugitive Paul Ceglia's Three Years on the Run," *Bloomberg*, November 10, 2018; and David Cohen, "Ecuador Won't Return Fugitive and Former Facebook Claimant Paul Ceglia to the U.S.," *Adweek*, June 25, 2019.

40 **"interesting approach to digital music":** Dan Moore wrote on machine learning and MP3s on *Slashdot*, April 21, 2003.

41 **They both moved on:** S. F. Brickman, "Not-So-Artificial Intelligence," *The Harvard Crimson*, October 23, 2003.

41 **Friendster:** The best account of Friendster is the two-part series on the *Startup* podcast that ran on April 21 and 28, 2017. Seth Feigerman's "Friendster Founder Tells His Side of the Story" (*Mashable*, February 3, 2014) gives Abrams's point of view. There are also good summaries in Angwin's *Stealing MySpace* and Kirkpatrick, *The Facebook Effect*.

42 **postmortem podcast:** "Friendster 1: The Rise," *Startup*, April 21, 2017.

43 **Buddy Zoo:** "AIM Meets Social Network Theory," *Slashdot*, April 14, 2003.

45 **Chris Hughes:** In addition to personal interview, Hughes tells his own story in *Fair Shot: Rethinking Inequality and How We Learn* (St. Martin's Press, 2018).

46 **"People would just spend hours":** Interview with Sam Altman, Y Combinator, "Mark Zuckerberg: How to Build the Future," August 16, *Zuckerberg Transcripts*, 171.

47 **steam coming from the suite's bathroom:** Interview with Y Combinator, "Mark Zuckerberg at Startup School 2013," October 25, 2013, *Zuckerberg Transcripts*, 160.

47 **his first notice:** S. F. Brickman, "Not So Artificial Intelligence," *Harvard Crimson*, October 23, 2003.

47 **"a bitch":** The online journal cited here, and first published by Luke O'Brien in the online Harvard alumni journal *02138* in "Poking Facebook," would become notorious in the movie *The Social Network*. The breakup scene in the film, however is purely screenwriter Aaron Sorkin's creation.

51 **"disciplinary probation":** Luke O'Brien's notes from the court documents, shared with me.

52 **an editorial in the *Crimson*:** "Put Online a Happy Face," *Harvard Crimson*, December 11, 2003.

53 **Steamtunnels:** Nadira Hira, "Web Site's Online Facebook Raises Concerns," *Stanford Daily*, September 22, 1999.

53 **"This has been on everyone's":** David M. Kaden, "College Inches Toward Campus-Wide Facebook," *Harvard Crimson*, December 9, 2003.

54 **"I was pretty screwed":** Interview with Y Combinator, "Mark Zuckerberg at Startup School 2013," October 25, 2013, *Zuckerberg Transcripts*, 160.

CHAPTER THREE: Thefacebook

56 **ConnectU:** There is an entire subgenre of journalism (and cinema!) devoted to the dispute between Zuckerberg and his classmates. Some of the most reliable accounts came under oath in depositions unearthed first by Luke O'Brien, "Poking Facebook," *02138* magazine, November–December 2007. Ben Mezrich's book *The Accidental Billionaires: The Founding of Facebook* (Doubleday, 2009) has a number of firsthand documents. Nicholas Carlson's reporting in *Business Insider* yielded

hitherto unknown instant messaging and emails, as well as valuable reporting. Kirkpatrick's *The Facebook Effect*, as always, was solid on the issue.

57 **"It was clear to him":** Shirin Sharif, "Harvard Grads Face Off Against Thefacebook .com," *Stanford Daily*, August 5, 2004.

57 **instant messages:** Nicholas Carlson, "At Last—The Full Story of How Facebook Was Founded," *Business Insider*, March 5, 2010; and "EXCLUSIVE: Mark Zuckerberg's Secret IMs from College," *Business Insider*, May 17, 2012.

58 **Aaron Greenspan:** Besides personal interviews, Greenspan's story is drawn from his book *Authoritas: One Student's Harvard Admissions and the Founding of the Facebook Era* (Think Press, 2008); John Markoff, "Who Founded Facebook? A New Claim Emerges," *New York Times*, September 1, 2007.

60 **"unfazed":** Matt Welsh blogged, "How I Almost Killed Facebook," February 20, 2009.

61 **Harry Lewis:** Alexis C. Madrigal, "Before It Conquered the World, Facebook Conquered Harvard," *The Atlantic*, February 4, 2019.

65 **"There was nothing like that":** Interview with Y Combinator, "Mark Zuckerberg at Startup School 2013," October 25, 2013, *Zuckerberg Transcripts*, 160.

65 **come from Microsoft:** Interview with Y Combinator, "Mark Zuckerberg at Startup School 2012," October 20, 2012, *Zuckerberg Transcripts*, 161.

67 **Saverin kicked in:** Information about Eduardo Saverin is drawn from Kirkpatrick, *The Facebook Effect*; Mezrich, *The Accidental Billionaires* (Saverin cooperated with the book); and Nicholas Carlson, "How Mark Zuckerberg Booted His Co-Founder Out of the Company," *Business Insider*, May 15, 2012.

68 **Hundreds!:** Alan J. Tabak, "Hundreds Register for New Facebook Website," *Harvard Crimson*, February 9, 2004.

69 **"I didn't really pitch him":** Seth Fiegerman, "'It Was Just the Dumbest Luck'—Facebook's First Employees Look Back," *Mashable*, February 4, 2014. A rich set of interviews with early employees here.

69 **Zuckerberg informed him:** Harvard University, "CS50 Guest Lecture by Mark Zuckerberg," December 7, 2005, *Zuckerberg Transcripts*, 141.

70 **lowest chance of success:** Interview with Y Combinator, "Mark Zuckerberg at Startup School 2012," October 20, 2012, *Zuckerberg Transcripts*, 9.

70 **"Would it be better":** Interview with James Breyer at Stanford University, October 26, 2005, *Zuckerberg Transcripts*, 116.

71 **"screws us in the long-term":** Phil Johnson, "Watch Mark Zuckerberg Lecture a Computer Science Class at Harvard—in 2005," ITworld, May 13, 2015.

71 **CC Community:** Christopher Beam, "The Other Social Network," *Slate*, September 29, 2010. Also, the Columbia *Spectator* gave considerable coverage to CC Community and its clash with Thefacebook.

71 **he urged CS majors:** Zachary M. Seward, "Dropout Gates Drops in to Talk," *Harvard Crimson*, February 27, 2004.

72 **prescient think piece:** Sarah F. Milov, "Sociology of Thefacebook.com," *Harvard Crimson*, March 18, 2004.

73 **"assholes":** Adam Clark Estes, "Larry Summers Is Not a Fan of the Winklevoss Twins," *The Atlantic*, July 20, 2011.

73 **"yea, I'm going to fuck them":** Another IM from the Carlson *Business Insider* collection.

73 **letter to a Harvard dean:** Email from Zuckerberg to John Patrick Walsh, February 17, 2004.

74 **blind tip:** Nicholas Carlson, "In 2004, Mark Zuckerberg Broke into a Facebook User's Private Email Account," *Business Insider*, March 5, 2010. Additional detail from personal interviews.

74 **"very novel idea":** Claire Hoffman, "The Battle for Facebook," *Rolling Stone*, September 15, 2010.

75 **"They blame me":** This IM is from Greenspan. On September 19, 2012, he published "The Lost Chapter" on his blog aarongreenspan.com, with a cache of newly discovered IM exchanges between him and Zuckerberg.

76 **copyright status:** *Adweek* staff, "Facebook Announces Settlement of Legal Dispute with Another Former Zuckerberg Classmate," *Adweek*, May 22, 2009.

76 **"This was a cool place":** Deposition in *ConnectU Inc. v. Zuckerberg, et al.*

77 **"the whiz":** Grynbaum, "Mark E. Zuckerberg '06: The Whiz Behind Thefacebook .com."

CHAPTER FOUR: Casa Facebook

79 **Sean Parker:** The Parker background is drawn from multiple sources, including a weeklong reporting trip I spent with him in 2011. Joseph Menn gives an excellent mini-biography, and besides *The Facebook Effect*, Kirkpatrick wrote a profile of Parker for *Vanity Fair* ("With a Little Help from His Friends," November 2010). Also see Steven Bertoni, "Agent of Disruption," *Forbes*, September 21, 2011. The Facebook chapter in *Valley of Genius*, an oral history of Silicon Valley compiled by Adam Fisher, provides great firsthand quotes from Parker and other early Facebook employees.

81 **Parker rubbernecked:** Hoffman, "The Battle for Facebook."

82 **"I don't think I said five words":** Adam Fisher did a series of podcasts based on his *Valley of Genius* interviews.

83 **an attorney would ask:** Zuckerberg deposition, *The Facebook v. ConnectU*, April 26, 2006.

83 **"You just had to turn the ignition":** Ellen McGirt, "Facebook's Mark Zuckerberg: Hacker, Dropout, CEO," *Fast Company*, May 1, 2007. McGirt's article was one of the first major magazine pieces on the young company.

85 **"I remember driving":** Personal interview with Zuckerberg, June 23, 2019.

90 **"Just don't fuck it up":** Sarah Lacy, *Once You're Lucky, Twice You're Good* (Avery; reprint edition, 2009), 154. Lacy's book is another valuable account of Facebook's early days.

90 **obsessed with movies:** *The Facebook Effect* has an extensive description of how early Facebookers used movie quotes. See 97–98.

95 **"We put a bullet in that thing":** M. G. Siegler, "Wirehog, Zuckerberg's Side Project That Almost Killed Facebook," *TechCrunch*, May 26, 2010.

95 **too old when you return:** Kevin J. Feeney, "Business, Casual," *Harvard Crimson*, February 24, 2005.

96 **landlady:** Mike Swift, "Mark Zuckerberg of Facebook: Focused from the Beginning," *Mercury News*, February 5, 2012.

96 **"Everyone else was like":** Feeney, "Business, Casual."

99 **"I maintain he fucked himself":** Nicholas Carlson, "EXCLUSIVE: How Mark Zuckerberg Booted His Co-Founder Out of the Company," *Business Insider*, May 15, 2012.

99 **5 percent:** That is the widely reported figure for the settlement. Brian Solomon, "Eduardo Saverin's Net Worth Publicly Revealed: More Than $2 Billion in Facebook Alone," *Forbes*, May 18, 2012. An SEC filing on March 17, 2012, reported that as of that pre-IPO date Saverin still had 53,133,360 shares, almost 2 percent of the pre-IPO company.

99 **Singapore:** Alex Konrad, "Life After Facebook: The Untold Story of Billionaire Eduardo Saverin's Highly Networked Venture Firm," *Forbes*, March 19, 2009.

CHAPTER FIVE: Moral Dilemma

102 **$12.7 million:** These figures are reported in Kirkpatrick, *The Facebook Effect*, 125.

103 **"I like Sprite":** James Breyer/Mark Zuckerberg interview, Stanford University, October 26, 2005, *Zuckerberg Transcripts*, 116.

103 **he was weeping:** The story was first told in Kirkpatrick, *The Facebook Effect*, 122–23.

104 **That month he painted:** Karel M. Baloun, *Inside Facebook* (Trafford Publishing, 2007), 22.

104 **FORSAN:** Rolfe Humphries 1953 translation. The celebrated 1983 Robert Fitzgerald translation is almost identical, flipping the words "even" and "this."

104 **"the original version of Facebook was a mess":** Matt Welsh blogged, "In Defense of Mark Zuckerberg," October 10, 2010.

105 **the heat in Facebook's server cages:** James Glanz, "Power, Pollution and the Internet," *New York Times*, September 22, 2012.

105 **entire inventory of fans:** Ryan Mac, "Meet New Billionaire Jeff Rothschild, the Engineer Who Saved Facebook from Crashing," *Forbes*, February 28, 2014.

109 **$600 a month:** Katherine Losse, *The Boy Kings: A Journey into the Heart of the Social Network* (Free Press, 2012), 71.

109 **David Choe:** Choe spoke of his experience on *The Howard Stern Show*, February 7, 2012.

109 **"I had to sit down":** Interview with Y Combinator, "Mark Zuckerberg at Startup School 2013," October 25, 2013, *Zuckerberg Transcripts*, 160.

109 **bailed halfway through:** Soleio Cuervo recounted that incident.

110 **"Domination!":** The "domination" call would be frequently cited in accounts of Facebook's early days, but first reported by Jessica E. Vascellaro, "Facebook CEO in No Rush to 'Friend' Wall Street," *Wall Street Journal*, March 3, 2010.

112 **"facebook group whores":** Katherine M. Gray, "New Facebook Groups Abound," *Harvard Crimson*, December 3, 2004.

113 **"the salon des refusees":** Michael Lewis, "The Access Capitalists," *New Republic*, October 18, 1993.

116 **freaked people out:** Zuckerberg deposition, *The Facebook v. ConnectU*, April 25, 2006, 214.

CHAPTER SIX: The Book of Change

119 **"limiting the retention period":** Josh Constine, "Facebook Retracted Zuckerberg's Messages from Recipients' Inboxes," *TechCrunch*, April 6, 2018.

120 **upping his service to the college world:** Interview with *Huffington Post*, "Mark Zuckerberg 2005 Interview," June 1, 2005, *Zuckerberg Transcripts*, 56.

124 **Chris Cox:** An excellent recent profile of Cox is Roger Parloff, "Facebook's Chris Cox Was More Than Just the World's Most Powerful Chief Product Officer," *Yahoo Finance*, April 26, 2019.

125 **a falling tree branch:** "Daniel Plummer, Cycling Champ, Scientist, Killed by Tree Branch," *East Bay Times*, January 4, 2006.

125 **"It was almost as if":** Noah Kagan, "The Facebook Story." The remark was made in an early (2007) version of what became Kagan's book *How I Lost 170 Million Dollars: My Time As #30 at Facebook* (Lioncrest, 2014).

133 **a newly hired engineer:** Kagan, "The Facebook Story," 24. Hirsch's departure is also discussed by former Facebook CFO Gideon Yu in Nick Carlson, "Industry Shocked and Angered by Facebook CFO's Dismissal," *Business Insider*, April 1, 2009.

133 **"Eight-thirty seems as good":** Sarah Lacy, *Once You're Lucky, Twice You're Good* (Avery; reprint edition, 2009), 165.

137 **He did not forget:** This sentiment is from Mark Zuckerberg's commencement address at Harvard on May 25, 2017.

142 **"Calm Down. Breathe":** The entire post is still available on Facebook/notes, September 6, 2006.

143 **Jeff Rothschild would later comment:** Adam Fisher, *Valley of Genius* (New York: Twelve, 2018).

144 **"Facebook could be hurt":** Stutzman's quote is in Rachel Rosmarin, "Open Facebook," *Forbes*, September 11, 2006.

CHAPTER SEVEN: Platform

149 **Jobs . . . graduation speech:** The text of Steve Jobs's Stanford commencement address on June 12, 2005, appears on the Stanford News website.

150 **Microsoft Five:** Not all of the five came from Microsoft; Charlie Cheever, one of the group, had been working at another Seattle company, Amazon.

156 **"We don't own the social graph":** I interviewed Zuckerberg for my *Newsweek* cover story, "The Facebook Effect," August 7, 2007.

157 **They bought ten:** Mark Coker, "Startup Advice for Entrepreneurs from Y Combinator," VentureBeat, March 26, 2007.

157 **F8 forever changed that perception:** David Kirkpatrick wrote the definitive article on Facebook's Platform, "Facebook's Plan to Hook Up the World," *Fortune*, May 29, 2007.

158 **Hadi and Ali Partovi:** Eric Eldon, "Q&A with iLike's Ali Partovi, on Facebook," VentureBeat, May 29, 2007.

158 **"In the history of computing":** Eric Eldon, "Q&A with iLike's Ali Partovi, on Facebook," VentureBeat, May 29, 2007.

158 **40,000 users:** Kirkpatrick, *The Facebook Effect*, 225.

161 **Zynga:** The best resource on Zynga is Dean Takahashi, *Zynga: From Outcast to $9 Billion Social-Game Powerhouse* (VentureBeat, 2011).

162 **People became obsessed:** *SF Weekly staff*, "FarmVillains," *SF Weekly*, September 8, 2010.

165 **"no way to maintain the business":** Partovi deposition, *Facebook v. Six4Three* (October 10, 2017).

165 **lead-gen advertisers:** Michael Arrington, "Scamville: The Social Gaming Ecosystem of Hell," *TechCrunch*, November 1, 2009.

166 **small Berkeley gathering:** Michael Arrington, "Zynga CEO Mark Pincus: 'I Did Every Horrible Thing in the World Just to Get Revenues,'" *TechCrunch*, November 6, 2009.

170 **"causing user pain":** Email from Sam Lessin to Mark Zuckerberg, October 26, 2012. This is from the "Note by Damian Collins, MP, Chair of DCMS Committee: Summary of Key Issues from the Six4Three Files," a cache of documents under seal that Facebook turned over to the courts in a lawsuit filed by a developer called Six4Three. The UK Parliament seized the documents from the Six4Three CEO, who just happened to have them with him on a trip to London. In December 2018, Collins released a selection.

173 **an internal email:** "Exhibit 48—Mark Zuckerberg email on reciprocity and data value," November 19, 2012, "Summary of Key Issues."

174 **cache of documents:** Another set of documents from the Six4Three seizure—around 7,000 pages—was leaked to journalist Duncan Campbell, who released them in November 2019. Facebook told Reuters that the documents were "taken out of context by someone with an agenda against Facebook."

174 **Xobni:** An email from Facebook executive Ime Archibong on September 9, 2013, identifies Zuckerberg as involved in the Xobni shutoff. From Six4Three files.

174 **Amazon Gifts:** June 2013 email exchange described in Six4Three files.

174 **"I am the only one":** Ilya Sukhar chat, October 15, 2013. From Six4Three files.

175 **"the switcheroo":** Sukhar seems to have given the API pivot this name in a January 31, 2014, chat. From Six4Three files.

175 **dating app Tinder:** "Exhibit 97—discussion about giving Tinder full friends access data in return for the use of the term 'Moments' by Facebook," March 13, 2015, "Summary of Key Issues."

CHAPTER EIGHT: Pandemic

182 ***three lines of code*:** "Facebook Privacy," Electronic Privacy Information Center website. The page is a virtual timeline of the company's privacy missteps. EPIC has been tracking Facebook for more than a decade and has filed some of the most significant complaints about the company to agencies and legislators.

184 **70 percent drop:** Kirkpatrick, *The Facebook Effect*, 242.

185 **She scorned the deal:** Kara Swisher, "15 Billion More Reasons to Worry About Facebook," *AllThingsDigital*, September 25, 2007.

186 **via her Facebook News Feed:** "5 Data Breaches: From Embarrassing to Deadly," *CNN Money*, December 14, 2010.

187 **"Christmas is ruined":** Ellen Nakashima, "Feeling Betrayed, Facebook Users Force Site to Honor Their Privacy," *Washington Post*, November 30, 2007.

187 **"Facebook has turned":** Josh Quittner, "R.I.P. Facebook?" *Fortune*, December 4, 2007.

187 **Facebook finally decided:** Dan Farber, "Facebook Beacon Update: No Activities Published Without Users Proactively Consenting," ZDNet, November 9, 2007.

188 **Stefan Berteau:** Juan Carlos Perez, "Facebook's Beacon More Intrusive Than Previously Thought," *PCWorld*, November 30, 2007.

188 **Beacon even gave:** Juan Carlos Perez, "Facebook's Beacon Ad System Also Tracks Non-Facebook Users," *PCWorld*, December 3, 2007.

188 **falsely assuring:** Brad Stone, "Facebook Executive Discusses Beacon Brouhaha," *New York Times*, November 29, 2007.

188 **privacy advocates:** Jessica Guynn, "Facebook Adds Safeguards on Purchase Data," *Los Angeles Times*, November 30, 2007.

188 **"Thoughts on Beacon":** Zuckerberg's note was published on Facebook, December 5, 2007.

CHAPTER NINE: Sheryl World

190 **"*Intimacy*":** Sheryl Kara Sandberg, "Economic Factors & Intimate Violence," Harvard/Radcliffe College, March 20, 1991.

190 **Sandberg's rise:** Excellent account of Sandberg's background in Ken Auletta, "A Woman's Place," *The New Yorker*, July 4, 2011.

191 **wedding toast:** Sheryl Sandberg, *Lean In: Women, Work, and the Will to Lead* (Knopf, 2013), 20.

191 **Not all:** John Dorschner, "Sheryl Sandberg: From North Miami Beach High to Facebook's No. 2," *Miami Herald*, February 26, 2012.

191 **Florida Gators sweatshirt:** Quote from Adam J. Freed, in Brandon J. Dixon, "Leaning In from Harvard Yard to Facebook: Sheryl K. Sandberg '91," *Harvard Crimson*, May 24, 2016.

191 **"I buckled down":** Sandberg, *Lean In*, 31.

192 **"Sheryl always believed":** Auletta, "A Woman's Place."

192 **"This is a rocket ship":** Ibid.

199 **WHAT BUSINESS ARE WE IN?:** Kirkpatrick, *The Facebook Effect*, 257.

200 **limiting the exodus:** Dan Levine, "How Facebook Avoided Google's Fate in Talent Poaching Lawsuit," Reuters, March 24, 2014.

202 **The Like button:** Besides personal interviews, I found the following accounts of the Like button's origin helpful: Clive Thompson, *Coders: The Making of a New Tribe and the Remaking of the World* (Penguin, 2019); Julian Morgans, "The Inventor of the Like Button Wants You to Stop Worrying About Likes," *VICE*, July 6, 2017; Victor Luckerson, "The Rise of the Like Economy," *Ringer*, February 15, 2017; and Jared Morgenstern's TEDxWhiteCity talk, "How Many Likes = 1 Happy," November 9, 2015. Interesting to note that in various stories, the inventor of the Like button may be Morgenstern, Pearlman, or Sittig.

203 **informal history:** On October 16, 2014, on Quora, Andrew Bosworth posted an annotated timeline responding to the question, "What's the history of the Awesome Button (that eventually became the Like button) on Facebook?"

205 **analyzed the data extraction:** Arnold Roosendaal, "Facebook Tracks and Traces Everyone: Like This!" *Tilburg Law School Legal Studies Research Paper Series* No. 03/2011. Later published as Arnold Roosendaal, "We Are All Connected to Facebook . . . by Facebook!" in S. Gutwirth et al. (eds.), *European Data Protection: In Good Health?* (Springer, 2012), 3–19.

205 **Bret Taylor:** Riva Richmond, "As 'Like' Buttons Spread, So Do Facebook's Tentacles," *New York Times*, September 27, 2011.

205 **"What people don't realize":** Ibid.

CHAPTER TEN: Growth!

208 **a heroic rise:** Palihapitiya has spoken numerous times about his background and his Facebook career. Most helpful were "How We Put Facebook on the Path to 1 Billion Users" (a lecture for a Udemy course on growth hacking); and Palihapitiya's appearance on the *Recode/Decode* podcast August 31, 2017. Evelyn Rusli's *New York*

Times profile, "In Flip Flops and Jeans, an Unconventional Venture Capitalist" (October 6, 2011) is an excellent one. Speeches by others in the Growth Circle were also helpful, especially Alex Schultz's talk at the Y Combinator/Stanford Startup School course. Overviews of the Growth team include Harry McCracken, "How Facebook Used Science and Empathy to Reach Two Billion Users," *Fast Company*, June 27, 2017; and Hannah Kuchler, "How Facebook Grew Too Big to Handle," *Financial Times*, March 28, 2019. I also found useful metric information in the growth section in Mike Hoefflinger's *Becoming Facebook: The 10 Challenges That Defined the Company That's Changing the World* (Amacom, 2017).

209 **"the most insipid, idiotic use of my time":** Palihapitiya, "How We Put Facebook on the Path to 1 Billion Users."

210 **"most people":** Ibid.

213 **"Everything stopped":** The executive is Alex Schultz, who would soon join the Growth team.

214 **"That was the only priority":** Noah Kagan, *How I Lost 170 Million Dollars: My Time As #30 at Facebook* (Lioncrest, 2014), 63.

215 **"Information Platform":** Toby Segaran and Jeff Hammerbacher, *Beautiful Data: The Stories Behind Elegant Data Solutions* (O'Reilly Media, 2009). Hammerbacher's essay is called "Information Platforms and the Rise of the Data Scientist."

216 **"It was turning from a place":** PandoMonthly interview with Sarah Lacy, "A Fireside Chat with Cloudera Founder Jeff Hammerbacher," *San Francisco*, March 22, 2015.

217 **"The best minds of my generation":** Ashlee Vance, "This Tech Bubble Is Different," *Bloomberg BusinessWeek*, April 14, 2011.

221 **"It is vital":** Moira Burke, Cameron Marlow, Thomas M. Lento, "Feed Me: Motivating Newcomer Contribution in Social Network Sites," *CHI '09 Proceedings of the SIGCHI Conference on Human Factors in Computing Systems*, 945–54.

222 **Kashmir Hill:** Hill's brilliant articles on PYMK include "Facebook Figured Out My Family Secrets and Won't Tell Me How," *Gizmodo*, August 25, 2017; "Facebook Recommended This Psychiatrist's Patients Friend Each Other," *Gizmodo*, August 25, 2016; "How Facebook Outs Sex Workers," *Gizmodo*, November 10, 2017; "How Facebook Figures Out Everyone You've Ever Met," *Gizmodo*, November 7, 2017; and "People You May Know: A Controversial Facebook Feature's 10-Year History," *Gizmodo*, August 8, 2018.

222 **"We do not create":** "House Energy and Commerce Questions for the Record," June 29, 2018. This is Facebook's response to follow-up questions from Zuckerberg's 2018 testimony before the committee.

223 **Backstrom:** His PYMK talk was given at the Society for Industrial and Applied Mathematics on July 7, 2010. The slide deck is currently viewable on graph analysis.org.

223 **Dunbar number:** Robin Dunbar explains his theory in *How Many Friends Does One Person Need?* (Harvard University Press, 2010).

229 **"It did not set up":** Lisa Katayama, "Facebook Japan Takes the Model-T Approach," *Japan Times*, June 25, 2008.

229 **Facebook Japan:** Statistics are cited from the global analytics company Statcounter.

231 **Internet.org:** Besides personal interviews, key sources for Facebook's program included Jessi Hempel, "Inside Facebook's Ambitious Plan to Connect the Whole World," *Wired*, January 19, 2016; her follow-up, Jessi Hempel, "What Happened to Facebook's Grand Plan to Wire the World?," *Backchannel*, May 17, 2018; and Lev Grossman, "Mark Zuckerberg and Facebook's Plan to Wire the World," *Time*, December 15, 2014. Hempel also generously provided me with access to her interviews.

231 **"Is Connectivity a Human Right?":** The white paper, "Is Connectivity a Human Right?," was posted to Facebook on August 12, 2013.

231 **he briefed me:** "Zuckerberg Explains Facebook's Plan to Get Entire Planet Online," *Wired*, August 26, 2013.

233 **Upon his return:** Casey Newton, "Facebook Takes Flight," *The Verge*, July 21, 2016.

233 **"techno-colonialism":** Grossman, "Facebook's Plan to Wire the World."

234 **Facebook now claims:** Hempel, "What Happened to Facebook's Grand Plan."

236 **farewell memo:** The memo is reprinted in Michael Arrington, "Facebook VP Chamath Palihapitiya Forms New Venture Fund, The Social+Capital Partnership," *TechCrunch*, June 3, 2011.

CHAPTER ELEVEN: Move Fast and Break Things

238 **150,000-square-foot:** Arden Pernell, "Facebook to Move to Stanford Research Park," *Palo Alto Online*, August 18, 2008.

238 **Analog Research Lab:** Background comes from personal interviews and David Cohen, "A Look at the Analog Research Lab, the Source of All of Those Posters in Facebook's Offices," *Adweek*, February 6, 2019; "Ben Barry Used to be Called Facebook's Minister of Propaganda," *Typeroom*, June 26, 2015; Steven Heller, "The Art of Facebook," *The Atlantic*, May 16, 2013; and Fred Turner, "The Arts at Facebook: An Aesthetic Infrastructure for Surveillance Capitalism," *Poetics*, March 16, 2018.

240 **in its original sense:** I helped circulate this definition by my own book *Hackers* (Anchor Press/Doubleday, 1984).

243 **younger people were . . . smarter:** Mark Coker, "Startup Advice for Young Entrepreneurs from Y Combinator," VentureBeat, March 26, 2007.

244 **he gathered the company:** Jessica E. Vascellaro, "Facebook CEO in No Rush to 'Friend' Wall Street," *Wall Street Journal*, March 3, 2010.

244 **"Working with Mark is very challenging":** Kirkpatrick, *The Facebook Effect*, 270.

245 **Quora was doomed:** Nick O'Neil, "Facebook Officially Launches Questions, a Possible Quora Killer," *Adweek*, July 28, 2010.

247 **a freelancer in Berkeley:** Kirkpatrick, *The Facebook Effect*, 133.

249 **a critical book:** Kate Losse, *The Boy Kings*.

250 **"a concerned parent":** Brad Stone, "New Scrutiny for Facebook over Predators," *New York Times*, July 30, 2007.

251 **traced the fake accounts:** Brad Stone, author of the *Times* article, says he can't remember how he vetted the source making the allegations.

253 **"nurse-in":** Benny Evangelista and Vivian Ho, "Breastfeeding Moms Hold Facebook Nurse-In Protest," *SFGate*, February 7, 2012.

256 **"I think many people forget":** Patricia Sellers, "Mark Zuckerberg's New Challenge: Eating Only What He Kills (And Yes, We Do Mean Literally . . .)," *Fortune*, May 26, 2011.

257 **"delicious vegan goodies":** Michelle Sherrow, "Mark Zuckerberg Only Eats Those He Kills," *PETA*, May 27, 2011.

258 **drove to downtown Palo Alto:** Besides personal interviews, the Twitter and Facebook meeting was drawn from Nick Bilton, *Hatching Twitter* (Portfolio/Penguin, 2013); and Biz Stone, *Things a Little Bird Told Me: Confessions of a Creative Mind* (Grand Central, 2014).

260 **EdgeRank:** At the 2010 F8 conference, Ruchi Sanghvi and Ari Steinberg of Facebook presented an explanation of the News Feed Algorithm (Jason Kincaid, "EdgeRank: The Secret Sauce that Makes Facebook's News Feed Tick," *TechCrunch*, April 22, 2010). The presentation also helped inform an explanation of the algorithm by Jeff Widman on edgerank.net.

261 **"congratulations":** I first reported this in "Inside the Science That Reports Your Scary-Smart Facebook and Twitter Feeds," *Wired*, April 22, 2014.

263 **"Gesundheit":** Eric Sun, Itamar Rosenn, Cameron A. Marlow, and Thomas M. Lento, "Gesundheit! Modeling Contagion Through Facebook News Feed," Proceedings of the Third International ICWSM Conference (2009).

264 **Chris Cox conceded:** Ryan Singel, "Public Posting Now the Default on Facebook," *Wired*, December 9, 2009.

264 **"Your profile is just":** Facebook posted an announcement, "Welcome to Facebook, Everyone," September 26, 2006.

265 **Terms of Service:** cwalters, "Facebook's New Terms of Service: 'We Can Do Anything We Want with Your Content. Forever,'" *Consumerist*, February 15, 2009.

265 **a contentious press conference:** Rafe Needleman, "Live Blog: Facebook Press Conference on Privacy," *CNET*, February 26, 2009.

265 **discard the idea:** Donna Tam, "The Polls Close at Facebook for the Last Time," *CNET*, December 10, 2012.

267 **trapped by convention:** Bobbie Johnson, "Privacy No Longer a Social Norm, Says Facebook Founder," *Guardian*, January 10, 2010.

268 **secret user IDs:** Emily Steel and Geoffrey Fowler, "Facebook in Privacy Breach," *Wall Street Journal*, October 18, 2010. Steel followed up with, "A Web Pioneer Tracks Users by Name," October 25, 2010.

269 **RapLeaf:** Earlier articles on RapLeaf include Stephanie Olser, "At Rapleaf, Your Personals Are Public," *CNET*, August 1, 2007; and Ryan Faulkner, "Can Auren Hoffman's Reputation Get Any Worse?" *Gawker*, September 18, 2007. Auren Hoffman responded on his company blog, "Startups, Privacy and Being Wrong," September 17, 2007.

271 **"privacy hairball":** Liz Gannes, "Instant Personalization Is the Real Privacy Hairball," *GigaOm*, April 22, 2010.

271 **"All Things Digital":** Kara Swisher and Walt Mossberg, "D8: Facebook CEO Mark Zuckerberg Full-Length Video," *Wall Street Journal*, June 10, 2010.

272 **"People have really gotten comfortable":** Ian Paul, "Facebook CEO Challenges the Social Norm of Privacy," *PCWorld*, January 11, 2010.

273 **a twenty-year period of oversight:** FTC Staff, "Facebook Settles FTC Charges That It Deceived Consumers by Failing to Keep Privacy Promises," November 11, 2011.

274 **"Verified Application":** Caroline McCarthy, "App Verification Comes to Facebook's Platform," *CNET*, November 17, 2008.

CHAPTER TWELVE: Paradigm Shift

277 **press was rhapsodic:** Pete Cashmore, "STUNNING: Facebook on the iPhone," *Mashable*, August 4, 2007.

278 **Apple wasn't moving fast enough:** Joe Hewitt blog, "Innocent Until Proven Guilty," August 27, 2009.

278 **released a statement:** Christian Zibreg, "Facebook Developer: 'Apple's Review Process Needs to Be Eliminated Completely,'" Geek.com, August 27, 2009.

281 **Cory Ondrejka:** The best early account of Facebook's journey to native applications is Evelyn M. Rusli, "Even Facebook Must Change," *Wall Street Journal*, January 29, 2013.

288 **"who want to join the company":** AllFacebook, "Mark Zuckerberg, Sarah Lacy SXSW Interview," March 10, 2008, *Zuckerberg Transcripts*, 16.

288 **Its top banker:** Background of Michael Grimes from Evelyn M. Rusli, "Morgan Stanley's Grimes Is Where Money and Tech Meet," *New York Times*, May 8, 2012.

290 **offering would go forward:** Nicole Bullock and Hannah Kuchler, "Facebook Chiefs Considered Scrapping 2012 IPO," *Financial Times*, August 9, 2017.

291 **cited this incident:** Ari Levy and Douglas MacMillan, "Morgan Stanley Case Exposes Facebook to Similar Challenges," *Bloomberg*, December 19, 2012.

291 **the IPO was slated:** The best overview is Khadeeja Safdar, "Facebook One Year Later: What Really Happened in the Biggest IPO Flop Ever," *The Atlantic*, May 20, 2013.

291 **scaling back its spending:** Sharon Terlep, Suzanne Vranica, and Shayndi Raice, "GM Says Facebook Ads Don't Pay Off," *Wall Street Journal*, May 16, 2012.

292 **the widow:** Safdar, "Facebook One Year Later."

292 **"an egregious fuck-up":** Hoffman made the remark to Sarah Lacy at a Pando Fireside Chat, posted online August 12, 2012.

293 **"They get married":** Rosa Price, "$19bn and Just Married . . . I Hope Mark Zuckerberg Got a Prenup, Says Donald Trump," *Telegraph*, May 20, 2012.

294 **"jokingly threaten":** Losse, *The Boy Kings*, 51.

CHAPTER THIRTEEN: Buying the Future

299 **Kevin Systrom recalls:** The section about Instagram's early days was drawn from a number of sources. My own interviews with Systrom, Krieger, and others were augmented by interviews generously shared with me by Jessi Hempel. Key published sources were Kara Swisher, "The Money Shot," *Vanity Fair*, May 6, 2013; Somini Sengupta, Nicole Perlroth, and Jenna Wortham, "Behind Instagram's Success, Networking the Old Way," *New York Times*, April 13, 2012; and Mike Krieger, "Why Instagram Worked," *Wired*, October 20, 2014.

304 ***Cathago delenda est:*** Antonio García Martínez, *Chaos Monkeys: Obscene Fortune and Random Failure in Silicon Valley* (HarperCollins, 2016), 287–89. Martinez's account of his experience at Facebook is a trenchant look at the company's culture.

305 **He summoned Systrom:** Background on the Instagram deal from Swisher, "The Money Shot"; and Shayndi Raice, Spencer E. Ante, and Emily Glazer, "In Facebook Deal, Board Was All But Out of Picture," *Wall Street Journal*, April 18, 2012.

305 **Amin Zoufonoun:** Background on Zoufonoun in Mayar Zokaei, "Lawyer and Musician Amin Zoufonoun Closes $1 Billion Instagram Merger for Facebook," *Javanan*, March 15, 2011.

306 **a review from the FTC:** Josh Kosman, "Facebook Boasted of Buying Instagram to Kill the Competition: Sources," *New York Post*, February 26, 2019.

307 **Snapchat:** In addition to interviews, I drew on Billy Gallagher's definitive book, *How to Turn Down a Billion Dollars* (St. Martin's Press, 2018). Also valuable was Sarah Frier and Max Chafkin, "How Snapchat Built a Business by Confusing Olds," *Bloomberg BusinessWeek*, March 17, 2016; J. J. Coloa, "The Inside Story of Snapchat: World's Hottest App or a $3 Billion Disappearing Act?" *Forbes*, January 6, 2014; and Sarah Frier, "Nobody Trusts Facebook, Twitter Is a Hot Mess, What Is Snapchat Doing?" *Bloomberg BusinessWeek*, August 22, 2018.

308 **"When Snapchat started out":** Brad Stone and Sarah Frier, "Evan Spiegel Reveals Plan to Turn Snapchat into a Real Business," *Bloomberg BusinessWeek*, May 16, 2015.

308 **he emailed Spiegel:** Alyson Shontell, "How Snapchat's CEO Got Mark Zuckerberg to Fly to LA for a Private Meeting," *Business Insider*, January 6, 2014.

309 **"I hope you enjoy Poke":** Gallagher, *How to Turn Down a Billion Dollars*, 84.

316 **"We hope to play a critical role":** Ingrid Lunden, "Facebook Buys Mobile Data Analytics Company Onavo, Reportedly for Up to $200M . . . And (Finally?) Gets Its Office in Israel," *TechCrunch*, October 13, 2013.

317 **"Project Voldemort":** Georgia Wella and Deepa Seetharaman, "Snap Detailed Facebook's Aggressive Tactics in 'Project Voldemort' Dossier," *Wall Street Journal*, September 24, 2019.

317 **a messaging company called WhatsApp:** In addition to personal interviews, I drew background on WhatsApp from a number of sources. Parmy Olsen's work for *Forbes* is best on its history up to the Facebook purchase: "EXCLUSIVE: The Rags-to-Riches Tale of How Jan Koum Built WhatsApp into Facebook's New $19 Billion Baby," February 19, 2004; and "Inside the Facebook-WhatsApp Megadeal: The Courtship, the Secret Meetings, the $19 Billion Poker Game," March 4, 2014. Other valuable stories were David Rowan, "The Inside Story of Jan Koum and How Facebook Bought WhatsApp," *Wired UK*, April 2014; and Daria Lugansk, "WhatsApp Founder: Most Startup Ideas Are Completely Stupid," *RBC*, September 8, 2015. Jan Koum has shared his story on several onstage interviews, all viewable on YouTube. Ones I found valuable included his appearances at DLD in 2016 and 2014; two sessions for the Y Combinator Startup School ("How to Build a Product," April 28, 2017; and with Jim Goetz on October 14, 2014); and with Alex Fishman at Startup Grind, March 1, 2017. Also, WhatsApp business head Neeraj Arora spoke at the Indian School of Business in an interview uploaded on February 18, 2015.

321 **"When we sat down":** "Why We Don't Sell Ads," WhatsApp blog, June 18, 2012.

322 **laid out the case:** Rowan, "The Inside Story."

323 **Onavo numbers told him:** The documents later released by UK Parliament in the Six4Three case mentioned above show multiple reports where Onavo tracked WhatsApp activity.

325 **"If partnering with Facebook":** "Facebook," WhatsApp blog, February 19, 2014.

327 **Oculus:** The definitive book on Oculus is Blake Harris, *The History of the Future* (Dey Street, 2019). It is particularly valuable for its wealth of original documents and emails. All the principals in the company, as well as those from Facebook, testified in the January 2017 *Zenimax v. Facebook et al*, trial in Texas, and I drew from those transcripts, as well as personal interviews.

328 **"Have you seen Oculus?":** The emails among parties in and out of Facebook were made public via the Zenimax litigation. All are reprinted in the Harris book.

328 **"This is really cool":** Zuckerberg's Zenimax trial testimony, January 17, 2017.

329 **had ordered a pizza:** Harris, *The History of the Future*, 328.

CHAPTER FOURTEEN: Election

334 **"Their tradecraft is superb":** Dmitri Alperovitch, "Bears in the Midst: Intrusion into the National Democratic Committee," *From the Front Lines* (CrowdStrike blog), June 15, 2016. This is one of a series of posts from CrowdStrike that broke ground in publicly exposing Russian involvement in the 2016 election. Other sources in addition to personal interviews include Michael Isikoff and David Corn, *Russian Roulette: The Inside Story of Putin's War on America and the Election of Donald Trump* (Twelve, 2018); and David E. Sanger, *The Perfect Weapon: War, Sabotage, and Fear in the Cyber Age* (Crown, 2018).

334 **alerted the FBI:** Nicholas Thompson and Fred Vogelstein, "Inside the Two Years That Shook Facebook—and the World," *Wired*, February 12, 2018.

335 **Alex Stamos:** Background on Stamos included Kurt Wagner, "Who Is Alex Stamos, the Man Hunting Down Political Ads on Facebook?" *Recode*, October 3, 2017; and Nicole Perlroth and Vindu Goel, "Defending Against Hackers Took a Back Seat at Yahoo, Insiders Say," *New York Times*, September 28, 2016.

335 **clashed repeatedly with his bosses:** Perlroth and Goel, "Defending Against Hackers."

336 **DCLeaks page:** The definitive source for the origin and operation of DCLeaks is the Mueller indictment, *United States v. Viktor Borisovich Netyksho, et al.* Filed July 13, 2018.

338 **"The results show":** Robert M. Bond, Christopher J. Fariss, Jason J. Jones, Adam D. I. Kramer, Cameron Marlow, Jaime E. Settle, and James H. Fowler, "A 61-Million-Person Experiment in Social Influence and Political Mobilization," *Nature* 489, September 12, 2012, 295–98.

338 **The study horrified observers:** Dara Lind, "Facebook's 'I Voted' Sticker Was a Secret Experiment on Its Users," *Vox*, November 4, 2014.

339 **Joel Kaplan:** Background on Kaplan included "Joel D. Kaplan, White House Deputy Chief of Staff for Policy," White House Press Office, April 24, 2006.

339 **"That was Joel's role":** Deepa Seetharaman, "Facebook Employees Pushed to Remove Trump's Posts as Hate Speech," *Wall Street Journal*, October 21, 2016.

340 **"Don't poke the bear":** Sheera Frenkel, Nicholas Confessore, Cecilia Kang, Matthew Rosenberg, and Jack Nicas, "Delay, Deny and Deflect: How Facebook's Leaders Fought through Crisis," *New York Times*, November 14, 2018. This is the explosive story that revealed much of the behind-the-scenes machinations in Facebook's policy world during and after the 2016 election.

340 **the CEO allowed it to stay:** Seetharaman, "Facebook Employees Pushed to Remove Trump's Posts."

341 **"news curator":** Michael Nunez, "Former Facebook Workers: We Routinely Suppressed Conservative News," *Gizmodo*, May 9, 2016. The best account of the Trending Topics debacle came from my colleagues Nicholas Thompson and Fred Vogelstein, "Inside the Two Years that Shook Facebook—and the World," *Wired*, February 12, 2019. The story in general provides a deep inside view of Facebook in 2016 and 2017.

341 **twelve-page response:** Facebook General Counsel Colin Stretch wrote to Hon. John Thune, Chairman of the Committee on Commerce, Science, and Transportation, on May 23, 2016.

342 **On the day:** Heather Kelly, "Facebook Ditches Humans in Favor of Algorithms for Trending News," CNN, August 26, 2016.

342 **"anti-Kelly fan fiction":** Abby Ohlheiser, "Three Days after Removing Human Editors, Facebook Is Already Trending Fake News," *Washington Post*, August 29, 2016.

343 **"All Lives Matter":** Jessica Guynn, "Zuckerberg Reprimands Facebook Staff Defacing 'Black Lives Matter,'" *USA Today*, February 26, 2016.

343 **the thought was:** Thompson and Vogelstein, "Inside the Two Years."

347 **"home run":** "Facebook CEO Mark Zuckerberg: Philippines a Successful Test Bed for Internet.org Initiative with Globe Telecom Partnership," *Globe Telecom*, February 25, 2014.

347 **Maria Ressa:** The definitive story on Facebook, the Philippines, and Maria Ressa is from Davey Alba, "How Duterte Used Facebook to Fuel the Philippine Drug War," BuzzFeed, September 4, 2018. Other useful sources besides personal interviews include Dana Priest, "Seeded in Social Media: Jailed Philippine Journalist Says Facebook Is Personally Responsible for Her Predicament," *Washington Post*, February 25, 2018; and *Frontline*'s documentary, *The Facebook Dilemma*, which ran on PBS on October 29 and 30, 2018.

350 **mini-documentary:** In a 2018 *Wired* story, Antonio García Martínez, a former Facebook ad executive, wrote about how the Trump campaign got more value from Facebook than the Clinton campaign, in "How Trump Conquered Facebook—Without Russian Ads," *Wired*, February 23, 2018. Facebook's Andrew Bosworth responded on Twitter with data that seemed to indicate the Trump campaign actually paid more per million views. This was disputed by Trump's digital director, who claimed that in some cases, Trump was getting a hundred times more value per CPM. Will Oremus in *Slate* ("Did Facebook Really Charge Clinton More for Ads than Trump?" February 28, 2018) found that while Bosworth might be technically correct, the larger advantage was Trump's because the Clinton campaign used less effective general interest ads while the Trump digital team ran more narrowly targeted "call to action" ads that indeed got more bang for the buck. As for the two-and-a-half minute ad, I was unable to locate that example but included it, as the source had firsthand knowledge of the episode. Later, Facebook conceded that the data showed the Trump campaign's superiority in this realm (Sarah Frier, "Trump's Campaign Says It Was Better at Facebook. Facebook Agrees," *Bloomberg Businessweek*, April 3, 2018).

351 **Parscale understood:** Useful sources for the Parscale campaign were Issie Lapowsky, "The Man Behind Trump's Bid to Finally Take Digital Seriously," *Wired*, August 19, 2016; Joshua Green and Sasha Issenberg, "Inside the Trump Bunker with Days to Go," *Bloomberg BusinessWeek*, October 27, 2016; Sue Halpern, "How He Used Facebook to Win," *New York Review of Books*, June 8, 2017; and Leslie Stahl (correspondent), "Brad Parscale," *60 Minutes*, October 18, 2017.

352 **Parscale began directing:** A good explanation of how Parscale used Facebook's tools comes from Martínez, "How Trump Conquered Facebook."

355 **Dave Goldberg was dead:** Sandberg's own account is in her book, co-written with Adam Grant, *Option B: Facing Adversity, Building Resilience, Finding Joy* (Knopf, 2017). She discussed the loss and its consequences in interviews including Belinda

Luscombe, "Life After Grief," *Time*, April 13, 2017; Jessi Hempel, "Sheryl Sandberg's Accidental Revolution," *Backchannel*, April 24, 2017.

356 **prone to yelling at subordinates:** This was described to me by multiple employees who worked with Sandberg.

356 **obsessed with her public image:** Besides personal interviews, Sandberg's image tending has been written about in the aftermath of Facebook's problems. See Nick Bilton, "'I Hope It Cracks Who She Is Wide Open': In Silicon Valley, Many Have Long Known Sheryl Sandberg Is Not a Saint," *Vanity Fair*, November 16, 2018. The aforementioned *New York Times* article, "Delay, Deny and Deflect," which portrays Sandberg as culpable in the post-election saga, was a turning point in the press's treatment of the COO.

356 **hostile prepublication article:** Jodi Kantor, "A Titan's How-To on Breaking the Glass Ceiling," *New York Times*, February 21, 2015.

356 **"Sandberg has co-opted the vocabulary":** Maureen Dowd, "Pompom Girl for Feminism," *New York Times*, February 23, 2013.

358 **the real newspaper:** Eric Lubbers, "There Is No Such Thing as the Denver Guardian, Despite That Facebook Post You Saw," *Denver Post*, November 5, 2016.

358 **"We've tried to do":** Laura Sydell, "We Tracked Down a Fake-News Creator in the Suburbs. Here's What We Learned," NPR, November 23, 2016.

358 **BuzzFeed tracked:** Craig Silverman and Lawrence Alexander, "How Teens in the Balkans Are Duping Trump Supporters with Fake News," BuzzFeed, November 3, 2016.

359 **"They only wanted pocket money":** Samanth Subramanian, "Inside the Macedonian Fake-News Complex," *Wired*, February 15, 2017.

359 **fake news stories on Facebook exceeded:** Craig Silverman, "This Analysis Shows How Viral Fake Election News Stories Outperformed Real News on Facebook," BuzzFeed, November 16, 2016.

359 **"The cover note was":** Roger McNamee includes the letter in his anti-Facebook polemic, *Zucked: Waking Up to the Facebook Catastrophe* (Penguin, 2019).

360 **Groups sprang up:** Blake Harris, *The History of the Future* (Dey Street Books, 2019), 442.

361 **"Sadly, News Feed optimizes for engagement":** Bobby Goodlatte posted on Facebook on November 9, 2016.

361 **"As long as it's on Facebook":** White House Press Office, "Remarks by the President at Hillary for America Rally," Ann Arbor, Michigan, November 7, 2016.

362 **diagnosed the problem:** David Remnick, "Obama Reckons with a Trump Presidency," *The New Yorker*, November 18, 2016.

362 **appearance in Berlin:** Gardiner Harris and Melissa Eddy, "Obama, with Angela Merkel in Berlin, Assails Spread of Fake News," *New York Times*, November 17, 2016.

362 **"a wake-up call":** Adam Entous, Elizabeth Dwoskin, and Craig Timberg, "Obama Tried to Give Zuckerberg a Wake-Up Call over Fake News on Facebook," *Washington Post*, September 24, 2017.

366 **the white paper:** Jen Weedon, William Nuland, and Alex Stamos, "Information Operations and Facebook," April 27, 2017.

CHAPTER FIFTEEN: P for Propaganda

368 **"He did not want it overly designed":** Cade Metz, "Facebook Moves into Its New Garden-Roofed Fantasyland," *Wired*, March 30, 2015.

370 **"My work is about connecting":** Mark Zuckerberg posted on Facebook on January 3, 2017.

371 **1862 speech:** Lincoln's concluding remarks to Congress on December 1, 1862.

372 **intelligence officials:** Massimo Calabresi, "Inside Russia's Social Media War on America," *Time*, May 18, 2017.

372 **Senator Mark Warner:** Tom LoBianco, "Hill Investigators, Trump Staff Look to Facebook for Critical Answers in Russia Probe," CNN, July 20, 2017.

372 **"I was pretty disappointed":** Warner's interview with *Frontline*'s James Jacoby was posted on May 24, 2018.

372 **"We have seen no evidence":** LoBianco, "Hill Investigators."

373 **"troll farm":** Adrian Chen, "The Agency," *New York Times*, June 2, 2015.

373 **Those pages posted:** One of the most thorough assessments of the IRA's work came in a report called "Tactics and Tropes of the Internet Research Agency," December 17, 2018, produced by New Knowledge on the request of the Senate Select Committee on Intelligence.

374 **"Woke Blacks":** *United States of America v. Internet Research Agency, et al.* Filed February 16, 2018.

379 **an anodyne report:** Alex Stamos, "An Update on Information Operations on Facebook," *Facebook Newsroom*, September 6, 2017. A major *New York Times* story first told about the Stamos draft, as well as uncovering other details in the 2017 Facebook saga, much of it tracking with the research I had been conducting, though several Facebook officials felt that the story did not reflect their motivations. Sheera Frenkel, Nicholas Confessore, Cecilia Kang, Matthew Rosenberg, and Jack Nicas, "Delay, Deny and Deflect: How Facebook's Leaders Fought Through Crisis," *New York Times*, November 14, 2018.

379 **directors were shocked:** In addition to personal interviews, "Delay, Deny and Deflect" provided background to the two days when Facebook's board heard about Russian state involvement.

381 **"We got to know":** Justin Weir, "Zuckerberg Pays Surprise Visit to Falls Family," *Vindicator*, April 29, 2017.

381 **war games:** Crystal Bui, "Mark Zuckerberg Meets Raimondo, Providence Students, Dines at Johnston Restaurant," NBC 10 News, May 22, 2017.

382 **$7.3 million:** Joanna Pearlstein, "The Millions Silicon Valley Spends on Security for Execs," *Wired*, January 16, 2019.

384 **changing Facebook's entire *mission*:** Zuckerberg posted his speech, "Bringing the World Closer Together," on his Facebook page, June 22, 2017.

385 **Tristan Harris:** The best background on Harris's crusade is Bianca Bosker, "The Binge Breaker," *The Atlantic*, November 2016.

388 **a dress of ambiguous color:** Cates Holderness, "What Colors Are This Dress?" BuzzFeed, February 26, 2015.

389 **fact-checking operations:** Facebook outsourced the authorization of the fact-checking organizations to the Poynter Institute. Some of its choices were controversial as they came to include conservative publishers like the alt-right Daily Caller.

390 ***Why are these people here*:** Benjamin Mullen and Deepa Seetharaman, "Publishing Executives Argue Facebook Is Overly Deferential to Conservatives," *Wall Street Journal*, July 17, 2018.

392 **reform Newark's schools:** The story of Zuckerberg's Newark donation is comprehensively chronicled by Dale Russakoff, *The Prize: Who's in Charge of America's Schools?* (Houghton Mifflin Harcourt, 2015).

392 **foundation shut down:** Leanna Garfield, "Mark Zuckerberg Once Made a $100 Million Investment in a Major US City to Help Fix Its Schools—Now the Mayor Says the Effort 'Parachuted' in and Failed," *Business Insider*, May 12, 2018.

394 **$7 trillion:** Jeremy Youde, "Here's What Is Promising, and Troubling, About Mark Zuckerberg and Priscilla Chan's Plan to 'Cure All Diseases,'" *Washington Post*, October 4, 2016.

395 **name be stripped:** Lauren Feiner, "San Francisco Official Proposes Stripping Mark Zuckerberg's Name from a Hospital," *CNBC*, November 29, 2018.

396 ***I'm very glad to be in Beijing*:** Vindu Goel, Austin Ramzy, and Paul Mozur, "Mark Zuckerberg, Speaking Mandarin, Tries to Win Over China for Facebook," *New York Times*, October 23, 2014.

396 **he even asked China's president Xi:** Loulla-Mae Eleftheriou-Smith, "China's President Xi Jinping 'Turns Down Mark Zuckerberg's Request to Name His Unborn Child' at White House Dinner," *Independent*, October 4, 2015.

396 **chose their own Chinese name for Maxima:** Mark Zuckerberg announced it in the "Happy New Year!" video on Facebook in 2016.

396 **"Every year this trip":** Mark Zuckerberg posted his trip at Tsinghua University, Beijing, China, on Facebook on October 28, 2017.

397 **"For those I hurt this year":** Mark Zuckerberg posted on Facebook on September 30, 2017.

398 **his 2018 resolution:** Mark Zuckerberg posted on Facebook on January 4, 2018.

CHAPTER SIXTEEN: Clown Show

399 **news of this broke:** Though there had been previous reporting, the Cambridge Analytica/Facebook story broke through on March 17, 2018, with simultaneous publication in *The Guardian/Observer* (Carole Cadwalladr and Emma Harrison, "Revealed: 50 Million Facebook Profiles Harvested for Cambridge Analytica in Major Data Breach") and the *New York Times* (Matthew Rosenberg, Nicholas Confessore, and Carole Cadwalladr, "How Trump Consultants Exploited the Facebook Data of Millions").

400 **Psychometrics Centre:** The best account of how the Cambridge Analytica scandal intertwined with the center is Issie Lapowsky, "The Man Who Saw the Dangers of Cambridge Analytica Years Ago," *Wired*, June 19, 2018.

400 **Kosinski:** Some background on Kosinski and his involvement in the Cambridge Analytica story came from a prescient story by Hannes Grassegger and Mikael Krogerus, "The Data That Turned the World Upside Down," *Motherboard*, January 28, 2017. It was originally published in German in *Das Magazin* in December 2016.

402 **creepiness of the discovery:** Michal Kosinski, David Stillwell, and Thore Graepel, "Private Traits and Attributes Are Predictable from Digital Records of Human Behavior," *PNAS* 110, no. 15, April 9, 2013: 5805.

403 **"Computer models need":** Wu Youyou, Michal Kosinski, and David Stillwell, "Computer-Based Personality Judgments Are More Accurate Than Those Made by Humans," *PNAS* 112, no. 4, January 27, 2015: 1037.

404 **had gotten a patent:** Facebook, Inc., Menlo Park, CA (US) got patent No. US 8,825,764 B2 with Michael Nowak, San Francisco, CA (US); Dean Eckles, Palo Alto, CA (US) as inventors. The date of patent is September 2, 2014. While it's unclear how this specific technique was employed, a detailed discussion of Facebook's data mining is found in Shoshana Zuboff, *The Age of Surveillance Capitalism: The Fight for a Human Future at the New Frontier of Power* (New York: Public Affairs, 2019).

404 **"Entity Graph":** This was described to me by Cameron Marlow, who was once head of Facebook's Data Science team.

406 **the most controversial study:** Adam D. I. Kramer, Jamie E. Guillory, and Jeff T. Hancock, "Experimental Evidence of Massive Scale Emotional Contagion Through Social Networks," *PNAS* 111, no. 24, June 17: 8788–90.

406 **"It was important to investigate":** Jillian D'Onfro, "Facebook Researcher Responds to Backlash Against 'Creepy' Mood Manipulation Study," *Business Insider*, June 29, 2014.

407 **"What many of us feared":** Reed Albergotti, "Furor Erupts Over Facebook's Experiment on Users," *Wall Street Journal*, June 30, 2014.

407 **"Facebook intentionally made":** Katie Waldman, "Facebook's Unethical Experiment," *Slate*, June 28, 2014.

411 **promotional copy:** This brochure was among a cache of documents that Wylie submitted to UK Parliament. Wylie also explains his background and involvement

with Cambridge Analytica in his book, *Mindf*ck: Cambridge Analytica and the Plot to Break America* (Random House, 2019).

411 **"We'll give you total freedom":** Carole Cadwalladr, "'I Made Steve Bannon's Psychological Warfare Tool': Meet the Data War Whistleblower," *Guardian*, March 18, 2018.

411 **his predecessor had died:** Wylie testimony to House of Commons, Digital, Culture Media and Sport Committee, March 27, 2018.

411 **The name came from Bannon:** Wylie testimony.

412 **Obama campaign:** Elizabeth Dwoskin and Tony Romm, "Facebook's Rules for Accessing User Data Lured More Than Just Cambridge Analytica," *Washington Post,* March 19, 2018.

412 **Graph API V1**: Facebook explained how Kogan's app took advantage of the Open Graph in its June 29, 2018, letter to the House Energy and Commerce Subcommittee, answering questions arising from Zuckerberg's testimony earlier that year.

414 **he simply used Google:** Wylie's explanation came in a document he submitted to UK Parliament after his testimony, "A Response to Misstatements in Relation to Cambridge Analytica Introductory Background to the Companies."

414 **With Stillwell and Kosinski out:** A solid account of the timeline of Kogan and SCL's experiment can be found in the FTC ruling, "In the Matter of Cambridge Analytica, LLC," released July 22, 2019.

415 **In a May email:** Matthew Rosenberg et al., "How Trump Consultants Exploited the Facebook Data of Millions."

416 **even more than the 50 million:** Dr. Alex Kogan spoke on "Big Data Social Science: How Big Data Is Revolutionizing Our Science" at a brown-bag lunch at the psychology department on December 2, 2014.

417 **headed to a party:** Brittany Kaiser, *Targeted: The Cambridge Analytica Whistleblower's Inside Story of How Big Data, Trump, and Facebook Broke Democracy and How It Can Happen Again* (HarperCollins, 2019), 147.

417 **a *Politico* article:** Kenneth Vogel and Tarini Parti, "Cruz Partners with Donor's 'Psychographic' Firm," *Politico*, July 7, 2015.

418 **stolen Facebook profiles:** Harry Davies, "Ted Cruz Using Firm That Harvested Data on Millions of Unwitting Facebook Users," *Guardian*, December 11, 2015.

418 **for months:** The internal email chain preceding and directly following the 2015 *Guardian* story was released in 2019 as a part of Cambridge Analytica civil litigation.

419 **Hendrix also contacted:** Kaiser, *Targeted*, 159.

419 **deleted the data:** In *District of Columbia v. Facebook*, the complaint cited the dates that Kogan and Cambridge Analytica affirmed that the data was deleted. In its response on July 8, 2019, Facebook conceded that those dates were accurate. The company has confirmed this to me directly.

421 **raw data in Cambridge Analytica's files:** Matthew Rosenberg and Gabriel J. X. Dance, "'You Are the Product': Targeted by Cambridge Analytica on Facebook," *New York Times*, April 8, 2018.

421 **Brad Parscale would later tell:** *Frontline*'s *The Facebook Dilemma* web page has extended interviews with sources including Parscale.

421 **"secret sauce":** Nicholas Confessore and Danny Hakim, "Data Firm Says 'Secret Sauce' Aided Trump; Many Scoff," *New York Times*, March 6, 2017.

421 **"data-driven communication":** Hannes Grassegger and Mikael Krogerus, "The Data That Turned the World Upside Down," *VICE*, January 28, 2017.

422 **Facebook's responses were misleading:** The characterization of Facebook's statements at this time as false and misleading are explicit in "Securities and Exchange Commission vs Facebook, Inc," July 24, 2019. The document presents yet another damning timeline of the Cambridge Analytica episode. Facebook paid $100 million to settle the SEC complaint.

422 **"Our investigation":** Mattathias Schwartz, "Facebook Failed to Protect 30 Million Users from Having Their Data Harvested by Trump Campaign Affiliate," *The Intercept*, March 30, 2017.

425 **"Several days ago":** VP & Deputy General Counsel of Facebook Paul Grewal, "Suspending Cambridge Analytica and SCL Group from Facebook," *Facebook Newsroom*, March 16, 2018.

426 **"I think the feedback":** Nicholas Thompson, "Mark Zuckerberg Talks to WIRED About Facebook's Privacy Problem," *Wired*, March 21, 2018.

428 **scrutinizing his non-hoodie apparel:** Vanessa Friedman, "Mark Zuckerberg's I'm Sorry Suit," *New York Times*, April 10, 2018.

429 **"Facebook is an idealistic":** The statement, and the complete transcript of Zuckerberg's hearing, is available at "Transcript of Mark Zuckerberg's Senate Hearing," *Washington Post*, April 10, 2018.

429 **In front of him:** Taylor Hatmaker, "Here Are Mark Zuckerberg's Notes from Today's Hearing," *TechCrunch*, April 10, 2018. AP photographer Andrew Harnick had enterprisingly captured the notes when Zuckerberg left his seat and failed to cover his talking points.

430 **he did that forty-six times:** Brian Barrett, "A Comprehensive List of Everything Mark Zuckerberg Will Follow Up On," *Wired*, April 11, 2018.

430 **69,000 apps:** Tony Romm and Drew Harwell, "Facebook Suspends Tens of Thousands of Apps Following Data Investigation," *Washington Post*, September 20, 2019.

CHAPTER SEVENTEEN: The Ugly

432 **streamed the entire spree:** Charlotte Graham-McLay, Austin Ramzy, and Daniel Victor, "Christchurch Mosque Shootings Were Partly Streamed on Facebook," *New York Times*, March 14, 2019.

434 **Arab Spring:** A firsthand account of social media in Arab Spring is found in Wael Ghonim, *Revolution 2.0: A Memoir* (Houghton Mifflin Harcourt, 2012).

434 **"Facebook had extraordinary power":** Tim Sparapani, "Frontline: The Facebook Dilemma," PBS, March 15, 2018.

436 **used Facebook as a weapon:** Human Rights Council (UN), "Report of the Detailed Findings of the Independent International Fact-Finding Mission on Myanmar," September 10–28, 2018. This 444-page report is devastating.

436 **"Rohingya terrorists":** Ibid., 170.

436 **Aela Callan:** Timothy McLaughlin, "How Facebook's Rise Fueled Chaos and Confusion in Myanmar," *Wired*, July 6, 2018. A firsthand account from entrepreneur David Madden trying to warn Facebook is in the uncut interview for *Frontline*'s *The Facebook Dilemma*, conducted on June 19, 2018.

437 **Facebook take major steps:** "Removing Myanmar Military Officials from Facebook," *Facebook Newsroom*, August 28, 2018. Damaging details are found in Paul Mozur, "A Genocide Incited on Facebook, with Posts from Myanmar's Military," *New York Times*, October 15, 2018.

438 **a firm called BSR:** BSR produced the report "Human Rights Impact Assessment: Facebook in Myanmar" in October 2018.

440 **the destruction of a watermelon:** Tasneem Nashrulla, "We Blew Up a Watermelon and Everyone Lost Their Freaking Minds," BuzzFeed, April 8, 2016.

440 **"We saw just a rash":** Jason Koebler and Joseph Cox, "The Impossible Job: Inside Facebook's Struggle to Moderate Two Billion People," *VICE*, August 23, 2018.

440 **some critics would attack Facebook:** Natasha Singer, "In Screening for Suicide Risk, Facebook Takes on Tricky Public Health Role," *New York Times*, December 31, 2018.

441 **twenty-eight-year-old man:** Daniel Victor, "Man Inadvertently Broadcasts His Own Killing on Facebook Live," *New York Times*, June 17, 2016.

441 **"The Ugly":** Bosworth's memo was first reported by Ryan Mac, Charlie Warzel, and Alex Kantrowitz, "Growth at Any Cost: Top Facebook Executive Defended Data Collection in 2016 Memo—and Warned That Facebook Could Get People Killed," BuzzFeed, March 29, 2018.

442 **"We've never believed":** David Ingram, "Zuckerberg Disavows Memo Saying All User Growth Is Good," Reuters, March 29, 2018.

442 **forty percent of divorces:** A 2012 Study from Divorce-Online-UK seems to be the source for this. According to *Divorce Magazine* (Daniel Matthews, "What You Need to Know About Facebook and Divorce," July 15, 2019), a UK firm called Lake Legal found the number to be 30 percent. The *Divorce* article quotes a high-volume attorney estimating 30 to 40 percent.

443 **had only been viewed live:** VP and Deputy General Counsel of Facebook Chris Sonderby posted "Update on New Zealand," *Facebook Newsroom*, March 18, 2019.

445 **came from academics:** There have been several deep studies of content moderators and policy by academics, notably Sarah T. Roberts, *Behind the Screen: Content Moderation in the Shadows of Social Media* (Yale University Press, 2019); Tarleton Gillespie, *Custodians of the Internet: Platforms, Content Moderation, and the Hidden Decisions That Shape Social Media* (Yale University Press, 2018); and Kate Klonick, "The New Governors: The People, Rules, and Processes Governing Online Speech," *Harvard Law Review,* April 10, 2018.

445 **subgenre of journalism:** One of the earliest and best accounts came from Koebler and Cox, "The Impossible Job." A deep look at setting policy for moderators came in "Post No Evil," *Radiolab*'s August 17, 2018, show.

446 **Dickensian elements:** Casey Newton's stories were "The Trauma Floor," *The Verge,* February 25, 2019; and "Bodies in Seats," *The Verge*, June 19, 2019.

447 **19.4 million pieces of content:** Facebook released its *Community Standards Enforcement Report* on May 2019, adding data for the time period of October 2018 to March 2019. For the first time, according to Facebook Transparency, they shared data on the process for appealing and restoring content to correct mistakes in their enforcement decisions. This is also the first time they reported on standards on regulated goods, covering firearms and drugs.

448 **1,400 pages:** Max Fisher, "Inside Facebook's Secret Rulebook for Global Political Speech," *New York Times,* December 27, 2018.

448 **cringe-worthy images:** Koebler and Cox, "The Impossible Job."

448 **"Men are scum":** A deep discussion of this case comes in Simon Van Zuylen-Wood, "'Men Are Scum': Inside Facebook's War on Hate Speech," *Vanity Fair,* February 26, 2019.

451 **Cognizant decided:** Casey Newton, "A Facebook Content Moderation Vendor Is Quitting the Business After Two Verge Investigations," *The Verge*, October 30, 2019.

453 **"the Conspiracy":** "Welcome to the AI Conspiracy: The 'Canadian Mafia' Behind Tech's Latest Craze," *Recode*, July 15, 2015.

453 **Facebook Artificial Intelligence Lab:** I wrote about Facebook's artificial intelligence efforts in "Inside Facebook's AI Machine," *Backchannel*, February 23, 2017.

455 **it blocked 2 billion attempts:** VP Integrity of Facebook Guy Rosen, "An Update on How We Are Doing at Enforcing Our Community Standards," *Facebook Newsroom,* May 23, 2019.

455 **"the vast majority":** Jack Nicas, "Does Facebook Really Know How Many Fake Accounts It Has?," *New York Times*, January 30, 2019.

455 **read the content of messages:** Viswanath Sivakumar, "Rosetta: Understanding Text Images and Videos with Machine Learning," *Facebook Engineering*, September 11, 2018.

456 **a Norwegian writer:** The story of the Napalm Girl on Facebook is told in detail in Gillespie, *Custodians of the Internet.*

459 **pulled the plug:** James Vincent, "Facebook Removes Alex Jones Pages, Citing Repeated Hate Speech Violations," *The Verge*, August 6, 2018.

459 **ban him as "dangerous":** Casey Newton, "Facebook Bans Alex Jones and Laura Loomer for Violating Its Policies Against Dangerous Individuals," *The Verge*, May 2, 2019.

CHAPTER EIGHTEEN: Integrity

462 **two kinds of CEOs:** Ben Horowitz, *The Hard Thing About Hard Things: Building a Business When There Are No Easy Answers* (HarperCollins, 2014), 224–28.

465 **advised against buying it:** Aisha Hassan, "These Brutal Reviews of Facebook's Portal Device Shows Why No One Wants It in Their Home," *Quartz*, November 9, 2018.

465 **Some major companies:** According to a "Note by Damian Collins MP, Chair of the DCMS Committee," Facebook had entered into white-listing agreements with certain companies, which meant that after the platform changes in 2014–15 they maintained full access to friends' data. It is not clear that there was any user consent for this, nor how Facebook decided which companies should be white-listed or not.

465 **"Jew haters":** Julia Angwin, Madeline Varner, and Ariana Tobin, "Facebook Enabled Advertisers to Reach 'Jew Haters,'" *ProPublica*, September 14, 2017.

466 **a firm called Definers:** Sheera Frenkel, Nicholas Confessore, Cecilia Kang, Matthew Rosenberg, and Jack Nicas, "Delay, Deny and Deflect: How Facebook's Leaders Fought Through Crisis," *New York Times*, November 14, 2018.

467 **"someone had just shot":** Nicholas Thompson and Fred Vogelstein, "15 Months of Fresh Hell Inside Facebook," *Wired*, April 16, 2019.

467 **exploited flaws:** Mike Isaac and Sheera Frenkel, "Facebook Security Breach Exposes Accounts of 50 Million Users," *New York Times*, September 28, 2018.

467 **"We have a responsibility":** Mark Zuckerberg stated on a Facebook post on March 21, 2018.

467 **"It's not enough":** Erin Durkin, "Michelle Obama on 'Leaning In': Sometimes That Shit Doesn't Work," *Guardian*, December 3, 2018.

469 **"Beep":** Nicholas Fandos, "Alex Jones Takes His Show to the Capitol, Even Tussling with a Senator," *New York Times*, September 5, 2018.

469 **her meeting with the Congressional Black Caucus:** Yamiche Alcindor, "Black Lawmakers Hold a Particular Grievance with Facebook: Racial Exploitation," *New York Times*, October 14, 2017.

471 **CTO Mike Schroepfer:** Cade Metz and Mike Isaac, "Facebook's A.I. Whiz Now Faces the Task of Cleaning It Up. Sometimes That Brings Him to Tears," *New York Times*, May 17, 2019.

472 **a poll taken by 29,000 Facebookers:** Deepa Seetharaman, "Facebook Morale Takes a Tumble Along with Stock Price," *Wall Street Journal*, November 14, 2018.

473 **a celebratory photo:** Mark Zuckerberg shared photos with his team on Facebook on May 15, 2017.

473 **"The thought process":** Mike Allen, "Sean Parker Unloads on Facebook: 'God Only Knows What It's Doing to Our Children's Brains,'" *Axios*, November 9, 2017.

473 **"bright dings of pseudo-pleasure":** Paul Lewis, "'Our Minds Can Be Hijacked': The Tech Insiders Who Fear a Smartphone Dystopia," *Guardian*, October 6, 2017.

473 **"I think we have created":** James Vincent, "Former Facebook Exec Says Social Media Is Ripping Apart Society," *The Verge*, December 11, 2017.

474 **clawed back his statement:** Palihapitiya posted his reversal on Facebook, December 15, 2017.

475 **Partner Categories:** It appeared on both the *Washington Post* and BBC. Drew Harwell, "Facebook, Longtime Friend of Data Brokers, Becomes Their Stiffest Competition," *Washington Post*, March 29, 2018. Jane Wakefield, "Facebook Scandal: Who Is Selling Your Personal Data?," BBC, July 11, 2018.

475 **"Businesses may continue":** Facebook no longer works with third-party data providers to offer their targeting segments directly on Facebook after April 2018. Facebook states its new data policy on the web page, "How does Facebook work with data providers?" under the "How Ads Work on Facebook" section in Facebook's help center.

476 **a mark it would actually reach:** "US Digital Ad Spending Will Surpass Traditional in 2019," *eMarketer*, February 19, 2019.

476 **around 60 percent:** Anne Freier, "Google and Facebook to Reach 63.3% Digital Ad Market Share in 2019," *Business of Apps*, March 26, 2019.

476 **more than two-thirds:** Khalid Saleh, "Global Mobile Ad Spending—Statistics and Trends," *Invesp*, March 31, 2015.

476 **second-quarter earnings:** "Facebook Q2 2018 Earnings," transcript on Facebook Investor Relations page.

477 **had lost $17 billion:** Bill Murphy Jr., "Mark Zuckerberg Lost Almost $17 Billion in About an Hour. Here's What Happened," *Inc.*, July 26, 2018.

478 **Kaplan's wife:** Laura Cox Kaplan's LinkedIn account reports her position and tenure at PwC.

478 **a report for the FTC:** The Electronic Privacy Information Center used a FOIA request to obtain PwC's "Independent Assessor's Report on Facebook's Privacy Program," April 12, 2017.

478 **found the company in violation:** Federal Trade Commission, "FTC Imposes $5 Billion Penalty and Sweeping New Privacy Restrictions on Facebook," July 24, 2019.

479 **Facebook reported:** Salvador Rodriguez, "Facebook Reports Better Than Expected Second-Quarter Results," CNBC, July 24, 2019.

480 **"Your fundamental problem":** Edward Docx, "Nick Clegg: The Facebook Fixer," *New Statesman America*, July 17, 2019.

481 **"If you're not":** Cook's remark was made on an MSNBC "Revolution" event in an interview with Kara Swisher and Chris Hayes, on April 6, 2018.

482 **Cook disagreed:** Matthew Panzarino, "Apple's Tim Cook Delivers Blistering Speech on Encryption, Privacy," *TechCrunch*, June 2, 2015.

482 **measure their worth with Likes:** Brian Fung, "Apple's Tim Cook May Have Taken a Subtle Dig at Facebook in His MIT Commencement Speech," *Washington Post*, June 9, 2017.

482 **After Cambridge:** Peter Kafka, "Tim Cook Says Facebook Should Have Regulated Itself, but It's Too Late for That Now," *Recode*, March 28, 2018.

482 **"extremely glib":** Ezra Klein, "Mark Zuckerberg on Facebook's Hardest Year, and What Comes Next," *Vox*, April 2, 2018.

483 **Facebook did withdraw:** Deepa Seetharaman, "Facebook Removed Data-Security App from Apple Store," *Wall Street Journal*, August 22, 2018.

483 **Those users also included:** Josh Constine, "Facebook Pays Teens to Install VPN that Spies on Them," *TechCrunch*, January 29, 2019. Constine's story was the apparent spur for Apple to take action.

484 **"Full-year 2018 revenue":** Facebook Q4 2018 earnings call transcript, Facebook Investor Relations Page, January 30, 2019.

CHAPTER NINETEEN: The Next Facebook

487 **"Privacy-Focused Vision":** Mark Zuckerberg, "A Privacy-Focused Vision for Social Networking," March 6, 2019.

492 **Puerto Rico:** Arjun Kharpal, "Mark Zuckerberg Apologizes After Critics Slam His 'Magical' Virtual Reality Tour of Puerto Rico Devastation," CNBC, October 10, 2017.

492 **Palmer Luckey problem:** Besides personal interviews, I drew on the primary documents and reporting in Blake Harris, *The History of the Future*.

493 **reporter thought otherwise:** Gideon Resnick, "The Facebook Billionaire Secretly Funding Trump's Meme Machine," *Daily Beast*, September 22, 2016.

494 **they were abandoning the platform:** Jeff Grubb, "Some VR developers Cut Ties with Oculus over Palmer Luckey Funding Pro-Trump Memes," VentureBeat, September 23, 2016.

494 **"We care deeply about diversity":** Cory Doctorow, "VERIFIED Mark Zuckerberg Defends Facebook's Association with Peter Thiel," *BoingBoing*, October 19, 2016.

496 **literally in people's heads:** Josh Constine, "Facebook Is Building Brain-Computer Interfaces for Typing and Skin-Hearing," *TechCrunch*, April 19, 2017.

497 **name of the feature:** Besides personal interviews, the background on Stories was informed by Billy Gallagher, *How to Turn Down a Billion Dollars* (St. Martin's Press, 2018).

499 **"I cannot stand Facebook":** "Miranda Kerr 'Appalled' by Facebook 'Stealing Snapchat's Ideas,'" *Telegraph*, February 7, 2017.

502 **"Respect for your privacy":** "Setting the Record Straight," WhatsApp blog, March 17, 2004.

503 **"I think everyone was gambling":** Parmy Olson, "Exclusive: WhatsApp Cofounder Brian Acton Gives the Inside Story on #DeleteFacebook and Why He Left $850 Million Behind," *Forbes*, September 18, 2018.

503 **$122 million:** Mark Scott, "E.U. Fines Facebook $122 Million over Disclosures in WhatsApp Deal," *New York Times*, May 18, 2017.

504 **resented the idea:** Kirsten Grind and Deepa Seetharaman, "Behind the Messy, Expensive Split Between Facebook and WhatsApp's Founders," *Wall Street Journal*, June 5, 2018.

505 **"I'm taking some time off":** Jan Koum, Facebook post, April 30, 2018.

506 **"I find attacking the people":** David Marcus, "The Other Side of the Story," Facebook post, September 26, 2018.

506 **ads began appearing on WhatsApp:** Jon Porter, "WhatsApp Found a Place to Show You Ads," *The Verge*, November 1, 2018.

508 **Zuckerberg instructed Growth leader:** Nicholas Thompson and Fred Vogelstein, "15 Months of Fresh Hell Inside Facebook," *Wired*, April 16, 2019.

510 **a service called Facebook Watch:** Fidji Simo, "Facebook Watch: What We've Built and What's Ahead," *Facebook Newsroom*, December 13, 2018.

511 **musical chairs:** Kurt Wagner, "Facebook Is Making Its Biggest Executive Shuffle in Company History," *Recode*, May 8, 2018.

513 **on the way out:** Nicole Perlroth and Sheera Frenkel, "The End for Facebook's Security Evangelist," *New York Times*, March 20, 2018.

513 **appeared at a *Wired* conference:** Alex Davies, "What's Next for Instagram's Kevin Systrom? Flying Lessons," *Wired*, October 15, 2018.

515 **long op-ed:** Chris Hughes, "It's Time to Break Up Facebook," *New York Times*, May 9, 2019.

516 **Elizabeth Warren:** Astead W. Herndon, "Elizabeth Warren Proposes Breaking Up Tech Giants Like Amazon and Facebook," *New York Times*, March 8, 2019.

517 **Facebook named the currency Libra:** The Libra Association, "An Introduction to Libra: White Paper," June 18, 2019.

519 **Marcus first revealed:** I wrote about the Libra launch (with Greg Barber) in "The Ambitious Plan Behind Facebook's Cryptocurrency, Libra," *Wired*, June 18, 2019.

519 **Marcus testified:** Daniel Uria, "Head of Facebook Libra Grilled by Skeptical U.S. Senators," UPI, July 16, 2019.

519 **The hearing began:** I wrote about the testimony in "Mark Zuckerberg Endures Another Grilling on Capitol Hill," *Wired*, October 23, 2019. One can view clips on YouTube or the entire day on the House Committee on Financial Services website.

Index

Abrams, Jonathan, 41–43, 81, 87
Abrash, Michael, 327, 492
Accel venture capital firm, 102–3, 132
Acton, Brian, 317–25, 438, 500–506
advertising
  and Beacon, 182–83, 186–88, 206, 212
  and Bosworth, 294–95
  and business model of FB, 170, 178, 199–200
  and Cabal group of Bosworth, 294–96
  and Cambridge Analytica, 399, 420
  Campus Flyers, 178
  competition for, 476
  and data brokers, 269–70, 475
  and data collection on FB, 207
  and fake news disseminated on FB, 359
  and FTC investigation and sanctions, 274
  Hammerbacher on, 217
  on Instagram, 477, 490, 508
  and Like button, 202
  and Lookalike Audiences, 352
  and Microsoft partnership, 179–80, 183–85, 198
  in mobile apps, 295–98
  and News Feed feature, 138
  number of engineers working on, 199
  and Pages, 182
  and Pages You May Like campaign, 295
  and Pandemic code name, 181, 185
  and personally identifying information (PII), 474, 476
  and privacy questions, 475
  questionable categories in, 465–66
  revenues from, 170, 178, 198, 275, 297, 477
  and Russian election interference, 372–76, 377, 378–79
  Sandberg's policies for, 199-200
  in sidebars, 181
  social advertising, 180–81, 183, 185
  success of, 198
  targeted ads, 181, 351–53, 399, 465, 475
  by Trump campaign, 351–54
  and WhatsApp, 320–21, 324–25, 504
  Zuckerberg's perspectives on, 201–2, 295–96, 474–75
  by Zynga, 167
African Americans, 343, 353, 374, 403, 469–70
Agarwal, Aditya, 105, 107
algorithms of Facebook
  amplifying effects of, 142
  charges of political bias in, 458
  engagement prioritized in, 385
  and fake news/misinformation on FB, 9, 11, 361
  privileging close relationships, 261, 391
  and ranking of posts on News Feed, 127–28, 163, 172, 260–61, 385
  and sharing of content, 401
Amazon, 293, 516
America Online (AOL), 28–29, 209
Analog Research Lab, 238, 368
Andreessen, Marc, 288, 327–28, 379
Andreessen Horowitz, 300, 327
Android platform, 172
Anker, Andrew, 357, 388–89
antitrust questions, 514–16
anti-vaccination movement, 346
Apple
  and antitrust investigations, 516
  campus of, 148
  and Cook's criticisms of FB, 481–83

Apple *(cont.)*
and Facebook app, 276–79
"full friend access" negotiated by, 175
and iPhones, 154, 276–79, 301
and Macintosh computers, 22
and Onavo Protect, 483–85
partnership with FB, 148
and platform for mobile phones, 172
application programming interface (API)
Graph API V1 ("the Friend API"), 171–72, 175–76, 271, 409, 412
Graph API V2, 175
initial version, written by Fetterman, 150–51
See also Open Graph
Applied Machine Learning (AML) team, 454
Aquila drone, 232–33
Arab Spring movement, 7, 434, 471
Arrington, Michael, 166
artificial intelligence (AI), 33, 452, 453–56

Backstrom, Lars, 223
Badros, Greg, 106, 201
Ballmer, Steve, 201, 239
Bannon, Steve, 411, 420
Barker, Brandee
and Beacon, 187
on growth emphasis of company, 235
and News Feed feature, 141–42
and Pandemic launch, 185
on Sandberg critics, 356
and Schrage's hire, 200
and Zuckerberg's public speaking, 156
Barry, Ben, 237–39, 241–43, 337
Beacon, 182–83, 186–88, 206, 212
Beck, Glenn, 343
Beluga, 313
Benchmark venture capital firm, 102
Berteau, Stefan, 188
Bickert, Monika, 340, 343, 432–38, 443–44, 448–49, 457–58, 480
Black Lives Matter movement, 342–43
Bloomberg, Michael, 256
Book of Change, 119, 122, 127–28, 144, 205, 527
Bosworth, Andrew
and ads engineering team, 294–95
background of, 126
and expectations Clinton would win, 350, 354
and fake news disseminated on FB, 349, 350
on FriendFeed, 203
and Like button, 204
management style of, 294
on "Napalm Girl" image, 457–58
and News Feed feature, 142, 391
recruitment of, 126–27
and solving FB's problems, 481
on Trending Topics feature, 341–42
and "The Ugly" internal memo, 441–42
as VP of Hardware, 495
and Zuckerberg, 473
Bowles, Erskine, 288, 379, 471
boyd, danah, 67
breastfeeding, 252–53, 254
Breitbart News, 391, 411
Brexit, 422
Breyer, Jim, 102–3, 133
Brin, Sergey, 289
Brown, Campbell, 389–90
Brown, Nat, 159, 164–65, 269
Buchheit, Paul, 203
Buddy Zoo, 43–44, 61
"Building Global Community" (Zuckerberg), 371, 383
Burma (later Myanmar), 11, 435–39, 526
business plan of Facebook
and advertising, 170, 178, 199–200
Callahan's model, 177–78
and Cook's criticisms of FB, 481, 482–83
and data collection on FB, 207, 524
and diversification, 198–99
and Kendall's manifesto, 180
and Like button on external websites, 206
and Microsoft partnership, 179–80
and Sandberg, 198, 199–200
Buttigieg, Pete, 381
buyout offers, 131–37
BuzzFeed, 262, 387–88, 390, 440, 442

Cadwalladr, Carole, 422–24
Callahan, Ezra
anticipation of FB's success, 98
and business plan for FB, 177–78
on Cohler's adult presence, 97
and Cox's recruitment, 124
and customer support, 246, 247
at La Jennifer house, 96–97
and Open Registration, 144
on redesign, 139
on Sandberg's management, 197
on status updates inspired by Twitter, 259
Callan, Aela, 436
Cambridge Analytica
banned from FB, 425
congressional hearings following, 427–30
and data deletion demanded by FB, 419, 420–21, 422, 424–25

FB's caution following, 464
and FB's digital currency bid, 520
FB's failure to alert users to issue, 421, 478
FB's response to news of breach, 425–27
FB's review of problem with, 418–19, 422
and *Guardian* article, 417–18
investigative journalism on, 422–24, 425
name of, 411
and political ads on FB, 11, 399, 411, 420
and SCL, 411, 415, 417
and Trump campaign, 399, 420, 421, 427
and user data supplied by Kogan, 399, 411–13, 414–19, 420–21, 422, 423–26
Wylie's role in, 410–15, 420, 422–25
See also Kogan, Aleksandr
Campus Flyers, 178
Carlyle Group equity fund, 113
Carmack, John, 326, 329, 491, 495
Casa Facebook (Palo Alto work space), 76–77, 82–84, 90–91, 95
Cathcart, Will, 170, 294, 507
Causes app, 155, 162, 164
Ceglia, Paul, 39
cell-phone numbers shared on FB, 71, 101
censorship, charges of, 457–59. *See also* content arbitration on Facebook
"Chain-Links" (Karinthy), 19–20
Chan, Priscilla
children of, 8, 393, 396
and confidence of Mark, 52
first date with Mark, 51
living with Mark, 255
in medical school, 106
philanthropy of, 391–95
and *The Social Network* film, 6
and texts discussing IPO, 290
wedding of, 292–93
Chancellor, Joseph, 408, 412, 417–18
Chandra, Arun, 451
Chan Zuckerberg Initiative (CZI), 392–94
Cheever, Charlie, 245
Chen, Adrian, 373
Chenault, Kenneth, 470
China, 256, 396–97
Choe, David, 109, 237
Christchurch massacre, 443–44, 471
civil-rights work of Facebook, 469–70
Clegg, Nicholas, 479–81
Clinton, Hillary Rodham, 9, 334, 374–75, 493
Clinton presidential campaign, 2016
and anti-Hillary ads, 353, 358–59
emails stolen from, 334, 336–37, 344, 353
and expectations of winning, 350, 359
and "Facebook Effect," 9
and low effort put into FB, 350–51, 354
Cloudera, 216
Co-Creation Hub (CcHUB), 1–3
Cohler, Matt
and buyout offers, 132, 136
on Campus Flyers, 178
on cash flow of FB, 178
criticisms of *Washington Post*'s funding choice, 101–2
departure of, 244
and fundraising pitch of FB, 89–90
and Instagram, 303
and News Feed feature, 138, 140, 141, 143
on Open Registration, 134
and Platform of FB, 161
recruiting computer science engineers, 105
role of, in FB, 97
and social advertising, 180
and venture capitalists, 101–4
and Zuckerberg's silences, 13
Columbia University, 70, 71
community/communities, 157, 371, 383–84, 398
Community Standards, 340, 437, 447–48, 460, 559n
Community Summit in Chicago, 383–84
Congressional Black Caucus, 469
connecting the world
and Course Match program, 46–47
and Growth organization, 235
as mission of FB, 14–15, 16, 119–20, 127, 231–32
and public company transition, 289
and rapid adoption of Thefacebook, 65
and "The Ugly" internal memo, 441–42
unanticipated consequences of, 524
and virtual reality (VR), 329–30
and Weinreich, 21, 522
Zuckerberg's ambitions for, 14–15, 127, 231, 289, 329–30, 370, 371, 524
ConnectU, 56–57, 60, 72–76
conservatives' meeting in Menlo Park, 342–43
conspiracy theories, 346, 362, 446
content arbitration on Facebook
and artificial intelligence, 452, 455–56
Bickert's role in, 433–34
and charges of censorship, 457–59
content inciting violence, 435–39
by Customer Support team, 246–50
developing standards for, 247–50, 252–54
difficult decisions in, 456–58

content arbitration on Facebook *(cont.)*
and Facebook Supreme Court, 460–61
and free speech, 459–60
in international markets, 436
and moderators, 252, 436, 444–51, 456
and Open Registration, 250
and pornography reports, 250–52
and reduction in questionable content, 477
and youth on FB, 250–51
See also hate speech/campaigns
Cook, Tim, 396, 481–83
cookies, 205
Cool, Jesse, 257
Costolo, Dick, 263, 303, 304, 305
Course Match programming project, 46–47, 61
Cox, Chris
and advertising, 181
background of, 124
and Community Summit, 383
departure of, 514
on increased visibility of user details, 264
and integration of apps, 511–12, 513–14
on international growth, 435
on misinformation on FB, 350
and News Feed feature, 124, 130–31, 142, 172
on news in FB, 387
and objectionable content on FB, 435
recruitment of, 124–25
and Sandberg's first day, 195
and sponsored stories, 180
and Stamos's security report, 364–65
and Twitter, 258
and Zuckerberg, 473
critics/criticisms of Facebook
and antitrust charges, 514–16
and congressional hearings, 429
and content moderation, 451, 457–59
Cook, 481–83
from current and former employees, 473–74
FB's attempts to impugn, 466
following the 2016 presidential election, 10–11, 395
as threat to democracy, 12
CrowdStrike security firm, 333–34
Crowley, Dennis, 310–11
Cruz, Ted, 417, 418, 421
CTRL-Labs, 496
Cuervo, Soleio, 106
culture of Facebook
and apologizing after the fact, 274
hacker culture, 240, 289–90
and headquarters, 254–55
and "Move fast and break things" motto, 6, 16, 106, 240–43, 427
and poster project of Barry, 241–43
and Sandberg, 237–38, 243
Cuomo, Andrew, 250, 251
customer support, 230, 246–54

D'Angelo, Adam
background of, 30
and Buddy Zoo, 43–44, 61
and Casa Facebook, 83
and Cox's interview, 124
and data mining, 215
departure of, 244
at Exeter with Zuckerberg, 30, 32
and Friendster, 41, 43
and Google internship, 76
and Instagram, 302, 303
and invisible keyboard project, 91
and Microsoft meeting, 184
and News Feed feature, 123–24
on Parker's influence on FB, 116
and Platform of FB, 150, 152, 158
and Quora, 245, 309
and Synapse-ai, 33, 38, 39–40, 41, 44
and third-party developers, 163
and Wirehog project, 92, 94
and Yahoo!'s attempt to acquire FB, 137
Daniels, Chris, 184, 506–7
"Dark Profiles" (people not on FB), 129–30, 205, 222–23
data breach of Facebook, 467
data gathering of Facebook
and business model of FB, 207
and D'Angelo's business card, 215
and email scraping, 215–16, 223
and Facebook Connect, 408–9
with FB apps, 409
and Graph API V1 ("the Friend API"), 171–72, 175–76, 271, 409, 412
and Like button, 206, 402–4, 416
in mobile Facebook via Onavo, 315–17, 323, 483–84
and Six4Three lawsuit, 465
on Snapchat use, 316–17
sources of data, 223–24
and Synapse, 40
and trust of users in FB, 464–65
and WhatsApp, 502–3
Data Science team
Chancellor's work with, 418
and emotional contagion study, 406–8
and Growth team's priorities, 337, 385, 405, 408

Hammerbacher on work of, 217
and People You May Know (PYMK) feature, 221, 223, 224
and viralized News Feed, 263
and voter behavior, 237, 337–38, 406
Dating feature, 464–65
Davenport, Ben, 313–14
Davies, Harry, 417, 418
DCLeaks, 336, 344–45
deactivated accounts, 274
deaths, protocols for dealing with, 125–26
Democratic National Committee, 334, 336–37, 344, 353
Denver Guardian fake-news outlet, 358
Desmond-Hellmann, Susan, 379
DeWolfe, Chris, 132, 160
digital currency, 516–21
DiResta, Renee, 346
"disputed content" labels, 389
Dorsey, Jack
congressional testimony of, 468, 469
and Instagram, 302, 303, 304–5
and origins of Twitter, 299
pushed out of Twitter, 258
and Trump's tweets, 340
and undercooked goat dinner, 15, 257
drones, Internet access delivered by, 232–33
Dropbox, 93, 95
Duterte, Rodrigo, 347–48, 435

Ebersman, David, 288, 290
Egeland, Tom, 456–57
elections
and charges of partisanship at FB, 337, 338, 349
and civic engagement of FB, 334–35
and Election War Room (2018), 485–86
and FB's impact on political behavior, 337–38
and "I Voted"/"I'm Voting" button, 237, 337–38, 406
Philippines' 2016 presidential election, 347–48
and political ad campaigns, 270
See also presidential election of 2016
Elman, Josh, 163, 165
email scraping, 215–16, 222
emotional contagion study, 406–8
employees of Facebook
and bonuses for living nearby, 109
boot camp for, 108
criticisms from former and current, 472–74
expectations for performance of, 108
and hacker culture, 240
interns, 77, 83
morale of, 137, 479–80
recruitment of, 104–7, 124–25
and Yahoo!'s attempt to acquire FB, 137
and Zuckerberg's staff meetings, 109–10
See also specific names
encryption, 438, 488, 500–501, 505, 513, 514
Everson, Carolyn, 201

Facebook, 502–3
board of directors of, 102, 133, 288, 470
and buyout offers, 131–37
co-founders of, 68–70, 96
conceptual predecessor of, 34–35
critics (*see* critics/criticisms of Facebook)
culture (*see* culture of Facebook)
data (*see* data gathering of Facebook; user data)
design and interface of, 62–63, 91, 113
domains for, 60, 104
early leadership of, 97
employees (*see* employees of Facebook)
features, 110–11 (*see also specific features, including* News Feed of FB)
founder (*see* Zuckerberg, Mark)
friends (*see* friends and "friending")
growth (*see* Growth Circle; growth of Facebook)
initial programming effort for, 60–61
launch of, at Harvard, 65, 67–68
mission (*see* mission of Facebook)
mobile (*see* mobile Facebook)
name change from "Thefacebook," 104
objectionable content (*see* content arbitration on Facebook)
offices (*see* work spaces of Facebook)
origins of, 6, 53
Platform (*see* Platform of Facebook)
positive impact of, 12, 16, 240, 371, 434
privacy practices of (*see* privacy)
redesigns of, 113, 138–39, 259–63, 525
reputation of, 11–12, 398, 484–85, 525
scandals (*see specific scandals, including* Cambridge Analytica *and* Russian interference in US presidential election)
security (*see* security measures of Facebook)
server space required for, 66, 67, 97–98, 100, 105, 115
Terms of Service agreements, 265, 369, 407, 414
users (*see* users)
values of, 111, 239–41, 289, 459

Facebook Artificial Intelligence Lab
(FAIR), 453–54
Facebook Basics, 8
Facebook Connect, 169–71, 173, 268, 297,
408–9
"Facebook Effect," 9
Facebook Effect, The (Kirkpatrick), 184
Facebook Guy, 62, 113
Facebook Home devices, 287, 297
Facebook Live, 343, 432, 439–41, 443–44,
468, 471
facebook program of Tillery, 34–35, 60
Facebook Research app, 483
facebooks of schools, 34–35, 48, 53, 59–60,
67, 76
Facebook Supreme Court, 460–61
Facebook Watch, 510
Facemash incident, 47–52, 53, 56, 58, 61,
64, 144
Faceweb, 281, 282, 283
fake accounts/users, 376–77, 455
fake news and misinformation
algorithms' amplification of, 9, 11, 361
and anti-Hillary ads, 358–59
and artificial intelligence, 455
data-based perspectives on, 349–50
and digital literacy, 436
and "disputed content" labels, 389
downranking of, 389
fake outlets of, 358
FB's policies on, 438–39, 519
and FB's reluctance to arbitrate truth,
337, 346, 357, 361
FB's response to concerns about, 345–46,
348–49, 357–58, 359, 389
FB's work to mitigate, 362–63, 370,
463, 484
and freedom of speech/expression,
363, 389
originating in Macedonia, 358–59,
364, 365
people profiteering on, 358–59, 365
and Philippines' 2016 presidential
election, 347–48
post-election reactions to, 361–62
and privacy in Next Facebook, 514
and Project P team, 365–66
spikes in, before election, 9, 358
Stamos's report on, 363–67
state-sponsored, 435, 438, 454
as threat to democracy, 362
Zuckerberg's "crazy idea" comment on,
10, 360–61, 370
Zuckerberg's perspectives on, 523
See also Cambridge Analytica
Family collection of apps, 511–12, 513–14
Fanning, Shawn, 79–80
Farmville app, 162–63
Farrakhan, Louis, 459
Faust, Drew, 382
Federal Trade Commission (FTC)
and antitrust charges, 515, 516
Consent Decree and sanctions of 2011,
11, 272–74, 478–79
and FB's acquisition of Instagram, 306–7
and $5 billion fine, 11, 479, 525
formal complaints about FB made to, 265
and WhatsApp, 503
"Feed Me: Motivating Contribution in
Social Network Sites," 221, 224
F8 developers conference, 154, 157–58
Feinstein, Dianne, 396
Ferrante, Danny, 217
Fetterman, Dave, 150–51, 155, 169
financials of Facebook
and costs of server space, 97–98, 100
and dilemma on funding choices,
100–104
disconnect between reputation and,
484–85
and first profit (2009), 256
and FTC's $5 billion fine, 479
initial capital for Thefacebook, 67
and initial public offering (IPO), 288–93,
297–98
and investments in security, 476
and Microsoft's funding, 185
and mission of FB, 524
and Parker's fundraising, 85–86, 89
and Pincus and Hoffman, 88
and pitch to venture capitalists, 88–90
and private funding rounds, 288
and public company transition, 287–93
and revenues from advertising, 138, 170,
178, 198, 275, 297, 477
and revenues from platform, 153–54, 168
and revenues from user data, 175
and revenues lost to mobile FB, 290
"seed round" funding from Thiel, 88–89,
90, 100, 101, 178
and stakes of co-founders, 96
and stock prices of Facebook, 291, 293,
297, 426, 477
and venture round with Accel, 102–3,
132, 178
and *Washington Post*'s investment,
100–102
"Find Friend" program, 215
Fishman, Ivan, 318
Flickr, 12, 114

Flixster third-party app, 161, 169–70
ForAmerica, 418
Forstall, Scott, 278
Foursquare, 309–11
Franklin, Rachel, 492
Free Basics program, 234, 347, 437
freedom of speech/expression
  and content moderation, 246, 248–49, 459–60
  and fake news disseminated on FB, 363, 389
  as founding ideal of Facebook, 111, 459
  and News Feed feature, 142
  and presidential election of 2016, 344, 357, 432
  and Russian election interference, 376
  Sandberg's emphasis on, 470–71
  standards for, 253–54, 340
  Zuckerberg's emphasis on, 111, 249, 254, 344, 357, 363, 459, 524
FriendFeed, purchase of, 203
friends and "friending"
  average number of, 223, 416
  and Dunbar number, 223, 225, 226
  in earliest FB release, 63, 65
  and "Find Friend" program, 215–16
  friends of friends (FoFs), 223, 268–69, 409, 412–13, 415, 416
  and "Friends Only" posts, 273
  and People You May Know (PYMK), 220–26
  user behavior trends in, 216
  and user retention, 220–21, 224–25
Friendster, 41–43, 59, 66, 69, 76, 86, 87, 91, 102

games on Facebook, 161–63
Gates, Bill, 14, 71, 95, 162, 184, 392, 394
Gehry, Frank, 368
Geminder, Katie, 108, 110, 122, 143
"Gesundheit! Modeling Contagion through Facebook News Feed," 262, 405
Gizmodo, 341
Gleit, Naomi, 217–18, 219–20, 235–36, 364–65, 383
Global Science Research (GSR), 412, 413
Goetz, Jim, 321, 325
Goldberg, Dave, 193, 208, 355
Goldman, Rob, 235, 354, 375–76
Goler, Lori, 239, 289
Goodlatte, Bobby, 361
Google
  and ad deal with FB, 183–84
  as advertising competitor, 476
  and content moderators, 445
  culture of, 243
  employees' migration to FB, 200–201
  and Google Plus, 304
  proposed acquisition of FB, 132
  Sandberg's time at, 192–93
  and Senate Judiciary hearing, 395
  and user profiles in search results, 219–20
  visibility of users' details to, 264
  and WhatsApp, 322, 323
Gowalla, 311
Graham, Don, 100–104, 171, 193, 288, 289, 524
Graham, Lindsey, 514–15
Graham, Molly, 239–40, 242, 285, 289
Graph API V1 ("the Friend API"), 171–72, 175–76, 271, 409, 412
graphics, messages embedded into, 455
Green, Joe, 45, 48, 50, 51, 52, 66, 155, 164, 196
Green, Joshua, 353
Greenspan, Aaron, 58–59, 60, 64, 75, 76
Grewal, Paul, 424, 426, 428, 430, 494, 505
Grimes, Michael, 288, 290, 322–23
Grossman, Lev, 233
Groups feature of Facebook, 112, 383–84
Growth Circle
  and aggressive pursuit of goals, 219, 224
  and algorithms driving News Feed, 385
  and dark profiles, 222
  engagement prioritized by, 385
  expanded scope of, 234–36
  formation of, 218
  and internationalization, 226, 229–31, 233, 235
  and Internet.org initiative, 347
  leadership of (*see* Olivan, Javier; Palihapitiya, Chamath)
  and Messenger, 314
  and mission of FB, 235
  and mobile Facebook effort, 281–82
  and Onavo acquisition, 315
  and People You May Know (PYMK), 220–21, 225
  and privacy defaults, 267
  and trust of users in FB, 235
  and WhatsApp, 322, 323
  See also Data Science team
growth of Facebook
  and Cambridge Analytica, 399
  and contact scraping, 215–16
  emphasis placed on, 214, 234–35, 399, 524
  expansion into high schools, 120–21

growth of Facebook *(cont.)*
expansion to other college campuses, 68, 70–72, 98
and first million users, 94–95
in international markets, 226–34, 235, 320, 323, 435–36
and Internet.org initiative, 4, 231–34
limited by server space, 97–98
and mission of FB, 235, 524
and mobile Facebook effort, 281–82
and Monthly Active User (MAU) metric, 213
and Open Registration, 119–23, 133, 137, 144, 157–58, 215, 250, 251
and Palihapitiya's proposals, 213–14
and People You May Know (PYMK), 220–26
and poster project of Barry, 242
and privacy defaults, 264, 267
problems that emerged from, 524
and recruitment of computer science engineers, 105–6
slow downs or stalls in, 133, 213, 507
and trust of users in FB, 235
and "The Ugly" internal memo, 441–42
and vulnerability to manipulation, 384
Zuckerberg's perspectives on, 525
See also connecting the world
GRU (Main Intelligence Directorate of the General Staff), 334, 336, 364, 366
Grudin, Nick, 387
Guccifer, 334

hacker culture, 74, 240, 289–90
Hammerbacher, Jeff, 214–15, 216
Hancock, Jeff, 407
Harris, Tristan, 385–86, 398
Harvard Connection (later ConnectU), 56–57, 60, 72–76
Harvard Facemash, 47–52, 53, 56, 58, 61, 64
Harvard University
and Course Match program of Zuckerberg, 46–47, 61
and facebook project of Zuckerberg, 53–54, 60–61, 67–68, 72
and Greenspan's houseSYSTEM, 58–59, 60, 64
indefinite leaves available to students of, 71, 95
and study group collaboration tool, 54–55, 61
Zuckerberg's acceptance to, 37–38
and Zuckerberg's class attendance, 45–46, 60
Zuckerberg's commencement speech at, 36–37, 72, 382
Zuckerberg's dropping out of, 71–72, 95–96
Zuckerberg's honorary degree at, 72
Hastings, Reed, 288
Hatch, Orrin, 429
hate speech/campaigns
in advertising categories, 465–66
and artificial intelligence, 454, 455
challenges of moderating, 448–50, 455
and encryption, 514
and Facebook Supreme Court, 460–61
and fake user accounts, 455
FB's delays in addressing, 434, 438
in Myanmar (previously Burma), 437–38, 526
and Next Facebook, 488
Ressa's warnings about, 347–48, 435, 486
and scope of moderators' work, 252
and Trump video on Muslim immigration, 340
Zuckerberg's perspectives on, 459–60, 523
Hegeman, John, 386, 388
Hemphill, Scott, 515
Hendrix, Allison, 418, 419
Hewitt, Joe, 276–79, 285
high schools, Facebook's expansion into, 120–21, 133
Hill, Kashmir, 222
Hinton, Geoffrey, 453
Hirsch, Doug, 114, 133
Hoffman, Auren, 269–70
Hoffman, Reid, 85–90, 161, 178, 220, 292
Hold 'Em Poker app, 162
Horowitz, Ben, 462–63
Hotmail, 184, 215–16
Hot or Not website, 48
houseSYSTEM of Greenspan, 58–59, 64
Hughes, Chris
and antitrust charges, 515
background of, 45
as co-founder, 69
departure of, 244–45
at Harvard, 45, 50, 72
and News Feed feature, 142
and Obama campaign, 334
role of, in FB, 97
on Wall feature, 111
Hug Me third-party app, 160

idealism, 16, 445, 524
iLike, 158–59, 164–65, 269
India, 233–34
InfoWars, 458–59

initial public offering (IPO), 288–93, 297–98
Instagram
advertising on, 477, 490, 508
and antitrust charges, 515, 516
benefits provided by FB, 509–10
and Dorsey, 302, 303, 304–5
Facebook's acquisition of, 303–7, 309, 508
filters of, 301–2
independence of, 489–90, 509
and Instagram TV, 510
leadership changes in, 512
messaging service of, 509
origins of, 300–303
popularity of, 5, 12, 302–3, 507–8
revenues from, 489, 508
Russian election interference via, 374, 489
Stories feature, 497–99, 507
Instant Articles feature, 387, 388
Instant Messenger
as inspiration for later apps, 43
and Saverin's termination, 99
and Wirehog project, 92
and youths of FB engineers, 28–29
Zuckerberg's damaging exchanges on, 46, 57, 73, 75, 119
Instant Personalization, 270–72, 409, 424
Integrity team, 448, 477, 485
international markets, 226–34, 235, 320, 323, 435–36
Internet access, 4–7, 14, 231–34
Internet.org initiative, 4, 231–34, 316, 347, 506
Internet Research Agency (IRA; troll farm), 373–76, 395
Internet start-ups, first wave of, 74
Iribe, Brendan, 326–27, 328–29, 493, 495
"Is Connectivity a Human Right?" (Zuckerberg), 231
Issenberg, Sasha, 353
"I Voted"/"I'm Voting" button, 237, 337–38, 406

Janzer, Paul, 140, 141, 247–48, 249, 252, 447
Japan, Facebook's entry into, 228–29
Jobs, Steve
disinterest in buying FB, 149
and Facebook app for iPhone, 277
Gates on, 184
and Hewitt's angry email, 278
at Indian ashram, 197
and iPhones, 154
keynotes given by, 156
Palihapitiya's obsession with, 286
and Zuckerberg, 481–82
Jones, Alex, 458–59, 469
journalism in News Feed, 386–87

Kagan, Noah, 214
Kaiser, Brittany, 421
Kamangar, Salar, 179
Kaplan, Joel, 339–40, 342–43, 344, 357, 360, 458, 467
Karinthy, Frigyes, 19–20
Katigbak, Everett, 237, 238
Kavanaugh, Brett, 467
Kelly, Chris
and customer support issues, 247, 249, 250
departure of, 266
and news of purchases shared on FB, 183
and Palihapitiya's Growth team, 219, 224
and Yahoo!'s attempt to acquire FB, 134–35
on Zuckerberg's personality, 256
Kendall, Tim, 178–79, 180, 183, 187, 189, 198, 199, 211
Kirkpatrick, David, 10, 244–45
Kogan, Aleksandr
background of, 399–400
banned from FB, 425
colleagues' concerns about, 415–17
consulting work with Facebook, 408, 419
critical *Guardian* article on, 417–18
and data deletion demanded by FB, 419, 420, 422, 424–25
and data for Cambridge Analytica, 399, 411–13, 414–19, 420–21, 422, 423–26
data gathering app of, 408–10, 412, 413, 414, 419
and FB's Terms of Service, 414, 415
and Global Science Research, 412, 413
and SCL, 410, 413, 417
Kordestani, Omid, 192, 193, 200
Kosinski, Michal, 400, 401–5, 410, 412–13, 414–17
Koum, Jan, 317–25, 501, 503, 504, 505
Kramer, Adam, 406
Krieger, Mike, 300–302, 304, 305, 489, 509, 512
Kushner, Jared, 351

Lane, Sean, 187, 189
Leahy, Patrick, 437
Leathern, Rob, 466
LeCun, Yann, 453–54
Lessin, Sam, 65–66, 173, 472
Levchin, Max, 88, 152, 160–61, 164

Levine, Marne, 200, 339
Lewis, Harry, 61–62
Li, Charlene, 186
Libra digital currency, 516–21
Like button, 202–7, 402–4, 416
LinkedIn, 86, 220, 269
location technology, 310–11
Lookalike Audiences, 352
Losse, Kate, 130, 136, 228, 246, 248–49, 294
Luckey, Palmer, 325–27, 492–95

Macedonia, fake news originating from, 358–59, 364, 365
machine learning, 452–53, 454
Marconi, Guglielmo, 19
Marcus, David, 315, 506, 517, 519
Marks, Meagan, 39, 64
Marlette, Scott, 105, 114
Marlinspike, Moxie, 500, 505
Marlow, Cameron, 224, 337–38, 408
Martin, Kevin, 339
Martínez, Antonio García, 466
Mayer, Marissa, 322
McCollum, Andrew
  on adding new features, 111
  and Casa Facebook in Silicon Valley, 76, 77, 83
  co-founder role of, 69
  at Harvard, 50, 54, 60
  at La Jennifer house, 97
  on Moskovitz's importance to FB, 96
  page design and logo for thefacebook, 62
  on Parker–Zuckerberg relationship, 93
  on personality of Zuckerberg, 70
  on servers, 98
  and Silicon Valley workspace, 77
  and Wirehog project, 94
  on work style of Zuckerberg, 92
McNamee, Roger, 134, 193–94, 211, 359, 473
Mechanical Turk, 409
"memorialization" protocols, 450
Mentions feature, 439
Mercer, Robert, 411, 417, 422
Messenger, 313–15, 477
Metcalfe's law (network effect), 67
Microsoft
  attempts to acquire FB, 179
  and contact scraping issue, 215–16
  investment in FB, 185
  and invitations to join FB labeled as spam, 184
  and MySpace ads, 179
  partnership with FB, 131–32, 183–85, 198
  and Platform of FB, 155
  Zuckerberg's respect for, 152
midterm elections of 2018, 485–86
Milgram, Stanley, 20
Miller, Ross, 30
Milner, Yuri, 288
Mini-Feed feature, 130–31, 139–40
misinformation. *See* fake news and misinformation
mission of Facebook
  to build communities, 384
  and commitment to remain independent, 137
  to connect the world, 14–15, 16, 119–20, 127, 231–32
  and consequences of early decisions, 16
  expansion of, 119–20
  and growth of FB, 235, 524
  Hammerbacher on, 216–17
  and post-elections problems of FB, 16
  and profit goals, 524
  revolutionary component of, 149
Mitchell, James, 443
mobile Facebook
  apps built in HTML5, 279–82
  and data collection via Onavo, 315–17, 323
  and Facebook smartphone, 284–87
  and Faceweb, 281, 282, 283
  under Growth's umbrella, 235
  and initial iPhone app, 275–79
  and IPO of FB, 287–88, 290, 293
  and lost ad revenues, 290
  and Messenger, 313–15
  native apps for, 282–84, 293
  on suspicions FB is listening, 475–76
moderators, content, 252, 436, 444–51, 456. *See also* content arbitration on Facebook
Monthly Active User (MAU) metric, 213
mood study, 406–8
Moran, Ned, 333–36, 372–73, 377
Morgan Stanley, 288–89, 291, 322–23, 324
Morgenstern, Jared, 204, 206, 215
Morin, Dave
  and advertising business model, 170
  and Amazon Kindle, 155
  on AOL's Instant Messenger, 29
  Apple experience of, 154
  background of, 147
  and Microsoft partnership, 179
  and People You May Know (PYMK), 224, 225
  and Platform of FB, 156, 159
  and privacy defaults, 266
  recruitment of, 147–49
  role of, in FB, 151

on Sandberg's management, 196–97
and third-party developers, 160, 163
Moskovitz, Dustin
background of, 68–69
and Callahan, 96–97
and Casa Facebook in Silicon Valley, 83
as co-founder, 68–69
and Cox's interview, 124
departure of, 244, 245
at Harvard, 45, 50, 59
at La Jennifer house, 97
and Morin, 148, 149
and News Feed feature, 142
and Platform of FB, 158
role of, in FB, 97, 108, 111
and Silicon Valley workspace, 77
stake in FB, 96
on time demands of growing infrastructure, 95
and Yahoo!'s attempt to acquire FB, 136
Mossberg, Walt, 271
Mosseri, Adam, 362, 364–65, 388–89, 390, 512
"Move fast and break things" ethic of Facebook, 6, 16, 106, 240–43, 246, 427
MoveOn.org, 188, 262
Mueller, Robert, 336, 374–75, 378
murders on Facebook Live, 441
Murdoch, Rupert, 132
Murphy, Bobby, 308, 311–12
music industry, 93–94
Musk, Elon, 5, 6, 7, 8, 88, 232, 506
Myanmar (previously Burma), 11, 435–39, 526
MySpace
acquired by Murdoch's NewsCorp, 132
advertising practices on, 199
and child predation settlement, 250
and iLike app, 158
images posted to, 113–14
market dominance of, 91, 132
obsolescence of, 297
and Open Registration at FB, 144
popularity of, 59
proposed acquisition of FB, 132, 179
and RapLeaf (data broker), 269
third-party apps on, 153–54, 155, 159–60

"Napalm Girl" image, 456–58
Napster, 80, 86, 93–94
Narendra, Divya, 56, 73, 76
NECO campaigns, 176
Netflix, 171, 175, 239
net neutrality, 233
network effect (Metcalfe's law), 67
News Feed of FB
advertisements in, 138, 181, 295–98, 475, 477
algorithms feeding and filtering, 127–28, 142, 163, 172, 260–61, 385, 391
and Cambridge Analytica, 399
and content publishers, 387–90, 391
criticisms of, 385–86
and customer support issues, 250
design and implementation of, 14, 123–31, 139–40
"disputed content" on, 389
and emotional contagion study, 406–8
and fact-checking operations, 389
fake news on (*see* fake news and misinformation)
and Instant Articles feature, 387, 388
and Like button, 202–7, 402–4, 416
and Mini-Feed feature, 130–31, 139–40
and monetization of FB, 180
and myPersonality survey, 401
need for changes in, 385, 386–89
news consumption on, 128, 386–90, 391
and new users, 224
and Next Facebook, 488
outside content appearing on, 307–8
power of, 142
and privacy questions, 137–38, 143–44
and propaganda, 346, 348
quality studies for, 391
sensational content amplified on, 399
and sharing of content, 401
and spam wars, 163–65
and sponsored stories, 180, 296, 476
and stories from newer, weaker ties, 225
testing of, 137
and third-party developers and apps, 163–65
toxic addictiveness of, 385–86
and Trump campaign's ads, 353
Twitter-inspired redesign of, 259–63, 525
and user activities reported from partner apps, 172
users' reactions to, 140–43
video content in, 390–91
and virality, 262–63
news sources on social media, 132, 347
nipples policy, 252–53, 254
Nix, Alexander, 410–11, 413–14, 420–23

Obama, Barack, 334, 339–40, 350, 361–62, 363
Obama, Michelle, 467–68
Obama presidential campaigns, 350, 352

objectionable content on Facebook, 247–54, 435, 456. *See also* content arbitration on Facebook
Octazen, 216
Oculus, 325–30, 488, 491–96
Olivan, Javier
background of, 226–27
and Growth team, 217
and Instagram, 508–9
and international growth, 227, 228, 229
leadership style of, 234
on Messenger, 314
and Project P team, 365
promotion of, 511
role of, in FB, 234–35
and Stamos's security report, 364–65
Olson, Billy, 45, 48, 50, 52
Onavo/Onavo Protect, 315–17, 323, 483–85
Ondrejka, Cory, 281–84, 286–87, 314–15
online bullying, 49
Open Graph
and data for Cambridge Analytica, 424, 554
developers' abuse of, 430–31
and friend-of-friend user data, 175, 267–70, 409
Graph API V1 ("the Friend API"), 171–72, 175–76, 271, 409, 412
Graph API V2, 175
introduction of, 171–72
and privacy issues, 175, 176
and social graph, 156–57
Open Registration, 119–23, 133–34, 144, 157–58, 215, 250, 251
Oracle, 107
"organic" traffic, 205
Orkut, 60, 106
Orriss, Iris, 229–30

Page, Larry, 193, 289, 323
Pages, 182, 202, 205
Pages You May Like campaign, 295
Palihapitiya, Chamath
background of, 208–9, 210
and Beacon, 188
criticisms of FB, 473–74
on "Dark Profiles," 222–23
departure of, 234, 236
and growth of FB, 213–14, 218–19, 225
handpicked team of, 217
and Hewitt, 278
and international market, 226–27
management style of, 196, 211–12, 218–19
and mobile Facebook effort, 284–86, 287
move to Facebook, 210–11
and People You May Know (PYMK), 225, 226
and privacy defaults, 267
role of, in FB, 194, 211, 212
on Snapchat, 308
and values initiative, 239
and Zuckerberg, 209–10
Pandemic. *See* advertising
Parakey, 276
Parakilas, Sandy, 242, 270
Parker, Sean
background of, 79
and business model of FB, 177
business start-up experience of, 85
at Casa Facebook, 82, 84
and Causes app, 155
and cocaine possession arrest, 115–16, 247
criticisms of FB, 473
early projects of, 80–81
and Fanning, 79–80
fundraising efforts of, 85–86, 89, 117
and Graham meeting, 101
involvement in Thefacebook, 84–85
at La Jennifer house, 97
management style of, 196
meeting with Zuckerberg and Saverin, 82
and Morin, 148
and Napster, 80–82, 93, 94
and Palihapitiya, 209
and Palo Alto office on University, 109
on personality of Zuckerberg, 84
and recruitment of computer science engineers, 106–7
removed as president, 115–16
role of, in FB, 97, 115–16, 194
stake in FB, 116–17
and *Washington Post*'s investment, 101–2
and Wirehog project, 93
and Zuckerberg's "Domination!" yell, 110
Parr, Ben, 140–41, 142
Parscale, Brad, 351–52, 353–54, 421
Partner Categories, 475
Partovi, Ali, 158–59, 165
Partovi, Hadi, 158–59
Path (social network), 225
Payne, Donald, 470
PayPal, 88, 315
Pearlman, Leah, 180, 202, 206
People You May Know (PYMK), 220–26
Perry, Todd, 32–33
personally identifying information (PII), 474, 476
Pesenti, Jérôme, 454

Peters, Gary, 475–76
Petrain, David, 31
Philippines' 2016 presidential election, 347–48, 435
phone numbers of users, 318, 502–3
photos, 52, 64, 113–15, 301–3
Pichai, Sundar, 468
Pincus, Mark, 79, 86–88, 161–62, 166–69, 178
Place app of Facebook, 310–11
Platform of Facebook
  and App Review rules, 413
  apps suspended from, 430–31
  benefits of, 153–54
  and Cambridge Analytica, 418
  and Causes app of Parker and Green, 155, 162, 164
  and developers who were potential competitors, 174
  and F8 developers conference, 154, 157–58
  and Facebook Connect, 169–70, 171, 173, 268, 297, 408–9
  and friend-of-friend user data, 175
  games on, 161–63
  idea of, proposed, 150
  and information sharing, 152
  interface for (*see* application programming interface)
  misbehavior of developers on, 165–67, 430–31
  and News Feed spam wars, 163–65
  number of developers writing apps for, 158
  and privacy questions, 151, 152–53, 176
  release of, 155–56, 157–58
  and rise of the Apple and Android platforms, 172–73
  rule changes for developers on, 163–65
  sharing of user information on, 171–72, 399
  and social graph, 156–57
  success of, 159
  and traffic on FB, 168, 170
  and user data, 169–70, 173–74
  See also Open Graph
Plaxo, 80–81, 84, 116
Plouffe, David, 392–93, 394
Plummer, Dan, 125–26, 214
Podesta, John, 334
Poke app of Facebook, 309, 311, 312–13
pokes in FB system, 64–65, 122, 160, 203, 227
political bias, charges of, 338–42, 343, 429, 458–59
pornography on Facebook, 251–52
Portal feature, 464
Potemkin, Aleksey Aleksandrovich, 336–37
presidential election of 2016, 333–67
  and accusations of FB's partisanship, 340–43
  and DCLeaks, 336, 344–45
  and expectations for Clinton's victory, 350, 354
  and "Facebook Effect," 9
  and fake news (*see* fake news and misinformation)
  interference in (*see* Russian interference in US presidential election)
  reaction of FB staff to outcome of, 9, 360
  and Sandberg's leadership of FB, 355, 357, 486
  and Trending Topics controversy, 341–42, 345, 346
  Trump elected president, 9, 360, 494
  voter suppression in, 353, 374
  See also Cambridge Analytica; Clinton presidential campaign, 2016; Trump presidential campaign, 2016
presidential election of 2020, 11–12
Pritchard, Marc, 474, 477
privacy
  and ads on News Feed, 475
  availability of user data to public, 402, 404
  and Cambridge Analytica, 427, 429
  and cell-phone numbers shared on FB, 71, 101
  changes in default settings for (2009), 263–67
  and congressional hearings, 429
  and damaging instant messages of Zuckerberg, 57
  and "Dark Profiles," 129, 205
  and data brokers' access to user data, 269–70
  discussed in Book of Change, 122
  in earliest FB release, 63–64, 68, 71
  and Facemash incident, 50, 53, 144
  and Friendster, 69
  and FTC investigation and sanctions, 11, 272–74, 402, 478–79
  Graham's questions about, 101
  and Graph API V1 ("the Friend API"), 271, 412
  and Instant Personalization, 271–72, 409, 424
  Lewis's caveats on, 62
  and Like button on external websites, 205
  and News Feed feature, 137–38, 143–44

privacy *(cont.)*
and Next Facebook, 487–88, 513–14, 519
and Onavo Protect, 316–17, 483–85
and Open Graph, 171–72, 176, 267–70, 409, 424, 430
and Open Registration, 122, 144
and opt-in requirement, 64
options that increased, 266
and origins of FB, 53
and People You May Know (PYMK), 221–24
and Platform of FB, 151, 152–53, 176
and purchases by users shared on FB (Beacon), 182–83, 186–89, 212
and removal of objectionable content, 248
and suspicions FB is listening, 475–76
and Synapse's data mining, 40
and User ID leak, 268–69
Zuckerberg's perspectives on, 40, 143, 267, 272–73, 487–88, 525
See also Kelly, Chris; Sparapani, Tim
profiles in Facebook, 64–65
programming projects of Zuckerberg
and ConnectU project of Winklevoss brothers, 56–57, 60, 72–76
Course Match project, 46–47, 61
Facebook as culmination of previous projects, 61
facebook project of Zuckerberg, 53–54, 60–61
Harvard facebook project, 53–54
Harvard Facemash, 47–52, 53, 56, 58, 61, 64
and houseSYSTEM project for Greenspan, 58–59, 60, 64
Rome of Augustus collaboration tool, 54–55, 61
and skills acquisition, 49
Synapse project, 33, 38–41, 44, 47
while at Harvard, 45–52, 54–55
Wirehog project, 92–95
and work habits of Zuckerberg, 77
Project P team, 365–66, 367, 369
propaganda, 346, 349, 364, 365, 376
public company, Facebook's transition to, 288–93
public figures, harassment of, 253
publishers, 387–90, 391
purchases by users announced on Facebook, 182–83, 186–89
Putin, Vladimir, 367

Qualtrex, 415
Questions feature, 245
Quittner, Josh, 187
Quora, 245, 309

Rabkin, Mark, 198, 199, 201, 294
RapLeaf (data broker), 269–70
Rappler, 347
regulation of FB, calls for, 11
relationship status field in FB, 64, 141, 150
Remnick, David, 362
reputation of Facebook, 11–12, 398, 484–85, 525
Ressa, Maria, 347–48, 435, 486
Risk and Response group, 442–43
Roberts, Sarah T., 445
RockYou, 159–60, 164–65
Roosendaal, Arnold, 205
Rose, Dan, 172, 179, 184–85, 293, 306
Rosen, Guy, 315–16, 448, 452, 455, 477, 480
Rosenberg, Jonathan, 200
Rosenstein, Justin, 180, 182, 202, 203, 206, 473
Rosensweig, Dan, 134, 193
Ross, Blake, 217, 245, 276
Rothschild, Jeff, 105, 124, 143, 243
Royal Bank of Canada, 176
Russian interference in US presidential election
and ad purchases, 372–76, 377, 378–79
briefing board of directors on, 379–80
and congressional hearings, 377, 378, 395–96
and DCLeaks, 336, 344–45
DNC emails stolen, 334, 336–37, 344, 353
engagement with FB users, 334
and evasion of FB's detection, 375
first signals of, 333, 334, 335
and free expression on FB, 376
going public on, 377, 378, 379
and inflammatory messages embedded in graphics, 455–56
and Instagram, 374, 489
and intent to sow conflict, 383, 398, 421–22
and Project P team, 365–66, 367, 369
proof of, 375
and propaganda, 364, 365
spear-phishing attacks, 334, 336
and Stamos's report, 364–67
and troll farms, 373–76, 395
and voter suppression, 374
white supremacism fueled by, 469
and WikiLeaks, 345
Zuckerberg's perspectives on, 526
Rust, John, 400, 401, 415–16

Sandberg, Sheryl
and author's research, 15
background of, 190–92
and business plan of Facebook, 198, 199–200
and Cambridge Analytica, 419, 425–26
and Congressional Black Caucus, 469–70
congressional testimony of, 468–69
criticisms of, 356
and culture of FB, 237–38, 243
and data collected by FB, 207
division of labor with Zuckerberg, 194, 255, 355
on fake news disseminated on FB, 346, 359
and FB's attempts to impugn competitors, 466
and "FOSS" term (friends of Sheryl Sandberg), 200, 339
and free speech, 470–71
frustrations of, 470–71
and FTC investigations, 479
and goals for the future, 469
at Google, 192–93
and husband's death, 355, 357, 469, 470
and initial public offering, 287–88, 290, 298
and Lean In movement, 467–68
loyalty to Zuckerberg, 473
management style of, 196–97, 200, 356–57
meetings with Zuckerberg, 193–94
and Palihapitiya's criticism of FB, 474
and Palihapitiya's growth proposals, 213, 214
and Pincus, 167
and presidential election of 2016, 9, 355, 357
recruiting efforts of, 200–201
recruitment of, 193–94
and Ressa's warnings, 435
role of, in FB, 194–95
and Russian election interference, 378, 380
and security team, 335–36
and Stamos's security report, 364, 366–67
and Stamos takedown, 380
and WhatsApp, 504
work style of, 191–92
Sanghvi, Ruchi, 105, 124, 127, 139, 140, 142
satellite launch failure, 5, 6–7, 8, 232, 233
Saverin, Eduardo
and advertising sales, 178
background of, 66–67
co-founder role of, 70
and incorporation of FB, 85
and Parker, 82, 98
partnership with Zuckerberg, 66–67
portrayed in *The Social Network*, 99
stake in FB, 96, 98, 99
terminated from FB, 98–99, 177
and Zuckerberg's profile in *Crimson*, 77
Schmidt, Eric, 192, 196
Schmidt, Sophie, 423
Schrage, Elliot
and anti-Semitism charges, 466
briefing board of directors, 379
and DCLeaks page, 344
departure of, 479
and Kaplan, 344
and Myanmar situation, 436
and "Napalm Girl" image, 458
and policy making, 357
and Ressa's warnings, 435
role of, in FB, 200
and Sandberg, 357
Schroepfer, Mike, 282, 454, 471
Schultz, Alex, 217, 219–20, 224, 234, 435, 455, 511
Scissors, Lauren, 408
SCL, 410–15, 417, 419, 421–24
Scrabble app, 162
search engine optimization (SEO), 219–20
search engines, visibility of users' details to, 264
security measures of Facebook
additional staff hired for, 476
and Election War Room (2018), 485
and encryption, 500–501, 513–14
Growth team's oversight of, 477
Stamos's leadership of, 335–36
Selby, Brad, 161, 169–70
Semel, Terry, 133, 134, 135
Senate Judiciary hearing, 395–96
server space for Facebook, 66, 67, 97–98, 100, 105, 115
sexual politics on Facebook, 249
shadow/dark profiles, 129–30, 205, 222–23
sharing dynamics of Facebook users, 171, 401, 405
Shavell, Rob, 205
Silver, Ellen, 440
Simo, Fidji, 390, 439, 510
Sittig, Aaron
and Book Reviews app, 155
and Like button, 203
and News Feed feature, 130, 140
and Photos feature, 114–15
recruitment of, 90–92

Sittig, Aaron *(cont.)*
and recruitment of computer science engineers, 106–7
and redesign of FB, 113
on Wirehog project, 92
sixdegrees, 19–22, 80, 87, 522
Six Degrees to Harry Lewis, 61–62
Six4Three lawsuit, 465
"61-Million-Person Experiment in Social Influence and Political Mobilization, A" study, 337–38
Slashdot, 40–41
Slee, Mark, 217, 259–60, 261
Slide, 152, 159–61, 164
slogans used at Facebook, 241–43
smartphone, Facebook, 284–87
Smith, Ben, 390
Snapchat, 307–9, 316–17, 496–99
Snowden, Edward, 501
social advertising, 180–81, 183, 185
social graph, 156–57. *See also* Open Graph
Social Network, The (film), 6, 7, 82, 99, 116
social networking
Abrams' perspective on, 87
anonymity in, 42
and Buddy Zoo, 43–44, 61
early versions of, 19–22
explosion of, 59–60
and finding one's tribe, 27
and Friendster, 41–43
Pincus and Hoffman's anticipation of, 86, 88
real names used in, 42, 63, 249
Sittig's perspective on, 91
and sixdegrees patent for, 22, 87–88
Social Reader app, 171, 172
Soros, George, 345, 466
SpaceX, 5, 6–7, 232
Sparapani, Tim, 266–67, 268, 434
spear-phishing attacks, 334, 336
Spiegel, Evan, 307–9, 311–12, 316, 496–99
sponsored stories, 180, 296, 476
Spotify, 171
Spratt, Ryan, 156
Stamos, Alex, 335–37, 345, 363–67, 372, 377, 378–80
Stanford University, 53, 72
status updates, 259
Steamtunnels website, 53
Stewart, Margaret, 294, 296
Stillwell, David, 400–405, 413, 414–16
stock price of Facebook, 291, 293, 297, 426, 477
Stories feature in Facebook, 507
Stremel, Jed, 216, 275–76, 277
Stretch, Colin
briefing board of directors, 379
on delay in releasing news to public, 377
on delay of sharing Russian ads with Congress, 378
and FTC investigations, 272
on privacy model, 267
and Russian election interference, 366, 374, 377
and Senate Judiciary hearing, 395–96
Students Against Facebook News Feed, 140–41, 142
Stutzman, Fred, 144
Su, Sarah, 436
suicides on Facebook, 440–41, 451
Summers, Lawrence (Larry), 72–73, 190, 192
SuperPoke! third-party app, 160
Super Wall third-party app, 160
Swisher, Kara, 185, 271
Swope, Allison, 440
Synapse-ai, 33, 38–41, 44, 47, 58
Systrom, Kevin
and advertising on Instagram, 508
background of, 299–300
and control asserted by Zuckerberg, 509–10, 512, 513
departure of, 491, 512, 513
and Dorsey, 302, 303, 304–5
and FB's acquisition of Instagram, 305, 325
FB's attempt to recruit, 106–7
on independence of Instagram, 489–91
and Instagram TV, 510
and messaging service, 509
and origins of Instagram, 300–303
and social media critics, 473
and Stories feature, 497–99
and success of Instagram, 304

Tayler, Alexander, 419
Taylor, Bret, 203, 205, 284, 286
Techlash, 482
Terms of Service agreements, 265, 369, 407, 414
terrorist content, 455
Thiel, Peter
and car for Zuckerberg, 97
on FB's board of directors, 102, 133, 288
and growth of FB, 94
party thrown by, 95, 275
and RapLeaf (data broker), 269
"seed round" funding from, 88–89, 90, 100, 101, 178
on shares and vesting schedule, 96

and Trump campaign, 494
and Yahoo!'s attempt to acquire FB, 136
thisisyourdigitallife app, 408
Threat Intelligence team of Facebook, 333–34, 336, 364, 366, 372
Tillery, Kris, 34–35, 60
Time's Person of the Year, Zuckerberg as, 8
Tinder, 175, 445
Tokuda, Lance, 159–60
Translate Facebook app, 227–30
Trending Topics feature, 341–42, 345, 346
troll farms, 373, 395
Trump, Donald
anti-Muslim posts of, 340, 457
elected president of United States, 9, 360, 494
and fake news disseminated on FB, 10
on marriage of Chan and Zuckerberg, 293
Ressa's comment on election win of, 348
and Russian election interference, 366, 378, 379
Twitter account of, 340
Trump presidential campaign, 2016
anti-Hillary ads run by, 353
and Cambridge Analytica, 399, 420, 421, 427
effort put into FB, 351–52, 354
targeted ads of, 351–53
voter suppression by, 353
trust
and Beacon, 187
and changes in default privacy settings, 265, 267
and Dating feature, 464–65
FB's efforts to restore, 484
following Cambridge Analytica crisis, 464–65
and growth of Facebook, 235
importance of, to FB's success, 121
and users' distrust of apps, 170
truth, Facebook's reluctance to arbitrate, 337, 346, 357, 361
Twitter, 257–64, 299, 304–5, 307, 395, 439, 445

Universal Facebook of Greenspan, 58–59, 60
US Congress, 377, 378, 427–28, 468–69, 474, 515, 516, 519–20
US Department of Justice, 515, 516
user data
availability of, to public, 402, 404
and Cambridge Analytica, 399, 411–13, 414–19, 420–21, 422, 423–26
and data breach of Facebook, 467
data brokers' selling of, 269–70
and deactivated accounts, 274
and Facebook Connect, 169–71, 173, 268
friends-of-friends data, 223, 268–69, 409, 412–13, 415, 416
and FTC investigation and sanctions, 273–74, 478–79
and Graph API V1 ("the Friend API"), 171–72, 175–76, 271, 409, 412
and Instant Personalization, 270–72, 409, 424
and Like button on external websites, 204–7
and mobile analytics of Onavo, 315–16, 323
and Onavo Protect, 316–17
and Open Graph, 171–72, 176, 267–70, 409, 424, 430
and personally identifying information (PII), 474, 476
policies for use of, 410
predicting users' traits with, 402–4, 416
safeguards for, 153
and User ID leak, 268–69, 270
user-generated content, 22, 86, 112, 221
User IDs, leak of, 268–69, 270
users
and anonymity on Friendster, 66
content created by, 64
and "Dark Profiles" (people not on FB), 129–30, 205, 222–23
and data collected via Platform, 169–70, 173–74
and elections on governance of FB, 265
engagement of, 202, 385
fake accounts/users, 376–77, 455
growing numbers of, 275, 297
(*see also* growth of Facebook)
integration of FB into lives of, 89
"Likes" pursued by, 205
limited audience of, 67
and Mini-Feed feature, 130–31
older generations of, 122
and Open Registration, 119–23, 133–34, 215
and opt-in requirement, 64
phone numbers of, 318, 502–3
and privacy setting defaults, 263–67
purchases by, shared on FB (Beacon), 182–83, 186–89, 206, 212
reactions to News Feed, 140–43
real names of, 42, 63, 66, 81, 147, 249
resistance to redesigns, 138–39
retention of, 220–21, 224–25
Russians' engagement with, 334

users *(cont.)*
safety of, 250–52
and usage patterns, 89, 90
and user testing, 122
and Workplace Networks, 121, 133

vacation days at Facebook, 85
values of Facebook, 111, 239–41, 289, 459
Van Natta, Owen
and buyout offers, 131–32, 134–35, 210, 211
demotion of, 211
management style of, 196
and Microsoft meeting, 184
and News Feed feature, 141
role of, in FB, 116, 194
"Verified Apps" program, 274
Vernal, Mike, 169, 170, 176, 280–81
video content in News Feed of FB, 390–91
violence incited on Facebook, 11, 234, 435–39, 473, 526
virality, 159, 262–63, 388, 391, 405
virtual private network (VPN) of Onavo, 316
virtual reality (VR), 325–30, 491–96
voting button and study, 237, 337–38, 406

Wall feature of Facebook, 111–12, 160, 227–28
Warner, Mark, 372, 395
Warren, Elizabeth, 516
Washington Post, FB's funding agreement with, 100–104
Web 2.0, 12, 86
Wehner, David, 476, 484
Weinreich, Andrew, 19, 20–21, 66, 87, 522–23
Welsh, Matt, 60, 104
WhatsApp, 499–507
advertising on, 477
and antitrust charges, 515, 516
and congressional hearings, 520
FB's acquisition of, 322–25
origins of, 317–22
popularity of, 12
used to incite violence, 11, 437–38, 473
WikiLeaks, 345
Williams, Evan, 258–59, 299
Williams, Maxine, 469, 470
Willner, Charlotte, 252
Willner, Dave, 250, 252, 253–54, 447, 457
Winklevoss, Cameron, 56–57, 64, 72–76
Winklevoss, Tyler, 56, 64, 72–76
Wirehog project, 92–95
Words with Friends app, 162
Workplace Networks, 121, 133
work spaces of Facebook
author's access to headquarters, 15
bonuses for staff living nearby, 109
consolidation of, in Palo Alto, 194
and culture of FB, 105, 254–55
Los Altos house, 96–97, 100
and non-disclosure agreements, 74
Palo Alto headquarters on California Ave., 238, 254, 368–69
Palo Alto house on La Jennifer (Casa Facebook), 76–77, 82–84, 90–91, 95
Palo Alto office on Emerson Street, 100, 105
Palo Alto office on University, 109
whiteboards in, 118, 369
Wu, Tim, 515
Wylie, Christopher, 410–15, 420, 422–25

Xobni, 174

Yahoo!
and MySpace ads, 179
proposed acquisition of FB, 133, 134–37, 210, 211, 239, 312, 508
and sixdegrees patent auction, 87–88
Yale University, 72
YouTube, 12

Zoufonoun, Amin, 305–6, 311, 329
Zuckerberg, Donna (sister), 24, 28
Zuckerberg, Ed (father), 22–25, 28, 38, 51, 329
Zuckerberg, Karen (mother), 22–24, 27–30, 38, 51, 329
Zuckerberg, Mark
annual challenges/resolutions of, 256–57, 369–70, 396
author's meetings with, 12–14, 368–71, 397, 522, 523–27
and author's research, 15
birth of, 22, 24
"Building Global Community" essay, 371, 383
car of, 97
children of, 8, 393, 396
community building emphasis of, 371, 383–84
and Community Summit, 383–84
competitive nature of, 27, 260, 525
confidence of, 52, 59
on connecting the world, 14–15, 127, 231, 289, 329–30, 370, 371, 524
copycat products after failed acquisition bids of, 309–11

and culture of FB, 240
damaging instant messages of, 46, 57, 73, 75, 119
decision-making of, 472–73
and departures of leaders, 244–45
and "Domination!" yell, 110
education of, 29–35 (*see also* Harvard University)
engineering mindset of, 5, 7
and fake news problem, 10, 357, 359, 360–61, 370
food choices of, 256–57
and free speech, 111, 249, 254, 344, 357, 363, 459, 524
and FTC investigations, 272–73, 274, 479
idealism of, 445, 524
"Is Connectivity a Human Right?" essay, 231
leadership role of, 97, 117, 194–95
leadership style of, 91, 137, 244–45, 462–63, 508
and mission of FB, 137, 384
and "Move fast and break things" ethic, 16, 106, 240–41, 243
notebooks of, 118–19, 122, 197, 527 (*see also* Book of Change)
note of atonement posted by, 397–98
Obama's meeting with, 362, 363
partner of (*see* Saverin, Eduardo)
on personal goals, 77–78
personality of, 25, 59, 256–57, 397
perspectives on FB's problems, 523–27
philanthropy of, 391–95
programming (*see* programming projects of Zuckerberg)
risk-taking of, 526
security detail of, 382
on sense of purpose, 382
silences in interviews/meetings of, 13, 101, 255–56, 463
and *The Social Network* film, 6
stake in FB, 117
and stress of public speaking, 156, 157
travels of, 1–8, 197–98, 396–97
on trust, 464–65
United States tour of, 380–84, 397, 398
vision for FB, 129
wedding of, 292–93
wife (*see* Chan, Priscilla)
work style of, 255–56
and Yahoo!'s attempt to acquire FB, 134–37, 312, 508
youth of, 24, 25–35
See also specific apps, events, and features
Zuckerberg, Randi (sister), 24, 27, 28, 37, 214, 329
Zynga, 161, 162, 166–69, 516

About the Author

Steven Levy is the editor at large at *Wired* magazine. He was formerly senior editor and chief technology correspondent for *Newsweek*, and contributor to many publications. He is the author of seven previous books, including *Hackers*, *Artificial Life*, *Insanely Great*, *Crypto*, and *In the Plex*. Levy lives in New York City with his wife, the writer Teresa Carpenter.